Remaking the News

Inside Technology

edited by Wiebe E. Bijker, W. Bernard Carlson, and Trevor Pinch

A list of the series appears at the back of the book.

Remaking the News

Essays on the Future of Journalism Scholarship in the Digital Age

Pablo J. Boczkowski and C. W. Anderson, Editors

The MIT Press
Cambridge, Massachusetts
London, England

This book was set in ITC Stone Sans Std and ITC Stone Serif Std by Toppan Best-set Premedia Limited. Printed and bound in the United States of America.

Library of Congress Cataloging-in-Publication Data

Names: Boczkowski, Pablo J., editor. | Anderson, C. W. (Christopher William), 1977- editor.
Title: Remaking the news : essays on the future of journalism scholarship in the digital age / edited by Pablo J. Boczkowski and C. W. Anderson.
Description: Cambridge, MA : The MIT Press, [2017] | Series: Inside technology | Includes bibliographical references and index.
Identifiers: LCCN 2016041365 | ISBN 9780262036092 (hardcover : alk. paper)
Subjects: LCSH: Journalism--Technological innovations. | Electronic newspapers--Technological innovations.
Classification: LCC PN4784.T34 R46 2017 | DDC 070.4/3--dc23 LC record available at https://lccn.loc.gov/2016041365

10 9 8 7 6 5 4 3 2 1

To our respective mentors, Trevor Pinch and Michael Schudson, for teaching us how to think about technology and journalism, and for embodying the virtue of scholarly dialog.

P. J. B. and C. W. A.

Contents

Acknowledgments ix

Introduction: Words and Things 1
Pablo J. Boczkowski and C. W. Anderson

I Revisiting Theoretical and Methodological Debates 13

1 Scholarship on Online Journalism: Roads Traveled and Pathways Ahead 15
Pablo J. Boczkowski and Eugenia Mitchelstein

2 From Heterogeneity to Differentiation: Searching for a Good Explanation in a New Descriptivist Era 27
Rodney Benson

3 Rediscovering the News: Journalism Studies' Three Blind Spots 47
Victor Pickard

4 Newsroom Ethnography and Historical Context 61
C. W. Anderson

Commentary: Reflections on Scholarship in the Study of Online News 81
William H. Dutton

II Rethinking Key Concepts 89

5 Digital News as Forms of Knowledge: A New Chapter in the Sociology of Knowledge 91
Rasmus Kleis Nielsen

6 On the Worlds of Journalism 111
Seth C. Lewis and Rodrigo Zamith

7 The Whitespace Press: Designing Meaningful Absences into Networked News 129
Mike Ananny

Commentary: Remaking Events, Storytelling, and the News 147
Zizi Papacharissi

III Interrogating Occupational Culture and Practice 155

8 Helping Newsrooms Work toward Their Democratic and Business Objectives 157
Natalie Jomini Stroud

9 Journalism Ethics and Digital Audience Data 177
Matthew Hindman

10 Reinventing Journalism as an Entrepreneurial Enterprise 195
Jane B. Singer

Commentary: Blurring Boundaries 211
W. Russell Neuman

IV Foregrounding Underexamined Themes 215

11 Check Out This Blog: Researching Power and Privilege in Emergent Journalistic Authorities 217
Sue Robinson

12 A History of Innovation and Entrepreneurialism in Journalism 235
Mirjam Prenger and Mark Deuze

13 A Manifesto of Failure for Digital Journalism 251
Karin Wahl-Jorgensen

Commentary: The Journalism Studies Tree 267
Michael Schudson

Postscript: The Who, What, When, Where, Why, and How of Journalism and Journalism Studies 273
Michael X. Delli Carpini

Contributors 289
References 291
Index 345

Acknowledgments

We thank the support received from Northwestern University's School of Communication through the Departmental Excellence Funding Initiative administered by the Department of Communication Studies. Dean Barbara O'Keefe and Department Chair Ellen Wartella generously funded the conference that took place at Northwestern University on April 11, 2015, which brought together most of the contributors to this volume for a day of fruitful and productive conversations. We also thank Brad Hamm, Dean of the Medill School of Journalism, Media, Integrated Marketing Communications, for providing additional funding, and David Gerstner, Chair of the Department of Media Culture at City University of New York, for the support and encouragement he provided during this volume's gestation. Northwestern's doctoral student Amy Ross went above and beyond the call of duty to making sure that everything ran smoothly before, during, and after the conference.

We are also very grateful for the feedback we received on the introductory chapter from Jane Singer, Rodney Benson, and Michael Delli Carpini, and the helpful and constructive comments on the full manuscript from two anonymous reviewers. Northwestern's doctoral student Jacob Nelson assisted with copy-editing matters in an effective fashion. Katie Helke of the MIT Press provided able editorial guidance throughout the process. Inside Technology series co-editor Trevor Pinch gave sage advice with his usual diplomatic flair, dry humor, and understated effectiveness.

Finally, we want to acknowledge the phenomenal effort of the contributors to this volume. Many of them are our scholarly contemporaries, and others are the academics whose work we grew up admiring. We were blessed to consider all of them as friends when we started this project, and our friendships have only been strengthened by our joint work in it. Collective endeavors of the kind represented by this volume can become tortuous and the process quite convoluted. But the contributors to this volume were a pleasure to deal with: they fully engaged with the idea behind this project, put their best work forward, responded well to feedback, and delivered their manuscripts in a timely manner. They even put up with our occasional misspelling of their last names in various drafts and with noisy hotel rooms with a healthy dose of humor. All this made working with them a most rewarding experience. We owe them our heartfelt thanks.

Pablo J. Boczkowski, Evanston, Illinois; and C. W. Anderson, Brooklyn, New York, July 1, 2016

Introduction: Words and Things

Pablo J. Boczkowski and C. W. Anderson

In the beginning was the Word.
—John 1:1

During most of the evolution of scholarship on news and journalism, accounts almost always began with the word. And they usually ended with it, too. Pick up a book or an article written during the twentieth century about how the news is made, what it consists of, how it circulates in society, and what effects it has on politics and culture. Most likely you will find an account of words, the people who wrote or said them, and the organizations they belonged to and interacted with. Even works of journalism history often tended to center on this "holy trinity." Virtually ignored were the transformations in the material conditions of production, distribution, and reception of information that have marked the past few centuries, in particular the previous one. Emanating from the texts that embodied our culture of inquiry was a world devoid of artifacts. Absent from the scholarly foreground—and often even from the background, too—were the very things, tools, machines, hardware, software, and other types of technology that were used not only to write and broadcast the news, but also paradoxically to produce the inquiries that made them invisible. Could the ethnographer proceed without pen and paper? The statistician without calculators and computers? The content analyst without newsprint, footage, and tapes? The publishers without printing presses? Even the most gifted observers of newsmaking, the scholars who left us the seminal texts that have informed our discussions for decades, like Gaye Tuchman in *Making News* (1978) and Herbert Gans in *Deciding What's News* (1980), managed to depict key practices and processes while stripping them of their material culture.

Then came the 1990s. After a long build-up that escaped the notice of most industry analysts and scholars alike, a confluence of technological, economic, policy, and cultural developments that led to the commercialization of the World Wide Web took the world of news by storm. Since then, "the digital" has slowly but steadily moved to the center of attention of media organizations and their reporters and editors, with the concomitant realization that technology is a central and indispensable element of their enterprise. In a fairly dramatic reversal of fortune, most contemporary discussions among journalists begin and end with the intersection of words and things. Focusing on the word alone has become almost unthinkable for those charged with making news.

A somewhat parallel process has taken place among scholars. When Boczkowski decided in 1996 to write a dissertation on making online news, scholarly treatments of the role of technology in journalism were few and far between, and often marginalized within the academic community. A decade later, when Anderson began his dissertation on the rebuilding of the newsmaking ecosystem, technology had gained a central place within scholarly discussions about journalism. Today, if you flip through the pages of any journal or browse through the catalog of any book publisher that focuses on media, news, or journalism, you will be hard pressed not to find many texts that begin and end with the intersection of words and things.

The foregrounding of this intersection in scholarly consciousness has not only destabilized the intellectual regime that characterized the making of scholarship about the news but also reinvigorated it. To return to the case of ethnographic studies of news production, the long moratorium (Klinenberg 2005) that followed the publication of the now-classic texts by Tuchman, Gans, and others in the 1970s and early 1980s was brought to an end with a vengeance through a flurry of accounts written by emerging scholars who almost invariably centered on some aspect of the intersection of words and things.

Thus, the title of this volume, *Remaking the News*, is a double entendre about these twin changes in the object of study and the process of inquiry. Taking Tuchman's canonical text as a stand-in for a particular era of journalism and for the scholarship about it, *Remaking the News* signals both an acknowledgment of the past and a break from it in terms of how the news is made. It also evokes both an homage to an intellectual tradition about the role of journalism in society and a departure from one of its foundational tenets. The rationale for this collection is found in yet another set of parallels between industry and the academy. Much as journalists have experienced the past couple of decades as a whirlwind, we have experienced what seems to us a whirlwind in research about journalism. A feverish level of activity has characterized the scholarly study of news. Missing amid this ferment has been an opportunity to collectively reflect on what scholarship has achieved so far and, especially, what we might want to add moving forward.

The initial impetus for this volume originated during the lunch break of an event at the University of Pennsylvania's Annenberg School for Communication in Philadelphia, organized by Daniel Kreiss and Joseph Turow. It was held in spring 2013, and both of us were in attendance. During that conversation emerged not only the idea for this project but also the guiding principle that both the assessment of the road traveled and, in particular, the necessary programmatic work would benefit from community-oriented conversations rather than individual or small group efforts. This quite naturally led us to consider starting with that old academic standby—a conference—as the launching platform for this project.

From the start our goal was to facilitate a series of conversations that had a significant degree of internal coherence while at the same time respecting diversity of viewpoints. We decided that participation would be by invitation only and focused on a relatively short list of colleagues who had already made an impact on the study of news and whose work was concerned with the relationship between journalism and technology broadly conceived. We wanted to gather scholars who were at a point in their careers when they could enjoy the opportunity to take a step back from their own empirical research. We asked them to write essays on a particular aspect of their own choosing about the relationship between technology and journalism, and to reflect on how it might build pathways moving forward. A second goal was to place these papers and ideas within larger conversations happening within the study of communication and media. To that end, we invited five scholars to comment on groups of the papers in a way that wove themes across the papers and showcased their resonance with those larger conversations.

The conference took place on April 11, 2015, at Northwestern University. Prior to the event, we provided feedback to most of the paper writers. During the event, groups of three or four participants, organized thematically, presented their precirculated papers, followed by respondent commentary from our group of senior scholars, and then by open question-and-answer sessions. The conference concluded with commentary by Michael X. Delli Carpini of the University of Pennsylvania, fittingly closing the circle that had started at his home institution two years before. Shortly after the conference wrapped up, the editors asked for second drafts based on the feedback and exchanges during the event. Once these second drafts were submitted, we circulated them to the section respondents and to Delli Carpini, who turned their contributions into section commentaries and the postscript for this volume, respectively.

The book now in your hands is thus the result of an extensive and elaborate process to build community through conversation. The pages that follow feature a series of chapters and commentary sections offering an array of interlocking arguments about potential paths forward in the study of news and its role in politics and culture, as well as connections to scholarship on other objects of inquiry. As you will see, there are at times significant disagreements among the chapter authors and also with their respective section commentators. We take these disagreements as an accomplishment of this volume in that they are signs of a healthy culture of scholarly debate. Thus, we thank the contributors for engaging with sometimes conflicting ideas in a respectful and productive manner, both during the conference and in their texts. However, we also think that underneath these disagreements there are key commonalities cutting across the various contributions in a way that endows the collective conversations, and the volume as a whole, with a robust identity. We want to conclude this section by highlighting two of them.

First is the realization that the journey from what we have termed a focus on words to a focus on the intersection of words and things might entail a fundamental transformation in what it is that we might want to know about journalism and the news and how to go about knowing it. In other words, the cumulative outcome of the work undertaken over the past couple of decades seems to be less an addition of technology to a somewhat stable canon of accounts about words, people, and organizations, and more an invitation to rethink the canon altogether. This does not mean starting *de novo,* but it does mean the de-reification of analytical building blocks long taken for granted. Paraphrasing Geertz's reflections about the blurring of genres of explanation in the social sciences and humanities in the 1960s and 1970s, the collective work included in this volume might represent, "or will if it continues, a sea change in our notion not so much of what knowledge is but of what it is we want to know" (1980, 34).

Second, and related to the first, is a sense of destabilization about both the news as an object of inquiry and the processes of studying it. Are the stories that a company like Narrative Science produces about stock performance or baseball games the result of newsmaking or computer programming? Or both? Or is it an entirely new category better suited to apprehending what is novel and what is not novel about algorithmic authoring? Is Facebook blending editorial and distribution functions with such a force that it obliterates traditional enactments of both? Should we revisit the very foundations of the editorial and distribution processes to make sense of how news is curated, appropriated, and retransmitted in the contemporary media ecosystem? Similarly, how can one conduct an ethnography of algorithmic work—not the work of producing an algorithm, which is a related but different study—or an account of "personal influence" (Katz and Lazarsfeld 1955) via second-order editing and distribution in social media? Questions like these, which emerge naturally from the essays that populate this volume, would have been unimaginable at the tail end of the twentieth century. But after several decades of apparent intellectual stability in the study of news, these questions now open generative spaces of inquiry by "restoring to our silent and apparently immobile soil its rifts, its instability, its flaws; and it is the same ground that is once more stirring under our feet" (Foucault 1994 [1966], xxiv).

Overview of the Volume

This book is composed of four sections, each section in turn comprised of three or four chapters followed by a commentary that takes a broader look at the theme of each section. The book concludes with a postcript by Michael X. Delli Carpini. The three-layer structure of chapters, commentaries, and postscript was devised to increase the interconnection across the arguments advanced by the individual pieces. Sometimes the authors address each other directly, while other times the argument emerges in a less direct yet equally powerful fashion.

In the four chapters and the commentary that make up the first section, the authors primarily address theoretical and methodological issues. In chapter 1, Pablo J. Boczkowski and Eugenia Mitchelstein first summarize the current state of research on digital journalism. Then, on the basis of this overview, they suggest that while online journalism scholarship has drawn extensively from other fields, there has been comparatively less effort to articulate what this scholarship might mean for academics who belong to other fields and/or focus on other objects of inquiry. This, in turn, has had the effect of hampering larger attempts at theory building and also rendering it difficult to determine which findings are unique to journalism and which ones are actually signs of wider societal developments. Thus, they propose to adopt a two-way street of knowledge circulation, illustrating this proposal with examples of studying news (a) as information brokerage and (b) as a variety of social movement activism—two concepts readily familiar to sociologists but little remarked upon by mainstream journalism scholars.

In chapter 2, Rodney Benson provides another angle of intellectual criticism by mounting a frontal assault on what he calls the "new descriptivism" that, according to him, characterizes a sizeable part of scholarship on digital news. For Benson, the new descriptivism amounts to an obsession with thick description, single case studies, and irreducible contingencies of both journalistic structures and actions. What media scholars ought to be focusing on, he contends, are "the common 'causal mechanisms' (e.g., field logics, habitus, social struggles for distinction, discourses) that seem to be operative across a class of social phenomena (Steinmetz 2004) [in this case, the news industry.]" For Benson, the solution to this problem lies in research that can provide what he terms "good explanations" by emphasizing the comparative and causal instead of the local and descriptive.

In chapters 3 and 4, Victor Pickard and C. W. Anderson, in very different ways, build on the previous two chapters by proposing revisions to key elements of the theoretical and methodological foundations of scholarship on the news. In chapter 3, Pickard draws from a political economy perspective concerned with the journalism–technology intersection to contend that studies of news need to correct for three major "blindspots"—to address normative questions, to concern itself more with journalism policy, and to reengage with the economics of (now digital) news production. What would the comparative, empirical, causally directed, publicly engaged scholarship on the news that Benson advocates for look like? In no small part, it would look like the program Pickard outlines.

In chapter 4, Anderson posits an alternate agenda by embracing a more temporally sensitive, culturally minded approach. Anderson is interested in moving the study of newsrooms away from their local orientation both geographically and temporally. To study newsrooms—as well as the journalistic values embedded in those newsrooms—in a way that can claim some explanatory purchase, Anderson turns to history and the "long view" of technological change. To this end, he foregrounds science studies' basic insight that causation stems as much from the "temporal saturation" of a distributed network as it does from structural determinants. It is an approach that draws on the humanities and the social sciences by placing contemporary ethnographic insights within the extended temporal patterns that are the hallmarks of historical accounts.

In summing up and probing the merits of these different arguments, in his commentary William H. Dutton takes the time to ponder the values, as well as the drawbacks, of exploratory research, while emphasizing that journalism research is inevitably interdisciplinary. But although in the early phases of scholarship into online news it might have been sufficient to describe the nature of the technological innovations in online journalism, the sharpest current research goes beyond the description of technological innovation in specific ways. As interdisciplinary research, Dutton argues, digital journalism scholarship must not only meet the demands of the disciplines it draws on; it must also be interesting. Although "being interesting" might seem like a low bar, it is different from the criteria for research that sits more comfortably in traditional disciplines, where advancing a particular field-specific theory is often enough to be intellectually praiseworthy. Digital journalism research, Dutton contends, is almost inevitably problem-based research. This contention parallels Delli Carpini's concluding arguments in some interesting ways, as we shall see.

Each of the chapters in the second section problematizes a major concept that has been used to make sense of news and journalism: knowledge, (art) worlds, and silence. Rather than seizing onto the latest theoretical trends, the respective authors propose going back to the social science canon to find older resources that can help answer new questions about journalism today. In chapter 5, Rasmus Kleis Nielsen argues that we can get a better sense of general changes in journalism by considering it as a form of knowledge and returning to a distinction (first advanced by Chicago School sociologist Robert Park) between systematic "knowledge of" and the more unsystematic "acquaintance with." For Nielsen, thinking about the news this way serves several explicit and implicit goals. Explicitly, considering news as a form of knowing capable of providing both a deep familiarity with a subject matter and a more casual "acquaintance with" or "orientation towards" that material helps us realize that digital developments have not only increased the amount of bite-sized, headline-oriented news content but also seem to have made a greater amount of systematic news available as well. Implicitly, this analysis counteracts the widespread tendency among media scholars to dismiss the more unsystematic forms of news as somehow less worthy than deeper, more in-depth types of news content. On a normative level, both Park and Nielsen contend that more casual, "acquaintance with" forms of news themselves serve important public purposes.

Seth C. Lewis and Rodrigo Zamith draw on the work of Howie Becker and his theory of "art worlds" in chapter 6. They argue that the notion of an "art world"—the idea that a consideration of the sociological status of art must encompass not only the artist and the art work but also the entire ensemble of actors and things that go into making art—can be usefully applied to journalism as well, particularly toward an understanding of the relationship between journalism and technological artifacts. The authors suggest that the idea of "worlds"—artistic, journalistic, and so on—can act as convenient analytical shorthand through which to understand the increasingly deinstitutionalized, decentralized, and technologically oriented forms of journalism that are emerging in the twenty-first century.

In chapter 7, Mike Ananny draws on the metaphor of "journalistic whitespace" for rethinking speech and silence in the media. Ananny analogizes the use of newspaper whitespace (the parts of a newspaper where there is no text or other graphical material) to invoke various conceptions of silence, listening, reflection, or quiet contemplation—the journalistic values, in short, of not speaking. We do not often consider silence as a journalistic value. This is partly because of the negative rights framework often used to analyze the First Amendment and partly because we traditionally think of journalistic affordances encouraged by digital technologies as increasing the number of people who can speak, share, or participate. But there is a deeply rooted, alternate pathway of legal analysis to bolster this perspective on the importance of silence, and Ananny fully draws out its implications for the networked news ecosystem in the digital age.

In her commentary on part II, Zizi Papacharissi tackles the relationship between journalistic events, facts, and stories. The three chapters in part II each deal with the relationship between facts, values, and narrative stories in different ways, Papacharissi argues, and each of them makes it clear that while technology is not the only actor involved in shifting news values, it is certainly a central one. Emerging news values of instantaneity, ambience, and hybridity each help lend form to a genre of news that Papacharissi calls "affective," which can blur the distinction between experience, opinion, and emotion—yet another frame that gets problematized in the digital age.

The third section moves from theory and methods to the ontology of journalism as an object of inquiry. In different ways, each of these chapters grapples with questions of occupational culture and practice, and interrogates the role of the scholar vis-à-vis an occupation in a period of transition economically, ethically, and politically. In chapter 8, Natalie Jomini Stroud draws upon pragmatism to argue against a critical studies tradition that sees cooperation with the journalism industry as a betrayal of scholarly independence. She problematizes the divisions between business and democratic ends of journalism by showing that there are moments when the pursuit of profit produces positive democratic outcomes. Similarly, she notes that there are moments in which a scholar can be both critical and administrative within the same project. Thus, helping media organizations achieve their business and democratic ends does not necessarily entail giving up the critical dimension of the scholarly enterprise.

Picking up where Stroud leaves off, Matthew Hindman tackles in chapter 9 a question of major importance to both practicing journalists and democratic theorists: how do changing journalistic understandings of audience, facilitated by the increasing accuracy of digital metrics, affect journalism ethics? How does "what journalism ought to do" get affected by "what journalists know" of the people who consume their work? Distinguishing among a traditional ideal (journalists seek to produce content of maximum civic importance), a commercial ideal (journalism strives to maximize revenue and audience attention), and "ethical metric use," Hindman argues journalists ought to embrace the ethical use of metrics. He concludes by outlining a scholarly research agenda that can assist journalism in uncovering this ethical sweet spot.

Digital technology has thus had a disruptive effect on the news industry in both general ways (how ought journalists and the academy work together during this time of occupational turmoil) and specific ones (how does knowledge of the audience affect journalistic norms). A third practical development of importance has been the emergence of so-called "entrepreneurial journalism," and Jane B. Singer addresses this topic in chapter 10. Singer considers the numerous ways in which the culture and rhetoric of "entrepreneurial journalism" affect the practice and research of news in the digital age. With some important caveats, Singer is generally supportive of these enterprising endeavors, and she counsels her fellow researchers not to ignore them the same way scholars broadly overlooked the first wave of digital outlets in the late 1990s and 2000s.

W. Russell Neuman ties these three chapters together by exploring a second common theme they share, beyond their considerations of occupational culture and practice—the manner by which each grapples with the question of blurring boundaries. In *Common Knowledge* (Neuman, Just, and Crigler 1992), he and his fellow researchers were "drawn to ask the journalists what feedback on the stories they had received from viewers and readers and accordingly how their audience had reacted to the piece. The overwhelming response from our journalists was that they sought feedback from colleagues and editors but none from readers other than the comment of an occasional neighbor or airplane seatmate. In fact they seemed to think our question a bit strange" (x). Bringing up this earlier work, Neuman reminds us that the issues explored by Stroud, Hindman, and Singer are not new ones but are embedded within a long tradition of research concerned with the news media's relationship to the public sphere.

Each of the chapters in part IV addresses what their authors see as a major void in current scholarship on the news: inequality, history, and failure. In chapter 11, Sue Robinson makes the claim that recent scholarship on journalism has been particularly ill equipped to analyze widening power differentials that have been fostered in the digital media environment, particularly those that manifest themselves along the classic cleavages of race, class, gender, and sexual orientation. Part of the problem, she argues, is that journalism studies has yet to adequately integrate productive discussions about journalistic authority with wider questions of journalism's social power. To Robinson, doing so involves rethinking the deeper theoretical relationship between power and authority per se, and applying this theorizing to empirical questions surrounding the deployment of digital news. She closes her essay by proposing a number of frameworks that could orient such a process in the future.

In chapter 12 Mirjam Prenger and Mark Deuze, like Singer in chapter 10, discuss entrepreneurialism and innovation in journalism, but from a different perspective than Singer. Prenger and Deuze are concerned with historicizing the concept of innovation in the news business and detaching it from its contemporary technological connotation. They do this by analyzing the emerging genre of current affairs television in the 1950s and 1960s in the United States, the United Kingdom, and the Netherlands. By understanding the emergence of a novel genre at a specific moment in history Prenger and Deuze not only historicize innovation in the news but also de-Americanize the concept of journalistic innovation. They also take steps toward a more critical analysis of entrepreneurialism. The two chapters on entrepreneurialism and innovation contained in this volume, when read back to back, nicely illustrate the different ways that the scholarship on the news might be mobilized to tackle a central empirical question.

Karin Wahl-Jorgensen, building off Robinson's call to more explicitly articulate the relationship between authority, power, and exclusion in digital news, argues in chapter 13 that studies of journalism need to do a better job investigating the marginal, the small-scale, and indeed the many failed projects that lay scattered in the wake of journalism's transformation over the past two decades. While failure has not been entirely ignored, the dominant discourse in much of journalism research—as well as in society at large—is one that tends to place undue emphasis on the large, the technologically sophisticated, and the successful. This chapter, coming near the end of the volume, also brings the book full circle by echoing Boczkowski and Mitchelstein's early arguments to turn the study of journalism into a two-way street through analysis that can be integrated into a larger theory-building exercise. Specifically, Wahl-Jorgensen notes that sociologists, organizational theorists, and political scientists have a long track record of studying the failure of everything from complex technical systems to banks. There is no reason, she contends, why scholarship on journalism cannot contribute directly to this larger intellectual enterprise.

Michael Schudson discusses each of these three chapters by weaving the metaphor of the "journalism studies tree," a field of scholarly investigation that barely existed a few decades ago and somehow grew large and healthy while the gardeners—sociologists, political theorists, anthropologists, and others—were mostly busy elsewhere. We as scholars should be careful, Schudson seems to be telling us, to tend and nurture this tree; we should also not take its unexpected existence for granted.

Coming on the heels of Schudson's essay, Michael X. Delli Carpini draws the volume to a powerful close in his postscript by reviewing the empirical evidence about what has actually happened in journalism over the past two decades, filtered through the classic "Five W's and an H" of journalism itself. The "who, what, where, when, why, and how" of digital news production and consumption have all shifted in recent years, Delli Carpini contends. He closes his essay with a powerful plea for scholarship on journalism that makes a difference in the world, focused on fostering a journalism that assists ordinary citizens in better understanding, navigating, and grappling with the complex reality around them.

I Revisiting Theoretical and Methodological Debates

1 Scholarship on Online Journalism: Roads Traveled and Pathways Ahead

Pablo J. Boczkowski and Eugenia Mitchelstein

How has the study of online journalism evolved over the past 25 years? What have been the main topics analyzed, and the theoretical and methodological strategies pursued to examine them? In this article we assess this scholarship and conclude that there has been remarkable growth in the volume of activity, the diversity of the topics examined, and the array of conceptual and methodological resources utilized. We also find that the dominant approach to conceptual and methodological work has been the incorporation of suitable ideas originating from across the social and behavioral sciences, and from the humanities. However, researchers have not engaged to the same extent with scholars from those various fields to undertake cross-disciplinary conceptual developments and compare relevant findings. In a nutshell, the connection has been mostly a one-way street. This has led to two limitations. First, it has made it difficult to ascertain which empirical trends are unique to online news and which ones might be shared across other domains of digital culture. Second, it has made it difficult to undertake conceptual exchanges with relevant theory-building endeavors in other areas of the humanities and behavioral and social sciences.

On the basis of our assessment of the main scholarly contributions summarized in the next section, we devote the following one to a proposal to alleviate the limitations of this one-way strategy of conceptual development. This proposal centers on building intellectual bridges that would allow the study of online journalism to engage in conversations with relevant scholarly partners who study different but related topics. To illustrate the potential of this strategy, we briefly describe two possible bridges built on defining features of journalism that have become particularly salient in the contemporary media environment. The first bridge focuses on the links between journalism and other instances of information mediation and brokerage. The second bridge elaborates on public service journalism online as a case of activism by drawing upon knowledge developed within the social movements literature. We conclude the chapter reflecting on how this bridging strategy might contribute to the development of studies of online news in the future.

The Road Traveled So Far

In this section we take stock of over 25 years of scholarship on online journalism by centering on five key topics: historical context and market environment; processes of innovation; practices of journalism; challenges to established professional dynamics; and relationships with audiences and the role of user-generated content.

Historical Context and Market Environment

Relatively limited attention has been paid to the history of online news. Some authors have called for more research (Barnhurst 2012; Mitchelstein and Boczkowski 2009), because "the application of a strong historical perspective to scholarship on online news is necessary to gauge the depth of any changes from and the strength of continuity with print and broadcast news" (Cawley 2012, 219). The limited historical scholarship has mostly focused on how mainstream media set up online operations due to concern from competition by new online players (Allan 2006; Dennis 2006; Nguyen 2008). Boczkowski (2004) argues that mainstream media innovation cultures reacted to moves by new entrants rather than pursued news paths; defended their turf rather than sought to conquer new ones; and centered on short-term profit making rather than invested in opportunities that might be fruitful in a more distant future.

Research indicates that few online newspapers were making a profit by the early first decade on the twenty-first century (Boczkowski 2004; Nguyen 2013; Singer 2003a). Some authors contend that lack of profitability may be explained by the absence of an adequate business model (Ahlers 2006; Cawley 2008; Chyi 2005; Garrison 2005; Greer 2004; Pavlik 2001). Scholarship also suggests that the decision not to charge for content led to audiences that are used to getting free online news (Bruns 2011; Chyi and Yang 2009; Geidner and D'Arcy 2013; Gentzkow 2007). Thus, research has found that online news operations have relied mostly on advertising revenue, at least until the end of the first decade of the twenty-first century (Bustamante 2004; Chan-Olmsted and Ha 2003; Gentzkow 2007; Herbert and Thurman 2007). Reliance on advertising has been related to two main concerns. First, news sites attract a small part of the online advertising expenditures (Chyi 2005; Filloux 2008; Garrison 2005; Greer 2004; Schudson

2010). Goyanes notes that news operations "are tiny players in a market dominated by large technology companies (Google, Amazon, Yahoo, etc.), where the top five account for 64 percent of all US digital advertising spending in 2012" (2014, 742). Second, scholars have noted an increased blurring of the wall between commercial and editorial content (Cassidy 2005; Cohen 2002; Hutchins 2007; Raviola and Norbäck 2013; Singer 2003a). Coddington finds that even journalists "who were defending the boundary felt the need to acknowledge that blurring the boundary could be a crucial element in maintaining news organizations' financial viability" (2015, 74).

However, during the past five years, more online news operations have experimented with paid subscription models (Geidner and D'Arcy 2013; Herbert and Thurman 2007; Pickard 2014). According to the Project for Excellence in Journalism, "paywalls helped the newspaper industry raise circulation revenue by 5 percent in 2012, the first gain in subscription revenue since 2003" (2014). Scholars have also looked into other revenue sources such as subsidies and philanthropy (Jian and Usher 2014; Pavlik 2013; Pickard 2014). However, donations account for less than one percentage point of all income (Project for Excellence in Journalism, 2014) and subsidies offer no major support for online-only news operations (Nielsen, 2014). Konieczna and Robinson examined 50 nonprofit news organizations and concluded that "many of these organizations (…) work to reconceptualize the industry for citizens, but depend upon a level of funding that might not be viable in the long term" (2014, 968).

Processes of Innovation

While some initial accounts of innovation in online journalism were technologically deterministic (Steensen 2011), most scholarship has focused on the social shaping of innovation (Barnhurst 2012; Boczkowski 2004; Conboy and Steel 2008; Deuze 2007; Hass 2005; Karlsson and Clerwall 2012). Studies have looked at the role of organizational and institutional contexts (Ashuri 2014; Boczkowski and Ferris 2005; Klinenberg 2005; Nielsen 2014; Singer 2004; Thurman 2008). Boczkowski (2010) found that two different units within a single online newsroom varied greatly in the rate of imitation in news production as a result of differences in their temporal and spatial organization, sourcing practices, and use of communication tools.

Scholars have also focused on how established journalistic operations have not fully realized the potential of new technologies (Cawley 2012; Deuze 2003; Garrison 2005; Larsson 2012; Quinn 2005; Steensen 2009). In his analysis of news sites up to 2010, Barnhurst notes that they "preferred content interactivity to interpersonal interactivity, continuing a long history of resisting innovation" (2012, 791). Scholarship has shown that these operations were often more reluctant to innovate than newcomers (Anderson 2013; Pavlik 2013; Ryfe 2012).

Researchers have noted that news sites have incorporated social networks into their sourcing and publication routines (Bruns 2011; Hong 2012; Revers 2014; Singer 2014). Dutton and colleagues look into how journalists use "social media as source material, filtering the best for a mass audience," thus "developing new skills and roles for curated or 'networked journalism' in the process" (Newman, Dutton, and Blank 2012, 15). However, Annany examines the policies of eight news operations and finds that they "are focused on mitigating the risks of their staff misusing social media, (and) their audiences misunderstanding social media activity" (2014, 948), which suggest a certain conservatism even when innovation in editorial work is pursued.

Practices of Journalism

Scholarship has also focused on four aspects of journalists' practices online: modifications in editorial workflow; alterations in news-gathering practices; acceleration of temporal patterns of content production; and the convergence of print, broadcast, and online operations.

Online news has increased the pressure on reporters to perform multiple tasks and combine several media formats (Cawley 2008; Deuze 2004; Hermans and Vergeer 2009; Karlsson and Clerwall 2012; Lawson-Borders 2006; Ursell 2001). Van Der Haak, Parks, and Castells argue that as "multimedia reporting and publishing have become the norm without comparable investments in training or new staff ... working conditions for many journalists have deteriorated, and their workloads have increased" (2012, 2924). Information-gathering practices have been altered (Boyer 2011; Millen and Dray 2000; O'Sullivan and Heinonen 2008; Pavlik 2001; Salwen 2005). For instance, Doudaki and Spyridou find that an increase in "secondary reporting, and lack of human and material resources lead to more copy–paste journalism and less original reporting" (2013, 918).

Researchers have also noted that online journalism has contributed to the collapse of the daily news cycle into continuous publishing (García 2008; Karlsson and Strömbäck 2010; Raviola and Norbäck 2013; Saltzis 2012; Williams and Delli Carpini 2011). This temporal acceleration has deepened over time. Boczkowski conducted research on the online operations of the *New York Times* in 1998 and found that editors posted stories toward the end of the day, mirroring the publishing cycle of its print parent company (2004). Boczkowski and De Santos examined Argentine online news sites the following decade and found that "an intense pattern of constant publication ha(d) ... become a dominant trend among Argentina's leading online papers" (2007, 171). Tenenboim-Weinblatt and Neiger analyzed two Israeli newspapers in 2012 and concluded that journalists are expected to produce more news in less time, and to adapt to a culture of immediacy and speed (2014). Scholarship has argued that increased temporal acceleration endangers fact-checking and in-depth reporting (Boczkowski 2010; Doudaki and Spyridou 2013; Quandt 2008).

Scholars have also examined media convergence through either mergers of different companies (Boczkowski and Ferris 2005; Deuze 2004; Klinenberg 2005; Lawson-Borders 2006) or the incorporation of multimedia content on news sites (Thornton and Keith 2009; Vobič 2011). The uptake of convergence has not been without tension due to different newsroom cultures (Dickinson, Matthews, and Saltzis 2013; Singer 2004). In her study of a newly "converged" newsroom, Robinson finds that a new power hierarchy privileges those with technological savvy and a willingness to work in a 24–7 digital culture and "force(s) a redefinition of what it means to be a 'connected,' 'successful' journalist" (2011, 1124).

Challenges to Established Professional Dynamics

The long-standing debate about who is a reporter is exacerbated by challenges to journalism´s occupational jurisdiction (Boczkowski 2004; Deuze 2007; Dutton 2007; Sousa 2006; Waisbord 2013). Picard contends that "the transformation under way is not only altering the methods of news production and distribution, but the functions of journalism itself" as "the traditional functions of bearing witness, holding to account opinion leadership, and shaming are no longer provided solely by the news media" (2014, 493).

Although early research pointed toward a "normalization" of journalistic work online (Quandt et al. 2006; Singer 2003b, 2009), the most recent scholarship suggests that the crisis in journalism and in mainstream media (Mancini 2013; Siles and Boczkowski 2012) is linked to alterations in key values of reportorial work, from a logic of control and objectivity (Tuchman 1972) to an ethics of transparency and openness (Domingo et al. 2015; Lewis and Usher 2014; Pavlik 2013; Robinson and DeShano 2011; Vergeer 2014). Lewis suggests that journalism's "ideological commitment to control, rooted in an institutional instinct toward protecting legitimacy and boundaries, may be giving way to a hybrid logic of adaptability and openness" (2012, 851). However, Anderson contends that "as long as the primary work task at these institutions (...) remains reporting ... most newswork in most places most of the time will struggle to be open and will struggle to be networked" (2013, 161).

The Relationship with Audiences and the Role of User-Generated Content

The challenge to journalistic control is signaled by the crisis of gatekeeping (White 1949) as the principal marker of journalism's occupational jurisdiction (Bro and Wallberg 2015; Gillmor 2004; Lowrey 2006; McCoy 2001; Ihlebæk and Krumsvik 2014; Robinson 2007). Scholarship has identified two main challenges to gatekeeping.

The first challenge is that of the audience actively engaging in content production (Deuze and Dimoudi 2002; Gurevitch et al. 2009; Wall 2005). Some researchers propose that citizen journalism emerged to fill a void left by mainstream media organizations, particularly at the local level (Deuze, Bruns, and Neuberger 2007; Robinson and DeShano 2011; Schudson 2010). The second challenge looks into the coincidences between user-generated content and partisan news production as the latter was undertaken before the establishment of journalism as an occupation (Anderson 2011; Hendrickson 2007; Lowrey 2006; Nerone and Barnhurst 2001; Russell 2001; Singer 2006).

Some authors have identified a change from a one-way model of communication to a dialogical kind of journalism (Benkler 2006; Boczkowski 2004; Dutton 2007; McCoy 2001; Tremayne 2007). However, others have noted that it might be early to anticipate the end of gatekeeping (Allan 2006; Hujanen and Pietikainen 2004; Schiffer, 2006). Scholars have found that news organizations are not often enthusiastic about allowing audience members to become coauthors of content (Domingo et al. 2008; Thorsen 2013; Thurman 2008; Williams et al. 2010).

Other researchers note that most members of the audience are not interested in either producing or accessing user-generated content (Boczkowski 2010; Chung and Nah 2009; Goode 2009; Thurman 2008). Boczkowski and Mitchelstein (2013) examined both the home pages and the most read stories at six American mainstream news sites and found that, whereas only 1 in 100 stories at the top of a site's home page were user-generated content, only 1 in 1,000 stories of the most viewed on these sites were from that kind of content. Furthermore, when members of the public participate in content creation, they do not engage with public affairs topics but dwell mostly on personal and entertainment topics (Gans 2007; Ornebring 2008; Trammell et al. 2006).

Differences between audiences conducting "random acts of journalism" (Lasica 2003) and reporters' work are often due to organizational factors (Carlson 2007; Lowrey 2006; Reich 2008; Rutigliano 2007). Citizen journalists lack formal training and access to sources which may lead them to getting their facts wrong (Pavlik 2013) or simply commenting on information originated in mainstream media (Annany 2014; Cawley 2012; Domingo et al. 2015; Grafström and Windell 2012). However, Meraz examines intermedia agenda setting between the blogosphere and mainstream news operations in the US and finds a "growing influence of the progressive political blogosphere in setting other media agendas, while resisting traditional media agenda setting" (2011, 187).

The second challenge to the gatekeeping function of journalists arises from what Napoli calls the "rationalization of audience understanding" (2010, 31). Scholars have examined the importance of audience metrics in defining what event should be covered and the placement of that coverage in news sites (Anderson 2011; Ashuri 2014; Barnhurst 2012; Coddington 2015; MacGregor 2007; Nguyen 2013; Picone 2011). In a survey of more than 300 editors, 58 percent said they monitor web traffic to scrutinize readers' behavior, and 31 percent use online metrics to plan content production (Vu, 2014). This change in the relative influence of editors and audiences suggest that we might be moving toward an environment in which, rather than media telling people what issues to think about, "people tell the media [what] they want to think about" (Chaffee and Metzger 2001, 375). Yet, Boczkowski and Mitchelstein find a gap between editors' and audiences' preferences in 20 leading mainstream news sites from seven countries, with consumers selecting significantly fewer public affairs stories than the ones journalists choose to publish (2013). This suggests news workers are still guided primarily by occupational rather than purely market logics.

Pathways Ahead

As the previous section shows, the past two and a half decades have seen a remarkable development of online journalism scholarship. Most of this scholarship has drawn from methodological and theoretical resources developed within and outside the study of media and communication. It has also framed its contributions in an inward-looking fashion, by targeting the online news research community almost exclusively, rather than casting a wider net of academic audiences. As noted in the introduction, this one-way strategy presents empirical and conceptual limitations. To help overcome them, in this section we outline two possible bridges that might enable analysts to turn the study of online journalism into a two-way street. These two bridges focus on two defining features of journalism as an occupation that have become particularly salient in the contemporary media environment: it is a mode of work centered on mediation and brokerage of information, and the outcomes of this work can contribute toward effecting major social and political change.

From a network analytic perspective, journalists occupy a bridging position between two worlds that are structurally unconnected: the sources of news and the members of the audience. Reporters in particular, and news media in general, act as bridges in that they usually are the only path between those two clusters for most audience members (Granovetter 1973). Because they span the distance between otherwise separate clusters of connections, they can be characterized as bridging a "structural hole" (Burt 1992). Network connections are instrumental to the diffusion of information and the generation of social capital (Granovetter 1973; Burt 1992; Burt et al. 2013). By controlling the information flow between newsmakers and their audiences, journalists and their media organizations derive benefits in the form of money, influence, prestige, and further access to information. This, in turn, strengthens their position as *tertius gaudens* (Obstfeld 2005; Simmel 1950), or the third who benefits from brokering between other players.

Journalists and their organizations are not the only workers and companies which profit from occupying a bridging position between two otherwise unconnected social worlds. Scholarship on a wide array of occupations and organizations, from stock brokers to scientists, and from boards of directors to research and development units, has yielded valuable knowledge that can inform future studies of online news. Drawing upon this knowledge could, in turn, position these studies for further contributions to wider conversations about structural holes and other network analytic issues. One possible topic is the issue of "bridge decay" (Burt 2002).

Because bridging ties require more effort to maintain, they show faster rates of decay over time than other kinds of relationships in a social network. Despite the fact that this is a well-established phenomenon, relatively limited research exists on the factors that affect the dynamics of bridge decay. "Despite the known performance implications of bridges, we know almost nothing about their origins and decay" (Burt 2002, 342). The rate of bridge decay in news production is considerably slower than for other arenas of work due to a basic structural factor: it is difficult for most newsmakers to address the public directly, and, vice-versa, it is virtually impossible for members of the audience to have direct contact with them. Thus, inquiry into news production as information brokerage could potentially shed light on the role of structural determinants in general dynamics of bridge decay.

We believe this line of inquiry could be particularly fruitful because current technological and organizational transformations have weakened the privileged position of journalists and media organizations as bridges between newsmakers and the public, thus making visible the relative contribution of various factors that might become more difficult to ascertain during periods of greater stability. Up until the early 1990s, making appeals directly to citizens while bypassing intermediaries was a relatively rare strategy, available only to important figures such as the president of a country (Kernell 1997). However, since the mid-1990s, technology has increased for a greater number of actors the ability to reach large segments of the public. This has not been a playing field that all actors have equally taken advantage of, but the cumulative effect appears to have been an increase in the rate of decay of journalists' bridging position. Traditional media and professional journalists are no longer the only news brokers. There are more actors who compete to occupy a bridging role. Members of the audience may migrate some or all of their information-seeking practices to them. Across all content categories, sites of news sources and those that crowdsource information have become formidable competitors of journalistic organizations. For instance, in January 2015, the NFL Internet Group—which focuses on a single sport—virtually tied for site traffic with the New York Times Digital (comScore 2015). Also that month, the traffic of TripAdvisor—solely devoted to travel—was almost half of all Gannett sites, including *USA Today* (comScore 2015).

These transformations in the flow of information have created an interesting laboratory to understand the dynamics of bridge decay, which in turn should open a window for two-way exchanges of ideas. On the one hand, for studies of journalism, network analytic perspectives might provide a fruitful conceptual vocabulary and a set of empirical tools to track down transformations in sourcing, reporting, and publishing practices. On the other hand, for studies of networks, these transformations provide a potentially fertile setting to account for the variable role of technological and organizational factors in the dynamics of bridge decay. Most of the work on structural holes and bridges tends to assume scarcity of information and opacity of the interactions within clusters to outsiders. In that context, bridge decay has profound implications for the stability of social capital (Burt 2002). However, the technological and organizational changes that have marked the recent development of online news challenge the twin assumptions of scarcity and opacity, and invite us to rethink the competitive advantage of bridges in social structures in which multiple, sometimes redundant relationships span structural holes.

Journalistic work has long been seen as a possible driver of major social and political change. We propose that the public service mission of journalism could fruitfully be understood as a case of social activism, aimed at effecting major social and political transformations. To better understand this matter, scholarship could draw from the social movements literature, and in turn could help contribute to some of its contemporary developments.

Scholarship on public service journalism tends to present a dire picture of its present and its future due to structural factors such as funding, staffing, and the commercial orientation of most news media organizations (Beam et al. 2009; Downie and Schudson 2009; McNair 2000). The subsections devoted in this paper to the economic travails of news operations, the jurisdictional challenges to journalism as an occupation, and the potential editorial consequences related to access to granular and instantaneous information on audiences' preferences offer some perspectives on this critical juncture. These transformations, in turn, challenge three fundamental roles of the news media in democratic polities: acting as watchdogs of the powerful in the public and private sectors, setting a common agenda for the citizenry, and serving as a space for public deliberation (Boczkowski and Mitchelstein 2013).

These roles are tied to the notion of information as a public good, in that it is nonexcludable and nonrivalrous, and thus potentially obtained through collective action (Olson 1965). With information serving a relatively large group, problems of collective action and free riding in the production and distribution of information resonate with similar problems encountered by social movements. Tilly defines social movements as people making "collective claims to power holders in the name of a population living under the jurisdiction of those power holders by means of repeated public displays of that population's numbers, commitment, unity, and worthiness" (1999, 257). According to this author, social movements are better understood in the "relations of power, conflict, and solidarity" (1984, 747) they maintain with other actors.

The market structure and social organization of the news media during most of the twentieth century (limited competition; strong jurisdictional space of journalists; and low bottom-line risk associated with ignoring audiences' preferences) enabled leading news organizations to perform journalism's public service mission in a vibrant manner. However, as those conditions have gradually changed over the past few decades (increased competition; weakening of the jurisdictional space of journalism; and high bottom-line risk linked to ignoring audiences' preferences), providing the public goods of this mission has become more difficult.

In this context, several nonprofit news organizations—such as ProPublica, the International Reporting Program, and the Huffington Post Investigative Fund—have appeared to fill that perceived void (Carvajal et al. 2012; Lewis 2007). They have flourished in the past decade, and partnered with mainstream news organizations to publish their articles. Even though many are besieged by problems of fundraising and instability (Mitchell et al. 2013), some have managed to produce important investigative reporting at the national and local level "to stimulate positive change" (ProPublica 2015). They represent a case of journalism as activism for a social good and a possible solution for the collective action problem of how this outcome could be funded and distributed.

The focus on (nonprofit) news organizations appears to be a remnant from an era in which formal organizations were a fundamental condition of investigative reporting. But loosely networked informal organizations such as WikiLeaks can be highly consequential in the production of public service news. This parallels broader transformations in the world of social movements, which have also been transitioning from the "more centralized nongovernmental organizations to direct action networks," facilitated by "social technologies that combine online and offline relationship building" (Bennett, 2005, 203).

The challenges to established, for-profit news organizations, the rise of their nonprofit counterparts, and the greater role of informal organizations collaborating loosely in the search and dissemination of public affairs information offers a unique window to study the transition from a large and formal organizational model to a smaller, networked, and informal one, both in the production of public service journalism and in social activism more generally. This transition is not without obstacles. In the previous model collective identities—being a journalist from a mainstream news organization or a member of a known NGO—provided actors with resources as certifications of credibility. In the new model identities are more fluid, motives are tested repeatedly, and obtaining financing is an ongoing struggle. However, this new action model could also serve to empower individuals and promote social change from a micro to a macro scale (Anderson 2013). Either way, the production of accounts about contemporary challenges to fulfilling journalism's public service mission could provide a fruitful ground for conceptual renewal in studies of journalism and social movements.

Concluding Remarks

We began this paper by critically examining some of the main areas of development in the scholarship on online news over the past 25 years. Our analysis shows major growth coupled with a one-way approach toward conceptual and methodological development. We have argued that this approach presents empirical and theoretical limitations and therefore have proposed a two-way approach that could contribute toward overcoming these limitations. Our proposal is based on a cosmopolitan stance of knowledge creation that is at odds with the trend toward specialization in the academy. An instance of this specialization that is close to the study of online news is the emergence and institutionalization of journalism studies as a domain of inquiry. This can be seen in the launch of new journals, the creation of dedicated units within learned societies, the organization of specialized conferences, and an overall attitude among many scholars of identifying themselves and their work with this new subfield.

Like all such processes of academic specialization in different fields of knowledge, this institutionalization has had the benefit of defining and strengthening an intellectual core. But it has also come at the potential expense of descriptive precision, heuristic power, and theoretical ambition. How do we know which trends in online news are unique to it and which ones are shared with other aspects of digital culture, information labor, media consumption, and so on, if our findings are not put in the context of other relevant studies? How can we ascertain which among our findings are the most fruitful explananda and which variables are the appropriate combinations of explanans if our explanations are not examined from both insider and outsider perspectives? How can we engage in innovative and robust middle-range theorizing without fluid and sustained exchanges of findings and explanations with scholars dealing with different yet related objects of study?

We think the time is right to turn the connections of scholarship with online news with other domains of inquiry into two-way streets. In addition to the empirical, heuristic, and theoretical gains, this would also enable a process of reintegration of the study of news and journalism into larger disciplines, some of the very same disciplines from which it once emerged (Zelizer 2004). Thus, perhaps the pathway forward would actually mean a return to the beginning. Time, after all, does sometimes move in a circular way.

2 From Heterogeneity to Differentiation: Searching for a Good Explanation in a New Descriptivist Era

Rodney Benson

As our digital media systems are becoming ever more complex, our capacity to criticize, map, and explain them risks becoming progressively poorer. No theory is perfect, of course. A theory can be overly simplistic, decontextualized, reductionist, or mechanistic. A lot of energy is expended criticizing this kind of "mainstream," often highly quantitative research, and these are legitimate concerns. But an even greater problem, I would argue, is when complexity is taken as an end in itself, preventing any gesture toward systematic critique, patterned observation, and generalizable explanation. This is precisely what is happening today, with the rise of what I call the "new descriptivism."[1]

This "new descriptivist" media research is an eclectic collection of approaches from Bruno Latour's actor–network theory to Manuel Castell's network society to Jeffrey Alexander's cultural pragmatics, among others.[2] What these diverse approaches have in common is a close eye for empirical detail, all to the good, but an indifference or hostility to questions of normativity, variation, and causality. While the new descriptivists offer unique perspectives and insights, the questions they ignore are at least as if not more important and deserve more rather than less attention by scholars of digital journalism. In particular, I argue that in order to understand our increasingly socially fragmented media systems we need theories of systematic differentiation rather than chaotic heterogeneity.

In the discussion that follows, I discuss legacy as well as "new" media. If in fact, we live today with a "hybrid" media system (Chadwick 2013), it is often the case that few media outlets can be deemed either "old" or "new" but instead are mixes of the two.

Descriptivism, Old and New

How prevalent is description in news media research? In their comprehensive survey of online news research after 2000, Mitchelstein and Boczkowski

(2009) convey the breadth and diversity of a research field in formation. What is striking is the overwhelming descriptive focus of the research: "historical context and market environment, the process of innovation, alterations in journalistic practices, challenges to established professional dynamics, and the role of user-generated content" (562). The review article only mentions the word "explain" twice. In the first reference (569), the authors suggest that Klinenberg's (2005) ethnographic findings of extreme time pressures in online newsrooms may help explain Cassidy's (2007) survey results showing skepticism by online journalists about the "credibility" of online news. In the second reference, Mitchelstein and Boczkowski call for cross-national comparative research that would both compare "practice and professional dynamics" across locales and "explain the sources and patterns of variance" (577).

Updating this analysis and providing a very modest check on its contemporary generalizability, I categorized all the articles about journalism in the first summer 2015 issues of the journals *Journalism, Journalism Studies, International Journal of Press/Politics*, and *Journal of Communication*. Of a total of 23 articles, 14 were descriptive whereas 9 were explanatory. In order to be coded as explanatory, the article needed to only make at least one explanatory claim (whether in the hypotheses, findings, discussion, or suggestions for future research): a majority of these articles offered minimal explanation.

Now why is the prevalence of descriptive—or weakly explanatory—research a problem? Descriptivism has a long and often prestigious lineage in the humanities and social sciences, exemplified by legendary scholars such as Clifford Geertz (1973), Herbert Gans (1979) and Howard Becker (2008).[3]

Geertz (1973) provides the catchall justification for any attempt to "understand" rather than "explain," calling for interpretive "thick description" of the process of meaning construction in any given social milieu. Becker (2008) offers concepts for guiding research (cooperation, collective action, conventions, etc.), but his overall approach avoids closure. A "world" of cultural production is an "extendable, open space" and "an ensemble of people (including various support personnel) who do something together" in ways that "are never entirely predictable" (374, 379) (see also Dickinson 2008 for discussion in relation to journalism). Gans, of course, is well known for his distaste for grand theory and deductivism: the important thing is to get into the field and start observing.

Nikki Usher's (2014) excellent in-depth profile of the *New York Times* in the midst of its online transition fits squarely into this inductive mold (she dedicates the book to Gans). She takes as her starting point the classic newsroom studies of the 1970s and 1980s and asks what has changed. Clearly, a lot has changed, including the shift toward a more laissez-faire economic order as well as the rise of the Internet: she places the emphasis on the latter. Usher identifies the rise of three news values that the digital environment has either intensified (immediacy) or introduced (interactivity, participation) into the newsroom. Usefully clarifying terms that are often used interchangeably, Usher conceptualizes interactivity as the user–computer interaction via various forms of online multimedia and defines participation as reporters' use of social media to engage or collaborate with their audiences (or as is most common, to publicize their stories to their audiences).

One of her key discoveries is that despite a myriad of pressures pushing the *Times* toward digital, the print side still dominates. Just to take one example, the amount of time and energy that goes into deciding what should go on the print "Page One" still far exceeds that of the much more heavily trafficked web home page. This finding of "legacy" inertia echoes findings from other recent US newsroom ethnographies, notably Anderson (2013) and Ryfe (2012), with the notable difference that Usher is looking at a successful newsroom that has made a concerted digital effort. So what she finds is not a wholesale rejection of digital imperatives but rather a complex, still unfolding rise of new hybrid news values shaped by the "clash and combination of external, internal, professional, and normative pressures on journalists" (152) with the direction of the "causal arrow" difficult to discern (21). As is generally the case with a single case ethnography, the "how" effectively becomes the "why."[4]

Now, one could argue that high quality descriptive research of this type is entirely appropriate for our age, and for analyzing digital journalism. In a moment of rapid change and complexity, it makes sense that our first move should be to get the lay of the land. Detailed description *should* come first, before explanation or critique.

Theoretical but antinormative and anti-explanatory new descriptivists build on this long tradition both by widening and deepening description. They widen description by calling attention to dimensions of digital journalism previously underplayed (such as technology, networks, and culture) as well as by refraining from making a priori normative judgments that could bias the analysis. They deepen description by comprehensively documenting the agents and action that go into constructing and maintaining a particular case. I focus here on actor–network theory, network society theory, and cultural pragmatics, but many of my comments also pertain to journalism research influenced by various poststructuralist currents or classic descriptivists such as Becker and Geertz.[5]

Actor–network theory was the focus of the first three articles of a recent special theory issue of *Digital Journalism* (Domingo, Masip, and Costera Meijer 2015; Lewis and Westlund 2015; Primo and Zago 2015) and has been highlighted in a number of prominent studies of digital news media and political communication (see, e.g., Anderson 2013; Anderson and Kreiss 2013; Braun 2013, 2015; Hemmingway 2008; Nielsen 2012a, 2012b; Plesner 2009; Turner 2005). ANT's specific theoretical vocabulary of actant, network, mediator, intermediary, black box, translation, and the like (Callon 1986; Latour 1991, 2005; Law 2009) all are oriented toward a single purpose: encouraging us to see the world as highly contingent, unstable assemblages of human and non-human elements. These terms, deployed in detailed descriptive (usually) single case study accounts, assure some minimal "commensurability between accounts" (Sayes 2014, 142), though as I discuss below the accumulation of generalizable knowledge does not seem to be on the ANT agenda.

Most notably, ANT adds to sociocultural approaches by calling attention to non-humans, generally technical objects, as exerting agency in shaping outcomes. As Primo and Zago (2015, 39) insist, these material artifacts are not secondary but just as important "as any other actant in the ongoing process of news production, circulation, and consumption." Non-human objects have agency not by exhibiting consciousness but in the minimal sense that they can make "a difference in another entity or in a network" (Sayes 2014, 141). Similar to Becker's insistence on including everyone who helps produce the work of art, actor–network theory takes comprehensiveness as a point of principle: "Let all the actants be recognized, human and non-humans, their agencies, the associations they engage in, the traces they leave" (Primo and Zago, 43). "Widen[ing] the observation field, letting us see what was not before identified" is justified as a way to open up "new and innovative conclusions" (ibid., 49). So, what are these new and innovative conclusions in relation to digital journalism?

Domingo, Masip, and Meijer (2015) summarize some of the notable findings of previous studies: for example, Schmitz, Weiss, and Domingo (2010) studied content management systems and found that practices "prescribed" by the software clashed with journalists' preexisting practices; Hemmingway (2008) showed how personal digital production equipment at a regional BBC office increased flexibility in dealing with deadlines and thus "enabled" journalists to produce more human interest stories. In another oft-cited study, Plesner (2009) finds that emailing, googling, and phoning have become "seamless" parts of journalistic work practices. Nielsen (2012a) credits ANT for providing the kind of open-ended inquiry needed to notice the important role played by technologists as well as journalists and business managers in shaping the development of online news operations. Anderson and Kreiss (2013) show how sociotechnical objects like political maps and content management systems are "actants" that "hold networks in place" in particular ways, shaping a particular course of action.

There is something missing in almost all of these ANT accounts, in short, a clear answer to the reasonable question: why should we care? On the principle that it is important to know more rather than less about the world in which we live, we can thank ANT for making us more aware of the objects in our midst and the roles they play in greasing and maintaining our social relations. But the promise of many ANT studies to show how some seemingly "mundane" object makes a "highly consequential" difference (Anderson and Kreiss 2013, 366–367) is rarely made good: at the end of the analysis, the object, it turns out, is just as mundane as it seemed at the beginning. The interesting questions lie elsewhere, where ANT (in its pure version) does not go. For instance, in Anderson and Kreiss's study, the significance of the Obama campaigns' use of political maps to facilitate an "instrumental" model of delivering votes only really becomes clear in the road not taken. Why didn't the campaign pursue an alternative "organizing model" oriented toward creating "shared narratives, articulating moral claims, and generating emotional commitments among voters" (374), as advocated by some activists? Answering this question requires a non-ANT explanatory theory that can point to why some actors have more power and influence than others.

Manuel Castells' network society theory (1996, 2009) is similar to ANT in that it calls attention to the "networked" character of digital news media (see also Van Der Haak, Parks, and Castells 2012), although as a theory it has a much less developed account of its epistemology, ontology, and orientation toward normativity. Whereas actor-networks are seen as the building blocks of society and predate digital network technologies, Castells is clearly focused on the changes wrought by the Internet. Network Society theory offers an historical account of how a new kind of social configuration has emerged since the early 1980s, brought about by transformations in the world economy, the growth of new identity-based social movements, and the rise of the Internet, a uniquely decentered system of global communications. NS theory is also highly empirical and descriptive but has a more restrained sense than ANT of what is worth observing. Theorists of "network journalism" (see also Bardoel and Deuze 2001; Heinrich 2011; Russell 2011) are also much more willing to acknowledge the existence of power relations and often speak of the important role journalism plays in promoting democracy worldwide. They also generally paint an upbeat portrait of news start-ups and their contributions to democracy, even going so far as to deny that there is really a crisis in the quality of contemporary journalism (Van Der Haak, Parks and Castells 2012).

In NS theory, description goes beyond microdetail to encompass categories of digital phenomena, largely built up on an inductive, ad hoc basis. For instance, in a recent article on the future of journalism, Van Der Haak, Parks, and Castells (2012) identify and provide examples of "new tools and practices" of digital journalism that are contributing to the "adequate performance of a democratic society": networked journalism; crowdsourcing and user-generated content; data mining, data analysis, data visualization, and mapping; visual journalism; point-of-view journalism; automated journalism; and global journalism.

This is a useful cataloguing, but as an analysis of the contemporary digital moment it leaves much to be desired. For starters, it could state more clearly what's at stake: What is meant by democracy? Which kind of democracy? Second, going beyond a mere listing, one could search deeper for patterns of variation. Are these new tools and practices randomly distributed across a range of types of media organizations and other actors, or can we observe patterns in the forms of ownership, management, and funding? A casual glance at Van Der Haak et al.'s (2012) listing seems to overrepresent public service broadcasters (BBC, NOS, Arte), state broadcasters (Al Jazeera), government agencies (National Film Board of Canada), elite broadsheet newspapers shielded by formal or informal "trust" ownership forms (*Guardian, New York Times*), foundation-supported nonprofits, and small donor-supported media, whereas large privately held or publicly traded commercial companies are underrepresented. Lacking methodological rigor and explanatory curiosity, Van Der Haak et al. attribute their findings to the "open, networked structure of the Internet" (2934), when in fact most of the journalism that most people consume on the Internet has little to do with these exemplary democratic practices and tools (see, e.g., Hindman 2008). By not identifying the common threads uniting the positive examples of media performance they list (as well as the characteristics of those they do not list), NS theory authors obfuscate the actual possibilities and limits of what Williams and Delli Carpini (2011) call "democratically relevant, politically useful" media content on the web. NS theory often discusses what "can" happen with digital media, but has no way of sorting out when a particular outcome is more or less likely to occur.

Jeffrey Alexander's (2011, 2015) cultural pragmatics represents a third kind of new descriptivist current that makes room for explanation, but only as a broad "value-added" collection of factors: actor, collective representations, means of symbolic production, mis-en-scène, and social power.[6] In *Performance and Power* (2011), Alexander insists that his approach offers both a hermeneutic model of understanding and "a model of causality: each of the [elements in the model] is a necessary but not sufficient cause of every performative act." The incorporation of social power is a departure from his previous "strong program" in cultural sociology, which focused only on cultural elements. Social power allows Alexander to ask such questions as (2011, 32): "Who will be allowed to act in a performance, and with what means? Who will be allowed attendance? What kinds of responses will be permitted? Are there powers that have the authority to interpret performances independently of those that have the authority to produce them?"

And yet, to give one example of the associated research, one comes away from Alexander's beautiful descriptive accounts of Obama's electoral campaigns and policy-making not any closer to knowing why Obama has succeeded or failed: is it really mostly about the success or failure of his performance? Or were there other factors that mattered more—for example, the strength and determination of his opponents, the divisions within the Democratic party, the power of lobbyists and money in politics, the failure of the media to go beyond personalities and strategies to adequately convey to the public the issues and interests at stake? The "value-added" model makes it difficult to sort out these important questions. With cultural pragmatics, we have arrived at the most compelling and useful of the three forms of new descriptivism, as is evident in Jacobs and Townsley's (2011) complex, critically self-aware mapping of US "space of opinion" journalism. And yet cultural pragmatics still falls short of the kind of specific, generalizable explanations needed not only to make sense of but also to intelligently respond to the actual challenges posed by the new media landscape. It seems no accident that none of the new descriptivist theories emphasizes commercial pressures or systematic class inequalities: their ideological elective affinity and effective complicity with neoliberalism (or in the case of network society, anarchist social movements) inevitably produces an uncomfortable silence about progressive policy responses.

In sum, new descriptivist research has its virtues, but each of its strains fall short of providing the complete theoretical toolkit needed to adequately analyze contemporary digital journalism. Committed to comprehensiveness, its descriptions often lose sight of the (contextual) forest for the trees; highly observant of heterogeneity, it fails to see patterned variation; careful not to impose normative judgments, it refuses to draw obvious connections to real-world concerns and possible solutions. These shortcomings are most centrally those of ANT, but the indifference to systematic explanation also applies to network society theory and cultural pragmatics. My intention in offering these critiques is constructive. By calling attention to the questions ignored or inadequately answered by the new descriptivism, I hope to stimulate a renewed effort to provide adequate responses.

Responding to New Descriptivism's Unanswered Questions

What's at Stake?

Why does this matter? This is the fundamental question that too many new descriptivist accounts fail to answer. How might we come up with better answers?

Domingo, Masip, and Meijer (2015, 62–63) make a case for ANT's "problematization" of normativity. They argue that if researchers do not take firm a priori positions about what journalism "should" be, they will be more open to discovering how journalists (and other stakeholders) define journalistic excellence. This openness in turn can improve the dialogue between scholars and journalists, so that the latter do not feel they are simply being preached at rather than listened to and understood (Blumler and Cushion 2014).

Yet hampered by the injunction to downplay normative questions, ANT-inspired digital media researchers are at risk of reporting trivial findings or of simply relaying the narrow worldviews of the actors they are studying (see also Couldry 2008, cited in Anderson and De Maeyer 2015, 7). In an otherwise well-constructed research article, Braun (2015) falls into this trap. Braun (31) tells the "backstory" of Newsvine, an innovative news site where "citizen journalism published by users would be aggregated and discussed alongside professional news content." Newsvine developed a set of "heterogeneous" tools (social, technical, commercial, and policy—a "code of honor") to create what Braun describes as a "productive and commercially viable" (33) social news site. With the sale of Newsvine to NBC News and the reuse of Newsvine as a general commenting site, the civility that the founders sought to create broke down—threatening to alienate the core users attracted to the site in the first place.

Braun focuses on the organizational implications, noting how the case demonstrates the anarchy that can ensue when a "sociotechnical" system is effectively "stripped of the social" that originally allowed it to "function correctly" (in this case, via the massive entry of new users who were not aware of the original purpose of the website) (36). His ANT methodology helped him show in detail what actually happens when a large corporation swallows up a start-up: "rather than having a homogenizing effect on organizational cultures, vertical integration and corporate partnerships tend to make things more complex, populating news organizations with diverse organizational subcultures, each with their own heterogeneous strategies and assemblages of tools" (41). But what is at stake here? Why should anyone beyond the businesses involved and perhaps business school students using this as a case study care? Braun never directly speaks to the issue, even though it is there in his findings: questions of news quality and democratic civility. Without taking a particular stance, Braun might have *normatively signposted* this issue as one of the reasons why this case study bears repeating and connecting to other studies: How have other websites attempted to create "sociotechnical" assemblages to promote news quality and civility? How were these terms enacted into practices? What general properties of the assemblage stand out as crucial to the outcomes?

The principled relativism of actor–network theory need not be abandoned in favor of an ideologically rigid stance in judging findings. One need only move from an unconcerned relativism to an observant relativism, that is, a comprehensive linking of findings to the array of political and aesthetic positions potentially at stake. This approach moves beyond network society's vague commitment to democracy: it is most notably used by Myra Marx Ferree et al. (2002) when they assess German–US differences in news framing of the abortion issue in relation to four partially competing democratic normative models. I adopt a similar approach in my comparative study of French and US immigration news coverage (Benson 2013). I also do this in my contribution (Benson 2010) to a multimethod, comprehensive study of online news in the United Kingdom spearheaded by Natalie Fenton of Goldsmiths, University of London. I argue that claims of declining journalistic "quality" actually pertain to three distinct democratic models: a watchdog accountability ideal linked to elitist democracy, a diversity ideal linked to participatory democracy, and a civil discourse ideal linked to deliberative democracy. Given that any given news article or outlet is likely to "score" higher on some ideals than others, it makes no sense to make a singular claim about quality. Thus, both criticism and praise should be placed in the context of clearly specified multiple criteria of quality and the public good (see also Hesmondhalgh and Baker 2011).

New descriptivists willing to carry their theories lightly should have no problem engaging with normative questions in their work. Indeed, Anderson's exemplary ANT-influenced network ethnography (2013; see also Howard 2002) of the Philadelphia digital news media ecosystem simply ignores Latour's antinormative injunction. To his credit, Anderson makes clear in the opening pages the "central normative problem" it will tackle (3): "Local journalism's vision of itself—as an institutionally grounded profession that empirically informs (and even perhaps, 'assembles') the public—is a noble vision of tremendous democratic importance. But the unreflexive commitment to a particular and historically contingent version of this self-image now undermines these democratic aspirations." Actor–network theory usefully helps Anderson break away from a narrow institutional analysis that would have only focused on traditional journalistic organizations. But when it comes to making sense of and explaining his findings, he (like many other ANT-inspired researchers) is forced to abandon ANT and look to institutional theories, to be discussed further shortly.

Is There Systematic Variation? Is This Case Like Others?

Every sociotechnical assemblage, we are often reminded, is an uncertain concatenation of noncomparable elements. The Internet, from a network society vantage point, facilitates an incredible diversity of journalistic and activist practices and strategies. Performances, cultural pragmatics insist, may be composed of similar elements but always in different ways, and their likelihood of success is always contingent. These conclusions are presented as a kind of hard-won wisdom, but they are actually untested premises. There is no attempt to search for evidence that might prove otherwise.

In short, new descriptivist accounts tell us a lot about heterogeneity, multiplicity, and diversity, but very little about systematic variation. And the search for variation—linked to systematic comparison—is fundamental to any form of critical knowledge. In the words of philosopher Peter Osborne, "Strictly speaking, the incomparable is the unthinkable" (cited in Steinmetz 2004, 390).

If everything is a supercomplex, unique heterogeneous assemblage, then the only possible conclusion is that no case is like another. And indeed, no case is exactly like another. But some cases are enough alike that we might compare them, note significant similarities and differences, and draw some useful conclusions. To paraphrase C. Wright Mills, this is the "comparative imagination."

The search for variation (and explanation, as discussed in the next section) requires a move that is often anathema to the new descriptivists, especially those of a poststructuralist bent: categorization. ANT's injunction against a priori categories needs to be scrutinized more closely. If a priori is understood to mean ex nihilo, this is certainly a valid warning. But in actual research practice, categories are built from a combination of theory, previous research, and preliminary inductive research; they are often modified in the course of research. Purely inductive research is a myth. In other words, if we want to conduct good research, we cannot avoid the use of categories. The question is not whether to use them but how.

As Mitchelstein and Boczkowski (2009) acknowledge, international research leads us almost inevitably to comparative questions and the search for variation. Happily, there has been an efflorescence of international news media comparative research in recent years. But there is no reason to limit the investigation of variation to the cross-national level.

For example, two landmark studies in the making (Christin 2014; Petre 2015) test the extent to which the use of audience metrics, ostensibly agents of commensuration (transforming different qualities into a common quantitative scale), are actually leading to convergence across newsrooms. Christin (2014) compares a US and French website (the latter overtly modeled on the US site) that both use the Chartbeat audience measurement tool, thus allowing her to test competing hypotheses of convergence (due to overt imitation and use of a common commensuration tool) and difference (due to their location in different national journalistic fields). Effectively deploying a basic categorical distinction between editors and reporters, she finds that different dynamics are at work in the two groups. US editors use metrics to guide story development and placement; French editors often "steadfastly refuse to cut articles and sections that are not successful (according to metrics)" and are supportive of reporters whose stories are deemed substantively important but attract few page clicks (23–24). On the other hand, US reporters are mostly indifferent to metrics, whereas French reporters are obsessed with Chartbeat. Christin attributes these differences to distinct types of organizational power (sovereign in the US, disciplinary in France), manifested in different divisions of labor between editors and reporters, ultimately rooted in the unique historical trajectories of and the relationship to political and market power in the US and French journalistic fields. As one can see, these findings are historically specific and finely attuned to the particularities of the cases (note: the larger study of which this is a part examines two news organizations in each country) yet speaks to broader theoretical questions and makes generalizable claims that could be tested through further research.

Petre (2015) pursues similar questions through an innovative study that not only compares the variable uses of audience metrics by two news organizations with distinctly different business models (*New York Times* and Gawker) but also analyzes the process of constructing these metrics inside Chartbeat. Among many other important findings, Petre shows how traditional journalistic values, far from being necessarily opposed to audience metrics, are actually consciously incorporated into the technology by metrics corporations concerned with maintaining good relations with their news organization clients.

Christin and Petre build on an encouraging miniwave of comparative news ethnographies, most notably by Ryfe (2012), Anderson (2013), and Kreiss, Meadows, and Remensperger (2015). Kreiss et al. (2015) improve on cultural pragmatics by showing how the production of performance at US political conventions is crucially shaped by the intersecting and competing logics of the journalistic and political fields: as they emphasize, cultural pragmatics "lacks a meso-level appreciation for the different standards of evaluation that field actors and audiences bring to performances within the civil sphere" (580). Other important small-N comparative studies in recent years include Graves (2016), Graves and Konieczna (2015), and Konieczna (2014). Numerous studies call attention to the crucial shaping distinctions of types of ownership, funding, and audiences: publicly traded vs. privately held companies (Cranberg et al. 2001; Dunaway 2008; Edmonds 2004), public service vs. commercial audiovisual (Aalberg and Curran 2011; Benson and Powers 2011), and the influence on content of various types of audiences (Benson 2013; Gentzkow and Shapiro 2010). Few studies have yet attempted to systematically trace the differences among various types of nonprofit media or the differences between nonprofit and other forms of media ownership (but see Konieczna 2014).

To be fair, much of the recent international comparative research, summarized and critiqued in Hallin and Mancini (2004) and Esser and Hanitzsch (2012), is based on sizeable samples of nation-states and media organizations. This type of research, as with any large-N study, could be legitimately criticized for inadequately incorporating historical and social context. Even so, because it makes clear its categories of analysis, justification of cases, and hypotheses, it opens itself up to productive critique and refutation in a way that most new descriptivist studies avoid. Variation-attentive ethnographic and in-depth interview based studies can and should operate in partnership to help improve the quality and use of the categories deployed in such large-scale digital media research.

Scholars influenced by new descriptivist theorists can join the conversation about patterned variation in media, whether at the national, field level, or organizational level, but in order to do so they must break with self-defeating new descriptivist theoretical purity. Fortunately, many are willing to do so.

Jose Van Dijk (2013) actually makes more effective use of Castells' and Latour's concepts than does either of these theorists themselves in order to identify meaningful variation across types of digital media. She compares several social media sites, drawing on Castells to identify institutional variables and Latour to specify technological variables. It is a rather heterodox use of Castells and Latour, but it works. And it provides us with some real insights into how and why Facebook, Twitter, YouTube, and Wikipedia are different as well as similar.

Anderson (2013) is clearly influenced by ANT when he casts a broad net to incorporate the full range of amateur as well as professional journalistic online organizations at work in the Philadelphia area during the 1990s and 2000s; yet he goes beyond ANT when he pays attention to systematic variations across the professional–amateur divide, across types of funding (notably commercial vs. nonprofit foundation), and audiences. Anderson also finds that efforts to collaborate across the professional–amateur divide largely came to naught. Braun (2015), while not fully drawing out the implications, makes it clear that NBC Online News is not just a sui generis heterogeneous assemblage but is also representative of a broader type of contemporary media organization, what David Stark (2011) has termed a "heterarchy." In both of these instances, there is a welcome movement toward generalizability, an effort to specify what their case is a "case of."

Why?

Once one begins asking normatively significant questions and identifying systematic variation, inevitably one stumbles upon the "why" question. What is the explanation? According to the Stanford Encyclopedia of Philosophy (2014), an explanation provides an answer to a question about "why" things happen—"where the 'things' in question can be either particular events or something more general, for example regularities or repeatable patterns. …"

Some might take my call for explanation as a return to Robert Merton's (1949) "middle-range" sociological theories, situated between grand theorizing and pure description. While the descriptive work I am critiquing is closely allied to some versions of grand theory, there are other versions of grand theory (Bourdieu, Giddens, Raymond Williams) quite amenable to the kind of research I am advocating. What I am calling for is an integration of theory (grand or otherwise) with "good" explanations that respond to the legitimate and socially useful desire to go beyond the particular case, to draw conclusions relevant to other cases, in short, to understand in a way that can guide action in the world. I acknowledge that there are multiple forms of explanation that might achieve such a goal. While I admire certain elements of the strict research design and methods of "explication" specified by Chaffee (1991), my own position is considerably more open minded while emphasizing the basic elements of concision and generalizability.

By concision, I do not mean that the most parsimonious explanation is always the best explanation. Because the social world is an open system with over-determined outcomes, a good explanation has to be contextualized and complex. However, the basic demands of communication suggest that we should strive for concise over long-winded explanations. Moreover, concision goes hand in hand with effective incorporation of variation into the research design and ultimately explanation. In order to identify the factors that matter most, cases need to be selected that are similar in enough ways that the potentially crucial differences can be isolated. This of course is always a "construction,"[7] ignoring the full complexity of each case, but if done carefully can be heuristically useful, indeed indispensable, in order to identify and explain patterned variation.

Generalizability (what Martin 2011, 340, calls "transposability") is even more fundamental to any explanation that seeks to speak to a class of phenomena. If we are able to show that a given event or process is a "case of something" that reoccurs either spatially or temporally, then we can and should proceed to generalization. Generalizable claims usually refer to causal mechanisms, understood as acting relatively independently or as part of a larger conjuncture.

Some ANT proponents have argued that actor–network theory is not in fact wholly descriptive, that it does offer explanations. It's just that these explanations are rooted in specific historical conditions unique to each contingent case. The explanation of something lies in recounting all that preceded it; the explanation lies in temporality (Kreiss 2013). As Latour (1991) writes, explicitly addressing this question: "Explanation does not follow from description; it is description taken that much further (120). … If we display a sociotechnical network … we have no need to look for any additional causes. The explanation emerges once the description is saturated" (128).[8]

It is a truism that every particular event is the result of a temporally unique, overdetermined mix of other events. But usually our goal is not to explain in full a particular event (even if on some occasions we might have reason to try), but rather to draw out the common "causal mechanisms" (e.g., field logics, habitus, social struggles for distinction, discourses) that seem to be operative across a class of social phenomena (Steinmetz 2004). Thus, while comparative research most directly advances this project, the accumulating evidence of single case studies can also be harvested for explanatory claims (ibid., 391).

To be clear, I could not agree more with C. W. Anderson's call (see chapter 4 in this volume) for historicizing news research. However, I do not see an inherent opposition, as he seems to, between structural and historical approaches (Benson 2013a). Structures are historically produced and to be properly understood must be placed in their full historical context. History is a causal factor to the extent that institutions tend to exhibit "path dependent" behavior and resist change (Thelen 1999); such an understanding of history does not preclude the possibility of unforeseen events opening up new paths, but it does suggest that such "critical junctures" will be rare and that they must be strategically exploited to realize their full transformative potential. These conceptions do not seem incompatible with a "genealogical model" per se, though they do conflict with a voluntaristic conception of the social world at odds with what most social scientific research has actually shown. The point of paying attention to systematic variation is not to insist on the difficulty (I would never say impossibility) of change, but rather to precisely identify the ways that action can be effectively harnessed to produce particular outcomes.

In cross-national studies, a number of comparative approaches are on offer (Krause 2015). Hallin and Mancini (2004) create their typology of three media system "models" in reference to distinct conjunctures of the four causal mechanisms they identify (market structure, political parallelism, professionalism, and state intervention). These "dimensions" are somewhat less than causal mechanisms in that they do not systematically specify news content outcomes. Yet they meet my definition of a "good" explanation in that both the models and the dimensions are concisely stated and "potentially" generalizable (a lively debate has ensued over precisely this issue; see Hallin and Mancini 2011).

If we look at prominent studies that attempt to explain systematic temporal or spatial variation (or lack of variation)—even those that draw on actor–network theory or cultural pragmatics (see, e.g., Anderson 2013; Jacobs and Townsley 2011; Kreiss et al. 2015)—almost inevitably their explanations are provided by macro- or meso-level institutional or field theories (Benson 2014; Bourdieu 1993; Cook 1998; Fligstein and McAdam 2012). Why is this? It is because these are theories that capture the central social dynamic of complex large-scale societies: hierarchically organized institutional differentiation.

While sharing an emphasis with critical political economy (McChesney 2014; Pickard 2014) on economic power, these institutional theories better acknowledge the limited pluralism introduced by partially competing organizational fields (politics, science, arts, religion, etc.) and emphasize the variable forms these institutional configurations take across societies and time periods. Political economy often ignores or is baffled by capitalism's apparent "contradictions" in relation to media production: Fox's *The Simpsons*, HBO's *The Wire*, Vice's hard-hitting documentaries. Although Bourdieu paid little attention to large-scale commercial production (Hesmondhalgh 2006), in principle field theory calls attention precisely to those cross-cutting factors that can help explain these apparent paradoxes, such as class stratification of audiences (Clarke 2014) and the varieties of commercial ownership logics operative within contemporary capitalist "social formations" (Williams 1973).

As noted with Christin (2014), cross-national studies revealing variation in news practices or content can find explanations in the balance of power between different societal fields (see also Saguy 2013). In my own research comparing the production of French and US immigration news (Benson 2013b), I draw on field theory to explain differences in news content both *across* nation-states (differences in the position of the journalistic field in relation to competing state–civic and state–market poles of the field of power) and *within* nation-states (differences among news outlets related to their particular location in the field, as indicated by type of funding and size and class composition of audience). Likewise, a comparative study of the degree of "external pluralism" across the online versions of agenda-setting newspapers in Denmark, France, and the United States showed how institutional factors mediated Internet affordances (Powers and Benson 2014).

In single nation case studies, institutional theories have been used to account for variation and the lack of variation, either spatial or temporal. Kreiss et al. (2015) use field theory to offer meso-level explanations of differences across fields, specifically distinct media practices by actors rooted in the political and journalistic fields. Ryfe (2012) argues that field-shaped "constitutive" rules structuring social interactions both inside the journalistic field and with neighboring fields are the reason for the persistence of traditional notions of journalistic professional authority, despite dramatic technological and economic transformations seemingly undermining such authority. Anderson (2013, 162–163) goes beyond this argument by showing that change (as well as continuity) is institution driven: "Successful networked news ecosystems depend on the presence of strong institutions dedicated to building networks."

So-called "local" interorganizational comparative studies can be seen as complementing rather than competing with macro- or meso-institutional theories. For example, Boczkowski's (2005) comparison of the adoption of digital technologies by three US news organizations during the 1990s emphasizes local contextual factors and local "contingencies" such as the relationship between the online and print newsrooms and conceptions of the audience as inscribed in web interface designs. Microanalyses are also crucial in identifying the precise mechanisms of institutional reproduction and change (Hallett 2010; Kellogg 2009). But whether or not these factors are only locally contingent cannot be determined without reference to a broader sample. For this reason, macro- or meso-level institutional analysis needs to supplement and build on purely local approaches.

Conclusion

Outside of academia, there are diverse efforts afoot to make sense of and evaluate systematic variation in our digital media landscape. For example, journalist Michael Massing (2015a, 2015b), in a multipart series for *The New York Review of Books*, asks the question too few academics are asking: "Digital journalism: How Good Is It?" Massing sizes up the leading US news and opinion websites, including Huffington Post, Talking Points Memo, Slate, Salon, The Drudge Report, Politico, Pro Publica, Buzzfeed, Vox, and a host of smaller "narrowcast" websites on specific issues. He makes clear some of the important ideals at stake—such as the "range of voices" beyond the usual American elite commentators or the degree of "sustained investigations" into "systemic problems" such as "the composition, shape, and reach of the global oligarchy"—and finds that even the most highly praised digital media outlets fall short.

Massing also makes note, though not systematically, of variation in the ownership and financing of the various outlets and the size and composition of their audiences. In short, he signpoints normativity, observes variation, and points toward potential explanations. In similar fashion, numerous useful empirical and policy reports are also being produced by nonacademic or publicly oriented academic research organizations such as Pew, Nieman, Tow, and the Reuters Institute.

To put it bluntly, without theory, Massing and other nonacademic observers are going further and saying more than many academic studies being produced on digital media today. Theory, if it is to have any justification at all, must provide a way to go beyond this kind of intelligent lay analysis. It must not just add to the mass of detail, but make sense of and identify patterns in the detail; it should help us look beyond the obviously complex heterogeneous assemblages and find the less obvious strands of variation.

Stripped of their programmatic stances, new descriptivist theories might be hitched to such a project, but it's clear the heavy lifting will be borne by institutional approaches. We have all the conceptual tools we need—epistemological, methodological, normative—to make sense of this moment in history. We just need to be brave and lucid enough to use them well.

Notes

1. This chapter builds on my remarks as a discussant at the Qualitative Political Communication preconference of the International Communication Association, Seattle, Washington, May 22, 2014, available at https://qualpolicomm.wordpress.com/2014/06/05/challenging-the-new-descriptivism-rod-bensons-talk-from-qualpolcomm-preconference. I wish to especially thank Rasmus Kleis Nielsen, Daniel Kreiss, Andrew Perrin, Juliette de Maeyer, Fenwick McKelvey, Dan Hirschman, and Elizabeth Popp Berman for insightful online posts that helped me formulate and reformulate my arguments in this chapter. I am also grateful for the additional helpful comments I received from the editors and other participants at the Northwestern conference that led up to this book.

2. Other versions of descriptivism are resurgent and increasingly assertive across the academy, many of them linked to poststructuralist or postcolonial theory. See, for example, the program for "Description Across the Disciplines," a three-day conference (April 23–25, 2015) sponsored by the Heyman Center for the Humanities at Columbia University, http://heymancenter.org/events/description-across-the-disciplines.

3. Another type of descriptivism, also not entirely new, is highly quantitative social network or "big data" analysis. With recent advances in computation, these research approaches have gathered steam (see, e.g., Chris Anderson's [2008] essay on the "end of theory.").

4. Nikki Usher, Twitter communication to author, May 2015.

5. Gans, despite his protestations against theory, actually produces work that speaks well to broader theoretical debates. Both *Deciding What's News* and Usher's Gans-influenced *Making News at the New York Times* provide potentially "good" explanations, as I will define them—systematic, generalizable, normatively sign-posted claims—even if they aren't always overtly offered in this spirit.

6. This value-added model, though differing in the particulars, is similar in kind to that used by Gamson and Modigliani (1989) in their model for explaining the construction of social problems in the media; see Benson (2004) for a critique.

7. Actor–network theory and field theory actually share a constructionist conception of knowledge. The difference is that field theory is able to break from a debilitating relativism by privileging "scientific" construction, understood not as the accumulation of facts via scientistic methods but rather as an "epistemological break" with common sense knowledge facilitated by the semiautonomous reflexivity of the researcher. Obviously, this project of fully reflexive knowledge construction (not limited to the individual researcher, but caught up in the institutional conditions underlying autonomous production of knowledge) is never fully realized, but that does not invalidate the attempt.

8. Another reason why ANT is unable to offer concise explanations is that it keeps the world "flat" and does not privilege one account over another. ANT's self-professed interest in "controversy studies" is thus linked to the relativist "strict constructionist" (Latour and Woolgar 1979) wing of this subfield, oriented toward simply cataloguing the various accounts of the actors involved. "Contextual" constructionists (Best 1989), on the other hand, by necessity privilege some accounts over others in order to arrive at an explanation, for example, economic data provided by government agencies, dominant discourses mapped by the analyst him/herself, etc. Field theory can be seen as a strong version of contextual constructionism, indeed a "structural constructionism" (Bourdieu and Wacquant 1992, 239).

3 Rediscovering the News: Journalism Studies' Three Blind Spots

Victor Pickard

Over the past decade, journalism studies has become an increasingly prominent area of research within the field of communication. This ascendance, however, coincides with a moment of deep turmoil for journalism itself. Thus far, much research has focused on specific aspects of journalism's transformation, leaving other areas of inquiry less developed. These gaps in the research deserve closer attention, especially as we begin to imagine journalism studies' trajectory into the digital age, with a greater focus on online news. Such a project requires a healthy degree of reflexivity. An examination of how journalism studies makes sense of the profound changes—in some ways dissolution—of its primary research subject brings into focus paucities in specific perspectives and research agendas. In this essay, I suggest that considering these neglected traditions could potentially enrich journalism studies by bringing it into conversation with other subfields and increasing its relevance for constituencies beyond the academy.

The Ongoing Journalism Crisis

"Crisis narratives" for journalism vary, and whether "crisis" is even a proper way to describe what is happening to journalism is open for debate (Almiron 2010; Chyi, Lewis, and Zheng 2012; Pickard 2011; Siles and Boczkowski 2012; Zelizer 2015). Yet few would dispute that the dominant model for American journalism over the last 150 years—advertising-dependent, paper-based newspapers—is in gradual dissipation. Despite general consensus about the root cause of this decline, assessments of its implications and predictions for how it will ultimately play out vary widely. For some, journalism has entered its "postindustrial" phase (Anderson, Bell, and Shirky 2012) with the traditional news industry in a state of irreparable collapse. Many take the Internet's role in these changes for granted and downplay the commercial news model's endemic structural vulnerabilities that predate the web, specifically its overreliance on advertising support. While some scholars and pundits view this structural shift as a tragic loss for democracy and a once noble profession, for others it has been a source of great excitement. Indeed, a certain level of "digital exuberance" has emerged in recent discourses surrounding the future of news. Even the venerable Pew Research Center, in its 2014 State of the News Media annual report, emphasized the creation of some 5,000 digital journalism jobs in recent years as digital news start-ups expand (Jurkowitz 2014). While the Pew study took an optimistic view of these developments, it did not put to rest ongoing concerns about the sustainability of these start-ups (especially given their dependence on venture capital and other questionable means of support) and the disparity between the relatively low number of new digital jobs created by these companies and the tens of thousands of jobs lost from traditional newsrooms in recent years.

Nonetheless, rapid shifts within news industries have led to a proliferation of studies attempting to keep up with new permutations. Many researchers have clustered their efforts around that which is most visible, especially changes in news-gathering practices and journalistic routines resulting from the use of new technologies (Steensen 2014). To make sense of these phenomena, analysts have constructed new categories, discussing the impacts of "data journalism," "hacker journalism," or "networked journalism," among other variants. In heralding these presumably new forms of journalism, researchers often implicitly assume that new technological affordances enable the production of more and potentially better journalism with less time and money, and in ways that are inherently participatory and democratic.

It is understandable that these new trends have attracted so much attention. But there are other recent developments that deserve closer analysis, though they do not support the more optimistic narratives: the rise of native advertising and content farms, the use of news metrics, and the ongoing loss and casualization of journalistic labor, to name a few areas of concern. Indeed, despite a growing exuberance regarding new digital models and innovations, troubling questions remain: Can new digital start-ups fill the journalism vacuum? Can we assume that technology and market forces will combine to produce sustainable, profit-seeking models, despite little evidence that digital experiments can achieve long-term commercial viability? What happens as local journalism becomes increasingly reliant on crowdsourcing and other forms of low or unpaid labor? Can legacy media institutions innovate themselves out of this crisis or must American society rely on media billionaires and foundation-supported news institutions (with varying motives) for their news? Is it symptomatic of our neoliberal age that we look to charitable and entrepreneurial individuals—consisting of mostly white men (Bell 2014)—to save journalism?

This essay does not pretend to answer these questions, but rather seeks to identify some possible impediments that help explain journalism scholarship's failure to sufficiently address them thus far. This failure, I suggest, stems from several underdeveloped areas of inquiry, or "blind spots," in journalism studies. By highlighting these gaps, I hope to encourage more research in these areas as the subfield of journalism studies continues to evolve.

Three Blind Spots

Every area of research has patterns of selection, omission, and emphasis. For example, many subfields of communication research have consistently downplayed the importance of historical and global comparative research. Also often neglected are contributions and perspectives from traditionally underrepresented groups—like other fields, communication research has long been dominated by white men. Another perennial charge, especially among critical scholars, is that the field of communication unduly accommodates the commercial nature of the media industries that it seeks to understand. Some of these weaknesses might be discerned in the still-emerging subfield of journalism studies, but the areas to which I wish to draw attention are those that come into focus when we specifically examine "future of news" research. These include normative, policy-oriented, and economic questions about journalism's future.

Normative Foundations

The first of these lacunae relates to digital journalism's normative foundations. Of course, journalism scholarship often contains tacit normative implications, but usually they are not a central concern. Kreiss and Brennen (2016) note "a striking lack of a developed body of normative theory about journalism and democracy in the era of digital media to rival work on mass communication." More problematic than simply an absence of such considerations are normative assumptions that remain operative but are rarely explicitly identified as such. Kreiss and Brennen rightly observe that normative claims within contemporary analyses and discourses connected to journalism are often implicit and therefore go unexamined. For digital journalism in particular, they argue, these technological utopian claims tend to focus on participation, deinstitutionalization, innovation, and entrepreneurialism, and assume that digital media are inherently emancipatory.

A lack of reflexivity renders it difficult to tease out and scrutinize these normative assumptions. In general, many scholars treat journalism as a subject to be studied and described, while avoiding ethical evaluations, especially regarding questions about the role of media in a democratic society. A trickle of research has examined normative questions both in historical context, as in previous norm-setting texts like those associated with the Hutchins Commission (Pickard 2015a), and in contemporary contexts (Christians et al. 2009; Schudson 2008). One theory regarding the lack of attention given to more normative issues in communication research is that the dominant social science tradition within the field focused more on prediction and description, and less on the prescriptive (Craig and Tracy 1995). Researchers have also neglected thorny questions about advertising support for journalism (among other external influences that might compromise news media's democratic mission) and the role of government in supporting journalism (especially in establishing structural alternatives to legacy commercial models and incumbent media corporations). Perhaps these concerns are a better fit for the area of "media ethics," but journalism studies could benefit from including these questions within its own purview.

Similarly, despite long-standing media criticism in other subfields like political communication (see, for example, Bennett 2011), critical interrogations of commercial support for the news are less common within journalism studies than might be expected. For example, studies of emerging trends like paywalls (Pickard and Williams 2014), news metrics (Anderson 2011), native advertising, casualized labor, and other developments might encourage questions like: Are these trends good for a democratic society? Should we as citizens be concerned? If this is a social problem, should we as a society intervene? This line of inquiry connects to a growing area of comparative research that considers how normative assumptions and policies differ across the globe (see, for example, Benson 2013; Hallin and Mancini 2004; Nielson 2011).

Further research into normative questions related to journalism might focus on definitions of "freedom of the press," and how the assumptions related to this ideal have been historically situated. Indeed, there is a kind of American exceptionalism regarding press freedoms, a default ideology in the US that rests on a deeply libertarian position and market fundamentalist logic (associated with what I refer to elsewhere as "corporate libertarianism"; see Pickard 2015a). Critical historical analyses show that this position is neither inevitable nor natural, but rather the result of ideological struggles in which a more social democratic vision of the press was defeated (Pickard 2014b). Based on this longer historical arc, one might argue that this current libertarian position is based not on the vision of the nation's founders—whose policy designs for the first important communication network, the postal system, assumed a proactive role for government in designing media systems (John 1995). Rather, one might see this arrangement based on a more recent vision—and only after red-baiting and other political developments rendered less market-friendly arrangements off limits (Pickard 2015a).

Another way of exposing this logic is to consider the emphasis that American discourse places on negative freedoms (freedom from) as opposed to positive ones (freedom for). Historically, much of the discourse about freedom of speech is framed in negative terms, exemplified by the US First Amendment ("Congress shall make no law ... abridging the freedom of speech, or of the press ..."). This privileging of negative rights, which often benefit media owners instead of audiences, has arguably impoverished law and policy discourses regarding individual and collective speech rights in the US. The emphasis on the former bolsters libertarian assumptions about media's responsibilities to society and the role of the state in mandating democratic obligations.

However, there are also important—and often overlooked—traditions that draw instead from a positive rights discourse. Article 19 of the Universal Declaration of Human Rights, which codified the right to "receive and impart information and ideas through any media," is an international example. Even within the largely negative US policy discourse there are important exceptions, like the Supreme Court's 1969 *Red Lion* decision regarding the Fairness Doctrine, which determined that "It is the right of the viewers and listeners, not the right of the broadcasters, which is paramount." The positive liberties tradition tends to privilege the ideal of having diverse voices and viewpoints in the media system. It tends to be equally skeptical of media power concentration, whether power consolidates among corporate actors or governments. And it tends to focus on collective rights held by publics, audiences, and communities as much as the individual rights held by persons or corporations (Pickard 2015a). Alexander Meiklejohn (1948) captured this principle in his oft-repeated assertion that "What is essential is not that everyone shall speak, but that everything worth saying shall be said." In other words, society's freedom to have access to a rich and diverse information system is more important than an individual's freedom to speak. And this is especially true when those "individuals" are actually large media corporations.

One hazard with focusing on positive rights is that it can shade into reducing the public to mere consumers of news products. I expand on this "market ontology" below, but it is worth noting that the normative position delineated here takes for granted the existence of democratic concerns that are not reflected in market transactions and measures of consumer satisfaction. Treating audience members as consumers instead of citizens of an engaged polity may privilege concerns for the individual as opposed to society's benefit as a whole. Focusing on the rise of the "prosumer," as well as other combinations of audience activity and news production like user-generated content and citizen journalism, may only further obfuscate this distinction.

Journalism studies has a role to play in interrogating assumptions about "freedom of the press," as well as engaging with broader normative questions. By recovering alternatives to dominant discourses that treat the press as a naturally existing commercial enterprise, journalism studies can call into question assumptions that enshrine individual freedoms and shield media corporations from government and public oversight. Focusing on these implicit relationships raises related questions about the "social contract" that binds government, media institutions, and the public. These arrangements are codified in media policy, another blind spot within journalism studies.

Journalism Policy

"Journalism policy" is an uncommon phrase and concept. Journalism and media policy are only rarely brought into conversation with one another, a long-standing weakness in both areas of study. Part of this problem stems from the libertarian mythology that government has never been involved in journalism. However, history tells us otherwise. From early debates over the design of the postal system to the founding of each new medium from the telegraph to the Internet, government policy has always been inextricably linked with journalism and media institutions. For issues as diverse as copyright, spectrum management, subsidizing the early Internet's development, and the enforcement (or lack thereof) of antitrust laws, government policy plays a central role in how media systems are designed, owned, and operated. To pretend otherwise is a dangerous fallacy. It discourages public involvement in media policy debates and helps reify a media system that disproportionately favors corporate interests (Pickard 2015a).

Media policy's relationship to journalism should be of greater concern now than ever before. Laws and policies dealing with intellectual property, privacy, and net neutrality will lay the foundation for journalism's digital future. While questions pertaining to media conduits and content are often dealt with separately, the recent net neutrality debate reveals the artifice of these divisions. This is especially true as we consider journalism's digital future. Without net neutrality protections, a feudalized Internet could quickly become a hostile environment for all manner of journalists, from cash-strapped newspapers to individual bloggers and citizen journalists who cannot afford to pay off Internet service providers and thus are relegated to digital "slow lanes."

Lack of quality Internet access is another infrastructure problem with serious implications for the future of news media, but it is rarely addressed within journalism studies. The "digital divide" is still a major problem for American society, where nearly a fifth of all US households still lack broadband Internet (Federal Communications Commission 2015). Even for those with access, services are often subpar and costly compared to other democracies. These inequities are neither happenstance nor inevitable; they result from explicit policies that accommodate oligopolistic markets and corporate interests more generally (McChesney 2013; Crawford 2013). Because they directly impact the flow of online news media, these policy issues should be a central concern for journalism studies.

The role of policy in directly sustaining journalistic institutions is a related concern regarding questions about digital infrastructure and content regulation. A public policy approach to promoting positive press freedoms might consider support mechanisms like various kinds of subsidies for public media. The argument for subsidizing public service journalism enjoyed a brief discursive opportunity from 2009 to 2011 when a number of journalism-related reports came out in support of preserving public service journalism (Pickard, Stearns, and Aaron 2009; Downie and Schudson 2009; Knight Foundation 2009; Federal Trade Commission 2010). However, calls for state intervention soon fell out of favor with the return of the "nervous liberals" (Pickard 2015b) whose evasion of addressing market failure (Pickard 2014a) is exemplified by the FCC's "Information Needs of Communities" report (Waldman 2011). Nonetheless, arguments for subsidies are no less relevant now than they were then, and the debate around their deployment was unfortunately closed down prematurely.

To be sure, the empirical record of media subsidies in the US context is not as extensive as it is in other democracies, and there are potential limitations to their implementation. But we can look internationally and historically to find numerous examples where they were successful to varying degrees (Pickard 2011). Although such interventions are not certain fixes for all that ails journalism—and to what extent they change audience preferences over time would have to be closely studied—policy reforms aimed specifically at reducing market pressures on media institutions could allow them to become more focused on adversarial reporting and more accountable to diverse communities. Reinventing journalism might involve subsidies for an expanded public media system, tax incentives to encourage struggling media institutions transition into low- and non-profit status, and government-sponsored research and development efforts for new digital models, including public/private hybrids (I discuss potential sources for these subsidies and offer examples, including a plan for transitioning post offices into community media centers, in Pickard 2015a, 212–231). Together, these initiatives could remove or reduce distortionary market pressures and help restore journalism's public service mission. A two-pronged approach would require subsidizing public journalistic alternatives in conjunction with structural interventions aimed at reining in monopolistic Internet service providers. This would require enforcing public interest protections like net neutrality and encouraging competition by allowing communities to offer their own broadband services. The question of structural alternatives to commercial journalism and the relationship between government and media institutions brings us to the third blind spot: economic theories of journalism.

Economics of Journalism

Much news media research understandably focuses on journalism's cultural and technological developments, since these are often the most interesting, dynamic, and visible areas for research. However, economic questions—beyond the obligatory nod toward funding mechanisms and business models when noting the current state of news media—should be more central to the conversation about journalism's future. This shift in emphasis might bring into focus how news work is increasingly becoming a form of digital, immaterial, casualized, and precarious labor. Though this area of research is slowly gaining some attention, much more work is needed. Other economic issues surrounding the future of journalism that deserve closer attention include media monopolies, journalism as infrastructure, public/merit/information goods, and media market failure.

The argument that the information produced by news media should be treated as a public good has gained greater visibility in recent years, especially as the journalism crisis worsened (Baker 2002, 8; Hamilton 2006, 8–9;

Pickard, Stearns, and Aaron 2009, 1–9; McChesney and Nichols 2010, 101–103; Pickard 2015a, 212–231). Because public goods are nonrivalrous (one person's consumption does not detract from another's) and nonexcludable (difficult to prevent "free riders" from using it), they differ from other commodities, like cars or clothes, within a capitalistic economy (Samuelson 1954). Many public goods—artificial light, clean air, knowledge—also produce positive externalities (benefits that accrue to parties outside of the direct economic transaction) that are necessary for a healthy society. The information produced by journalistic practices is a public good that carries tremendous positive externalities because it confers social benefits beyond its revenue stream. As an essential public service, an ideal news media serves as an adversarial watchdog over the powerful, a forum for diverse voices and viewpoints, and a rich source of information regarding the important social issues upon which citizens will vote.

As a public good exhibiting positive externalities, most journalism derives only limited support from direct market transactions. Print journalism in particular has almost always been subsidized to some degree (for the profitability of broadcast media news, see Socolow 2010). Since the late nineteenth century, this has usually come in the form of advertising revenues that cross-subsidize quality journalism. More so than other democratic countries (Benson 2013), American print media institutions have been especially reliant on advertising support. Advertisers sought the audiences that only newspapers could deliver, and in many ways the news itself was a byproduct—a positive externality—of this transaction. Especially for print journalism, this model is increasingly unsustainable as audiences and advertisers migrate to the Internet, where ads sell for a small fraction of the cost of print ads.

Although increasing, digital ad revenues are not offsetting the enormous losses in traditional advertising revenue. A 2012 Pew study found that the decline in print ad revenue, more than 50 percent since 2003, exceeded any gain in online digital revenue by an order of magnitude (Edmonds et al. 2012). Although by 2013 there was growth in digital advertising, Pew found that it still "does not come close to covering print ad losses (Edmonds et al. 2013)." Based on these and other data, advertising-dependent revenue models for journalism appear to be increasingly unviable, with no other commercial models—like online subscriptions (Pickard and Williams 2014)—filling the vacuum. The ongoing disinvestment in news production is demonstrated by drastic actions like cutting staff and reducing home deliveries of leading metro dailies such as *The Plain Dealer* in Cleveland and *The Times-Picayune* in New Orleans (Pickard 2014a).

Given this evidence, the elephant in the room for journalism studies (and discourse about the future of news media more generally) is that a commercially viable model for supporting news media might not exist—or at least not the news media that democracy requires. The inadequacy of commercial support for democracy-sustaining infrastructures reiterates what should be obvious by now: the systematic underproduction of vital communications like journalistic media and accessible, affordable, and reliable broadband Internet services qualifies as a clear case of market failure. "Market failure" generally refers to the market's inability to efficiently allocate important goods and services (Taylor 2007, 15), and typically occurs when private enterprise withholds investments in critical social services because it cannot extract the returns that would justify expenditures, or when consumers fail to pay for such services' full societal benefit. Although this analytical framework derives from neoclassical economic theory—where it is often treated as a rare deviation from the norm—market failure has also been a concern among critical political economists who focus on media's normative foundations and who observe a number of market failures specifically affecting media. For example, commercial media markets tend toward concentration and produce both negative and positive externalities that must be managed via government regulation (Freedman 2008, 8–9). Positive externalities are especially important when examining "merit goods": goods that society requires but individuals typically undervalue (they are unable or unwilling to pay for these goods), leading to underproduction in an unregulated market (Musgrave 1959, 13–15; Leys 2001, 97–98; Ali 2013).

The history of American media is in many ways a history of market failure (Pickard 2015a), yet these recurring patterns almost always go unrecognized in mainstream policy discourse. Due to the special nature of media goods and services, the market's failure to provide them is particularly deleterious for democratic governance. The "public good" qualities and other characteristics intrinsic to media result in a systemic market failure that cannot be entirely eliminated. However, this market failure can be reduced or compensated for via public policy that recognizes the tremendous public benefits associated with a healthy media system. According to textbook scenarios, cases of market failure normally necessitate public policy intervention. This necessity, however, is usually obscured by the previously mentioned "corporate libertarianism"—an ideological project that equates corporations' freedoms generally, and media firms' privileges specifically, with individual liberties like First Amendment protections. Such an arrangement renders state interventions on behalf of public interest protections a priori illegitimate. Therefore, we dare not talk about market failure because to do so legitimizes government intervention.

A number of market failures specifically affect the media industry. In addition to the lack of support for public goods, another kind of market failure that frequently occurs in the American media system is associated with structural flaws like oligopolistic concentration and profit maximization. Uncompetitive markets can lead to perverse incentives and the abuse of market power by media conglomerates, resulting in the degradation of the entire media system (lack of access, overcommercialization, misrepresentation of important social issues, low-quality information, etc.). In addition to problems with monopolies, public goods, and externalities, commercial media markets are seldom the ideal means to sustain vital communication infrastructures. For example, they often fail to ensure interconnection between networks and to provide communication services to all of society, which can lead to red-lining that privileges profitable markets and communities over disadvantaged ones, especially rural and poor urban areas (Cooper 2011). Media markets are also subject to other failures including information asymmetries and high first-copy costs that encourage natural monopolies by allowing firms to exploit economies of scale and scope (Baker 2002, 9).

While market failure in the American media system is increasingly visible, the vocabulary for describing these structural problems remains impoverished. Similar situations require state intervention to ensure public access to essential services and infrastructures not sufficiently provided by the market: public education, a standing military, and a national highway system. That the government provides these services is largely naturalized and rarely requires justification. Yet it generally remains counterintuitive that the same rationale can be used to justify enacting public policy to address failures in media markets. Although its relationship to market failure is rarely stated, the foundation of a public broadcasting system is an example of this logic put into practice. Similarly, the existence of "public interest" policies implicitly acknowledges endemic market failure in commercial media. Nonetheless, an explicit discussion of these political economic relationships rarely occurs. Whether appraising the lack of support for local journalism or deficiencies in providing universal access to affordable and reliable Internet service, a focus on market failure deserves more prominence in journalism discourse.

Once market failure is acknowledged at the discursive level, it can be treated as a social problem that warrants public policy intervention. However, whereas a neoclassical economist might simply foster more competition to create the desired outcomes, I argue that additional policy interventions are necessary to address what I refer to as "systemic media market failure" and to ensure that a media system's positive externalities are supported and enhanced. Potential remedies include structural antitrust interventions to break up monopolies and increase competition as well as significant investments in alternative, nonprofit, and noncommercial communication infrastructures to promote structural diversity. According to this view, a muscular policy approach is called for because oligopolistic media firms have little incentive to make the necessary investments to address structural problems like insufficient capacity in their broadband markets on their own (Pickard 2014a).

Transitioning to a new digital media system requires a paradigm shift away from libertarianism toward a framework that recognizes media's public good qualities and positive externalities, and embraces government's affirmative role in providing for society's communication needs. But before any of these reforms can occur, discourse about the future of news media institutions must first be reframed to acknowledge systemic media market failure. And before this can happen, we must come to terms with a number of "discursive inhibitors" enabling the market ontology that pervades key assumptions about journalism.

Challenging the Market Ontology in Journalism Discourse

Scholars rarely acknowledge a number of implicit assumptions about journalism's fundamental nature. These assumptions essentially treat the information produced by journalism as widgets with commercial properties similar to other commodities on the market. But of course, media products are not just widgets. As noted above, media contain special properties and serve important roles in society that cannot be reduced to their commercial value. Moreover, supply and demand in the unfettered free market do not always reflect accurate assessments of social value. First, we know that it is rarely, if ever, actually a "free market." Media markets are often dominated by monopolies that face little competition. Furthermore, the supply of commercial news does not just account for the interests of the consuming public, but also reflects the influence of advertisers and investors. And even if the public demand for quality journalism is lacking, unsexy stories about the bureaucratic goings-on at state legislatures, the condition of our core infrastructures like reservoirs, roads, and bridges, and the banalities of local public affairs are necessary for an informed polity. These kinds of stories often lose out to irresistible listicles, celebrity news, partisan bickering, and other forms of "click bait" that sell advertising and are cheap to produce. Nonetheless, democracy still requires public interest journalism, regardless of profitability and popularity. We would not discontinue primary education simply because there is not a clamoring demand for its services.

But these basic understandings often seem absent, especially in popular discourse about journalism. And in varying degrees, this lack of attention to the special economics of news media also constrains and distorts academic discourse about journalism. There has been a raft of important studies in recent years (see, for example: Anderson 2013; Boczkowski and Mitchelstein 2013; Herndon 2012; Patterson 2013; Ryfe 2012; Starkman 2014; Usher 2014). Most of these fine works at least touch on the normative and economic implications of traditional journalism's decline. But there is still, in my view, an underdeveloped theoretical framework for understanding what exactly is being lost (or, more accurately, was never there in sufficient quantity and quality), and what democracy actually needs from journalism that new digital technologies and economics are unlikely to provide. By not adequately addressing this structural critique, the resulting "policy drift" may perpetuate worsening communication inequalities in the US (Pickard 2015a).

Current commentary on the state of news media often presupposes a commercial relationship within which news readers are treated as consumers, not as citizens of a polity, and journalists are blamed for not being savvier about utilizing new technologies and monetizing their content. Often it is assumed that news workers are themselves responsible for enabling a disconnect between the journalism they want to produce as opposed to the journalism the public actually wants. This taken-for-granted assumption about journalism's inherently commercial nature implicitly suggests that if the market no longer supports public service journalism—that if people (or advertisers) do not pay enough for hard news to make it profitable—there is nothing to be done. But we also know that media value is measured in ways that are not only expressed in terms of dollars (Picard 2011, 6). In addition to having tremendous positive externalities, the products of the news media are also set apart from other goods because they are subject to both individual- and society-level preferences in ways that most other goods and services are not (Picard 2010, 17). Unfortunately these important distinctions are typically neglected in discussions about sustaining journalism.

This market ontology helps bolster a corporate libertarian paradigm that masks market failures and discredits government intervention. A long-standing American exceptionalism in journalism discourse assumes almost no legitimate role—aside from servicing business interests—for government intervention in media markets. Pervading this discourse are tacit assumptions that technological fixes, the charity of benevolent billionaires, and bold entrepreneurialism—with the market acting as the final arbiter—will combine to support the journalism that democracy requires. Journalism studies could play an important role in challenging these misconceptions.

Toward a New Focus

My modest proposal for journalism studies is that the growing subfield should focus more on normative, policy-oriented, and economic questions pertaining to the future of news media institutions. Such a refocusing might help correct the creeping "new descriptivism" that critical scholars like Benson (2014) decry, and also challenge the technological enthusiasm that colors a lot of the commentary about journalism's future. Furthermore, the critical political economic framework I outline here challenges the "market ontology" that constrains discussions about the future of digital journalism. Beyond questions of whether the market will organically produce new technologically driven commercial models for journalism, or whether we understand journalism's viability as hinging on consumers' desires and media firms' profit imperatives, larger considerations are at stake. Addressing these questions, however, requires a deeply normative and structural approach to understanding—and, when necessary, intervening against—digital journalism's future trajectories. Bringing these often-neglected areas of inquiry into clearer focus will only bolster journalism studies' intellectual robustness and relevance in the coming decades.

4 Newsroom Ethnography and Historical Context[1]

C. W. Anderson

Traces of the Past in Digital News Spaces

Scene One

An immersive ethnographic research project is underway at a small digital media company that is pioneering the use of interactive informational graphics in order to display news stories in creative ways. The newsroom researcher is studying the different ways this media company makes extensive use of publicly available datasets in order to collect, analyze, and visualize statistical information. Most of the employees working on these projects are in their 20s and 30s and are trained to use a variety of types of digital visualization software. Some even have a background in computer programming. The editors and business executives of this company make regular social media proclamations about the different ways in which their "data journalism" is breaking new ground. During her time at the company, however, the ethnographer also encounters one or two older journalists who tell her about something called "precision journalism," which apparently dates back all the way to the 1960s. They recommend she join something called the "NICAR listserv." Once on this listserv, our ethnographer begins to learn about an entire past history of data journalism, one that extends the conversation further into the past, back even to the 1910s and 1920s and the early history of news infographics. At the same time, however, her day-to-day ethnographic project churns along, with new information and organizational intricacies emerging every day. Far from the history-minded listservs, our ethnographer struggles to place the shop floor buzz and frenetic pace of the digital newsroom into context.

Scene Two

A team of researchers is looking at the way reporters and editors have begun to utilize online metrics and audience-tracking technologies to understand the impact of their journalism. Digital affordances seem to make measuring the audience for news easier than ever before, and newsrooms are increasingly pioneering innovative ways to analyze the effect news is having on the people that consume it. Both the academic literature about newsroom metrics and the workers in these newsrooms are adamant that these changes in audience measurement tools represent a new era in the relationship between the consumers and producers of news. Some even claim that these changes mark a step toward news becoming more democratic as companies increasingly cater to the needs and wishes of their audiences.

One member of the ethnographic research team, however, has a background in newspaper audience research from the 1980s and early 1990s. Unlike other members of her team, she is aware that newspapers did not suddenly "discover" their audience through Chartbeat, but rather have always had different ways of trying to quantify both the consumption of news as well as the preferences of news consumers. She mentions this older research to her colleagues, but the points cause some degree of consternation. There does seem to be a qualitatively different relationship between journalists and their audiences today, other team members argue. Besides, what relevance can the existence of these older methods have for ethnographic newsroom research today?

Scene Three

An ethnographer is studying newsroom innovation at a public radio station in a tiny northeastern city. One of his primary fieldsites is a so-called "innovation lab," which has been set up inside the station in order to spur creative thinking about journalism outside the normal constraints of day-to-day newsroom practice. Members of this lab are aggressively future oriented—to the degree they consider present-day practices at all, they do so in order to decry them, pointing to "newsroom culture" and the "newsroom workflow" as impediments to making necessary journalistic change. One evening over beers, this ethnographer encounters several station veterans who begin to crack jokes about the lab; "I wonder how this will be different from the last time these guys tried this?" one producer smirks. Intrigued, the ethnographer follows up, asking what they mean by "the last time?" "Don't you know," says another journalist, shaking his head dramatically, "they pull this stunt every few years around here. Some consultant comes in and makes a big speech and produces a lot of memos, and then one day they are gone and everything is more or less like it was before." Puzzled, the ethnographer asks his academic colleagues about these remarks. "Oh, don't you see?" says one, "this obsession with the failures of the past is one of the things you ought to be studying. These newsroom dead-enders rely on history as way to justify never having make any changes in what they are already doing. Things have actually changed a lot around that station, and will continue to do so for a while!" Now the young ethnographer is really confused. Lots of people are talking about the history of his newsroom, but are doing so in very different ways. And does he even have time for this? There is a business strategy meeting tomorrow morning that he really has to get ready for. For ethnographic purposes, what does it really matter what happened five or ten years ago?

As should be clear, all three of these invented anecdotes grapple with the question of the historicity of the newsroom, and the manner in which ethnography, as traditionally conceived, can best integrate a historical perspective into its often very presentist research. In response, my argument in this chapter is straightforward: ethnographic research on news needs to develop a stronger historical sensibility than it has (with a few exceptions) so far. We need to see the newsroom as a space and an organizational assemblage that travels through time. This sense of the historical context is particularly important in the digital age, an age in which both scholars and everyday media users seem to be living within a culture of technology without a history. As everyday media users, we are bombarded by new information, new social media start-ups, new apps, and new technological demands on our attention. As scholars of communication, we run the risk of moving our analytical gaze rapidly from one new device and software system to the next, often failing to consider the roots of different technologies or the genealogical connections between them. While there are important counterweights to this tendency in the field of communication history (for example, Abbate 1999; Dunbar-Hester 2014; Gitelman 2014; Marvin 1988), historical amnesia is particularly acute in journalism studies research, where much scholarship has been weighted toward analyzing "the next big technology" or "the current crisis in news" at the expense of uncovering the technological antecedents or larger social structures that play a role shaping how journalistic change plays out over the long-to-medium term.

What have been the reasons for this, and how might we adjust our thinking to better accommodate the apparent pace of newsroom change and innovation? This chapter begins by comparing three classic newsroom ethnographies—Herbert Gans's work in *Deciding What's News*, Gaye

Tuchman's *Making News*, and Pablo Boczkowski's *Digitizing the News*—in order to demonstrate that they each adopted a different perspective on the historicity of the newsrooms, perspectives which tied back to both the time period in which they were written as well as to the larger academic commitments of each scholar. I then turn to a discussion of my own past research of newsroom collaboration in Philadelphia, as well as to current research on the use of data in journalism, as examples of what a properly genealogical ethnography would look like in the digital age. The subject of ethnographic genealogy concludes the chapter insofar as it addresses the larger theoretical and methodological commitments I think are necessary to most productively study digital newswork in an era of supposedly rapid transformation. Ethnographic research, I argue, can only succeed if it maintains a critical yet engaged stance toward its object of study, and this stance depends on both slowing down the passage of digital time as well as tracing the transformation of different visions of technology, rhetorics, and public keywords as newsrooms gallop, headlong, into a postindustrial era of news production (Anderson, Bell, and Shirky 2012).

I also think that my discussion here of the relationship between history and ethnography sheds light on some of the arguments on causality and structure explicitly broached by Benson in chapter 2 of this volume). It is important to state that I think the distance between a scholar like Benson and myself is quite small; we are certainly closer to each other in our methods and research aims that either of us is to mainstream quantitative sociology and political communication in their disciplinary search for objective causal laws of social development. Nevertheless, I think differences in we how we understand the balance between stasis and change, and the larger role played by contingency in the development of journalistic technologies, systems, and editorial products, can be usefully seen through a close reading of our two chapters. To put it baldly, I think that Benson is more concerned with the comparative analysis of journalistic structures and the way history operates as a constraining force on those systems, and, through that analysis, determining how structure limits many forms of media development. I, on the other hand, am more open to the possibility of the contingency of existing media structures (at least as a methodological starting point), the way history demonstrates this contingency, and the very real difficulty of determining the "origin" of a particular historical tendency. Drawing on Daniel Kreiss (Kreiss 2013) who himself draws on Thomas Sewell, we might say the difference here lies in an "experimental" (Benson) and "eventful" (me) vision of the relationship between structure, causality, and temporality. Whether these finely grained theoretical distinctions are useful enough to be a difference that makes a difference will, of course, be up to the reader to decide.

Newsroom Ethnographies and History: The 1970s and 1980s

All in-depth ethnographies of newsrooms inevitably involve a discussion of how news organizations report on current happenings. Insofar as this is the case, every newsroom study is chock full of dates and events—a protest outside the White House after Richard Nixon announced his resignation (Gans 1979, 55); President Johnson's earlier announcement that he would not be running for President in 1968 (Tuchman 1978, 60); even Houston, Texas's, eighth annual Art Car Week in 1995 (Boczkowski 2004, 105). But the mere presence of dates and specific events does not signify historicity in the sense I mean here. We can conceptualize the perspective I am interested in documenting as "genealogical ethnography." By genealogical ethnography I refer to an ethnographic sensibility that sees newsrooms as both embedded within and subject to the passage of time, and that seeks to understand the way different technologies, material forms of evidence, organizational processes, cultural values, and deployments of rhetoric both evolve over time and are currently embedded in different spaces and institutional routines.

The section that follows is based on a contextual overview of a large number of classic and contemporary newsroom ethnographies (Epstein, 1973; Fishman, 1980; Schleisinger, 1978; Eliasoph, 1988, Born 2005, Kleinenberg 2005) and a close reading of three of them: *Making News: A Study in the Construction of Reality* (Tuchman 1978), *Deciding What's News: A Study of CBS Evening News, NBC Nightly News, Newsweek, and Time* (Gans 1979), and *Digitizing the News: Innovation in Online Newspapers* (Boczkowski 2004). The argument of this section can thus be summed up as follows. A close analysis of ethnographic newsroom research over the last five decades demonstrates that no newsroom ethnography is content to simply operate on the emic level—they all carry within them epistemological explanations that go beyond the small-scale. However, the larger aspirations in the newsroom ethnographies of the 1970s and 1980s are different than those of more recent vintage. In the first wave of newsroom ethnography there was a tendency to look "up," connecting field-level observations to structural and more widespread cultural tendencies. However, this focus on structure radically decontextualized the 1960s and 1970s newsrooms from their position in history. Looking at the newer wave of ethnographic research we can see the contrast even more clearly, insofar as there is now a sharper focus on organizational innovation, technological adoption, and institutional transformation in order to adapt to the waves of digital change now sweeping newsrooms. Newsroom ethnography has become far more conscious of the historical contingency of its findings, and in the process, has sacrificed some of the explanatory power that made the first generation of newsroom research so powerful and enduring. At the same time, it has opened itself up to an understanding of newsrooms that sees them as far more precarious artifacts than they might have once appeared.

At the risk of oversimplifying the relationship between these three works, we can posit the following chart as a summary of the studies I will then discuss in more detail. We might think of this as a kind of "meta-theory" of ethnographic research, an argument that different research foci often emerge from particular historical moments, and that the goals/disciplinary backgrounds of different ethnographic researchers also influence the perspective adopted in tandem with the times in which they emerge. One of the most fascinating things about ethnography, after all, is that it has many practitioners—from sociologists to anthropologists to business school case builders, and that each of these disciplines has permeated the rather porous field of media studies research. In so doing, these scholars have carried their different understandings of "what ethnography is for" with them. In the chart below, this sensitivity to research goals is gestured toward the distinction between the "focus on values" and the "focus on organizational processes." The context of the times is captured in the distinction between "periods of journalistic change" and "periods of journalistic stasis" in the news business. *Deciding What's News*, by Herbert Gans, focuses on values and was written at a moment of apparent newsroom stability. Both *Making News*, by Gaye Tuchman and *Digitizing the News*, by Pablo Boczkowski, on the other hand, are primarily concerned with organizational processes, with the major difference between them being that Boczkowski focuses on newsroom process during a period of rapid technological change while Tuchman studies how these processes remain stable even when challenged by outside social groups like the women's movement.

That a sense of historical contingency is largely missing from the classic newsroom ethnographies can be immediately seen in Gans's discussion of the fact that he engaged in two separate periods of fieldwork for *Deciding What's News*, but the gap in time between his research in the 1960s and 1975 was largely irrelevant. "After shelving two drafts of the book in the early 1970s," Gans writes, "I thought my data were too old. Consequently I spent another month at NBC, *Newsweek*, and *Time* in 1975, and interviewed

Table 4.1
Ethnographic Conceptions of Stasis, Change, Values, and Routines

	Periods of Journalistic Stasis	Periods of Journalistic Change
Focus on Organizational Processes	Gaye Tuchman, *Making News*	Pablo Boczkowski, *Digitizing the News*
Focus on Values	Herbert Gans, *Deciding What's News*	

at CBS. As it turned out, the world had changed in between the two fieldwork periods, but the way journalists work had not. The book, therefore, is based on both sets of fieldwork data and some last-minute interviews to update them, conducted in June 1978" (xii). This argument is indicative of Gans's overall argument in *Deciding What's News*: a certain cluster of journalistic values and journalistic work practices pervade the news industry regardless of time or technology; these values are largely consonant with the values of American society as a whole; these values are self-reinforcing insofar as they affect the overall content produced by journalists; and, minus an even larger shift in American society and culture, we are unlikely to see meaningful changes in either news values or news content. The most central topic heading in part I of *Deciding What's News* is, after all, "Enduring News Values," and the argument is made by Gans that the values he isolates have not only remained consistent across two decades but "are probably of a far more venerable vintage." (41).

The structure of the far longer section of *Deciding What's News*, "The Journalists," is similar. Once again, Gans acknowledges that his fieldwork took place over two separate time periods and that "the historical moments I did my fieldwork may therefore affect my findings." (74). But his discussions of story selection, news sources, story suitability, the relationship between objectivity and journalism, and the journalist–audience dynamics all document a fairly static conception of newsroom behavior. These processes are all, in a word, highly decontextualized. The one major exception to this ahistorical tendency can be seen in Gans's analysis of changes to story format—a process that does, he acknowledges, occur relatively regularly. Even this discussion, however, is hedged in several important ways: "the basic formats in news media are remarkably stable," "the evening news program has not changed fundamentally" since 1964, "most format changes are small," and "drastic format change is rare" (164). The driving force behind most format change that does occur, Gans argues, is economic distress: "when news media encounter serious economic difficulties ...

more drastic format changes may be instituted" (ibid.). Still, while Gans does provide a useful overview of how and why story formats change over time, this is the exception that proves the rule; the lack of a corresponding analysis of sourcing, deadlines, story selection, and news audiences demonstrates just how central the concept of newsroom continuity was to Gans's analysis.

The simplest explanation for this focus on continuity is that Gans was open to the possibility of newsroom change but simply did not observe empirical evidence of it. His claim that there was little difference between the journalism of 1965 and 1975 should probably be taken at face value. But even if it is true that news changed little in the decade Gans discusses, his focus is still overwhelmingly structural. Even ethnographic research conducted during a period of relative stasis might open itself more to the possibility of organizational change over time. Indeed, when we turn to Tuchman's ethnographic work from the same time period we can see a somewhat different sense of the relationship between structure, newsroom values, and history, despite the fact that she too is writing in a period of relative newsroom stasis. This difference stems from the fact that Tuchman—with her focus on coverage of the women's movement—appears slightly more sensitive to the fact that changing social values and technological infrastructures do in fact affect organizational processes. For Gans, on the other hand, relatively static cultural values can be empirically linked to static journalistic values that then reflect themselves in static organizational processes.

In contrast to Gans's largely de-historicized analysis of journalism, Tuchman dedicates an entire chapter—"Facts, Freedom of Speech, and Ideology"—to the incorporation of news history into her ethnomethodological analysis. In part, this different understanding of history emerges from a reversal of emphasis; whereas Gans sought to show how stable, society-level news values become routinized, Tuchman is more interested in demonstrating how particular news values intersect with organizational processes and routines. Perhaps because of her aforementioned sensitivity to the shifting fortunes of the women's movement, Tuchman remains more sensitive than Gans to the basic notion that news organizations are affected by the era in which they operate—that changing social values are part of dialectical relationship with relatively stable organizational routines and processes. And while her primary attempt to provide historical context in chapter 8 falls a bit flat, her discussion of news narrative in chapter 6 points the way toward how later news ethnographies would think of the embeddedness of newsrooms in history.

"The Cultural Arm of the Industrial Order," a subsection of chapter 8, "Facts, Freedom of Speech, and Ideology," is more or less a literature review of two journalism history works: Schudson's *Discovering the News* and Dahlgren's PhD dissertation on the relationship between journalism and the state. Tuchman introduces this history because, as she admits, her ethnography "does not permit historical or crosscultural generalization." Nor, she adds, does she demonstrate "that contemporary news frames develop in concert with other institutions and are historically linked to them" (157). It is Tuchman's hope that the overview of this history will allow her to reconsider news as a particular type of ideology and make the links to the present day clear. In retrospect, it seems clear to us that Tuchman succeeds admirably in her attempt, especially when her work is considered in concert with the newsroom ethnographies that emerged around the same time. This link, however, does not emerge from her actual ethnographic research. In the final two chapters Tuchman uses theory, particularly the sociologically oriented phenomenology of Schutz, Berger, and Luckmann, to draw the connections between the history of journalism, the social construction of reality, and the ideological functions of the news. But her news organizations themselves are never really placed into the flow of history. The historical chapter largely takes place without reference to the news organizations she analyzes in the other chapters.

The section of *Making News* with the most sensitivity to history is, not surprisingly, the section of chapter 6 that deals with the relationship between TV technology and narrative form (104–109). "News narratives have their own history," Tuchman declares. "The use of film and videotape in television has evolved over two decades ... [and] newspapers [embrace] shorter sentences and whiter pages." What then follows is a discussion of how narrative has evolved over time. Note, however, the technological angle Tuchman takes here, particularly her concluding claim that the shift in newspaper style emerged as an effort to combat the allure of television. History, in short, becomes relevant when technology makes it so. This focus on technological innovation will be one of the keys to explaining the turn toward historical contextualization that emerges when we turn our gaze to the newsroom ethnographies of the digital era.

To sum up, both Tuchman and Gans largely fail to see the newsrooms they studied as historically contingent—Gans explicitly so, Tuchman more as a consequence of her emphasis on present-day processes and routines. I think the main reason for this emphasis on structural stability arises from the context in which these works were written—moments when it appeared, on the surface at least, like there was little change occurring in newsrooms, at least in comparison to the larger changes that seemed to be occurring in other sectors of society. Although a focus on these larger changes (embodied in her focus on the women's movement) leads Tuchman to be slightly more sensitive to the historicity of news practices, this is a largely a matter of relative degree. Both Gans and Tuchman present a largely stable world of news production, with Gans focusing on the values that drive journalism and Tuchman focusing on the organizational processes that embody these values.

Digitizing the News: Innovation in Online Newspapers, by Pablo Boczkowski (2004), has been credited with reintroducing ethnographic research into the study of journalism after an absence of several decades (Stonebely 2014); for our purposes here, it is important to note that its appearance marked a fundamental shift in the manner by which ethnographic fieldwork was contextualized historically. The evidence documenting this shift is in plain sight: the subtitle of the book is, after all, ***Innovation** in Online Newspapers*, with "innovation" signifying or even necessitating some sort of chronological passage of time between one pre- and one postinnovative state of affairs. In the methods section as well, Boczkowski is clear that his work "combined ethnographic case studies of contemporary innovation efforts by online papers with archival research about consumer-oriented electronic publishing." Comparison with the previous wave (Cottle 2000) of ethnographic research drives home just how significant these changes are and makes clear the various shifts in research emphasis that were inaugurated by Boczkowski's work. Fundamental to these changes are a much sharper sense of the historicity of newsrooms and newsroom practices, an emphasis on the materiality of innovation, and a corresponding deemphasis on journalistic values.

Chapters 2 and 3 of *Digitizing the News* are almost entirely historical, and they act as a preface to the three case studies that follow in chapters 4, 5, and 6. For a student or scholar returning to *Digitizing the News* more than a decade after it was originally published, these second and third chapters are some of the most valuable parts of the book. They allow us to read *Digitizing the News* not simply as an ethnographic history of newsrooms in 1997, 1998, and 1999, but also as a history of the first two decades of what Boczkowski calls "consumer-oriented electronic publishing," a history that extends well back into the 1980s. In some ways, we can compare the opening chapters of this book to Tuchman's generalized history of news in "The Cultural Arm of the Industrial Order," but analysis in Bockzkowski's book is related more directly to the actual case at hand. What's more, Boczkowski does not simply relegate his historiography to its own stand-alone chapter, but includes a contextual account of organizational history at the start of each of his case studies. This background helps situate the discussion of the organizational innovation (or the failure thereof) inside the flow of history as opposed to seeing it as solely the outcome of macro-level structural forces.

Digitizing the News also marks the moment when journalism studies began to intersect with science and technology studies (STS), a process now well underway across the broader field of communication (Gillespie, Boczkowski, and Foot 2014). The book thus promises to focus on the previously neglected material-spatial aspects of newsroom work. "Despite [the] valuable contributions [of previous newsroom fieldwork,]" Boczkowski writes, "research in this area has been less successful in making sense of the material dimension of news production, an issue that has become particularly pressing in view of the computerization of newsrooms in the last few decades." (80). Drawing on STS and organizational sociology, *Digitizing the News* aims to rectify this oversight (although in contrast to Boczkowski, the earlier sections of this chapter make it clear that this is less an oversight than it is a genuine conclusion that the impact of technology on news production and news values was extremely limited). This merger of STS and journalism studies thus highlights a second major difference between the earlier round of newsroom ethnographies (with their roots in classic sociology) and *Digitizing the News*. Whereas the work of Tuchman, and especially Gans, focused on the values driving journalistic behavior, Boczkowski rarely discusses values. Take his concluding discussion of the "factors that shaped the online paths pursued in the three case studies" (177). These determining criteria are summarized as "relationship between print and online newsrooms," "user inscriptions," and "character of newsroom practices" (ibid.). While values might be extracted from each of these categories (for instance, the "inscription" of the user obviously implies a particular understanding of what users do with news and thus what news workers ought to value when they do their work), values are far from the primary focus of *Digitizing the News*. While Boczkowski would shift his focus somewhat in future work, it is clear that, insofar as values are largely unaddressed, they are thus not subject to the same historical change that the news organizations themselves are—or if they are, the history of these values lies largely outside the purview of the book. The news business, in short, travels though time—but news values themselves do not.

Digitizing the News thus represents a marked departure from earlier ethnographic work. Originating in science and technology studies and organizational sociology rather than mainstream sociology per se, it focuses far more explicitly on organizational processes and procedures and the material technologies that lie behind them. Its research question is no longer "why have newsrooms changed so little given the amount of apparent ideological tumult in the larger world?" Rather, it asks "why have newsrooms changed so little given the amount of internal experimentation in news organizations, especially considering their accompanying technological shifts?" Far more than earlier works, it is concerned with attempts to place technological evolution at newspapers in context, tracing their evolutions and genealogies, particularly the ways that these technologies meant different things at different times to different people, and especially the ways that they failed to instigate larger changes in newsroom practices. This focus on genealogical shifts in material practices and meanings is part and parcel of mainstream STS, and helps counteract some of the larger tendencies toward technological determinism present in the the broader literature on technology and the news. It remains, finally, firmly on the level of operational analysis, largely dispensing with the focus on news values so common to the earlier generation of newsroom ethnographers.

Genealogical Ethnography

In my own ethnography of the Philadelphia media ecosystem (Anderson 2013), I focused on newsroom values and routines (much as the original newsroom ethnographers from the 1970s did) but adopted Boczkowski's more subtle understanding of the role played by technology and his far more historical sense of newsrooms as embedded in time. Within this general framework I examined the evolution of several "digital era" journalistic values (Kreiss and Brennan 2015)—citizen participation, interorganizational collaboration, standards of digital evidence, ideas about the audience, and journalists' understanding of "the public"—as they were mediated by journalistic routines, economic forces, and technological changes between 1997 and 2008. In terms of the chart I outlined above, my own book was written in a period of journalistic change (like Boczkowski) but focused more on values than organizational routines (like Gans). An appropriate name for this focus, I argued above, might be genealogical ethnography.

To see more clearly how genealogical ethnography might be put into practice, I want to discuss, very briefly, the example of interorganizational collaboration (which I examined in *Rebuilding the News*) and the role played by the document in journalistic understandings of newsworthy evidence (part of my current research into the history of data journalism.)

Table 4.2
Ethnographic Conceptions of Stasis, Change, Values, and Routines

	Periods of Journalistic Stasis	Periods of Journalistic Change
Focus on Organizational Processes	Gaye Tuchman, *Making News*	Pablo Boczkowski, *Digitizing the News*
Focus on Values	Herbert Gans, *Deciding What's News*	C. W. Anderson, *Rebuilding the News*

Genealogies of Journalistic Collaboration in *Rebuilding the News*

The ability of news organizations to collaborate with other news organizations in order to produce valuable media content has been a stated journalistic value in the digital era. In popularizing the idea of newsroom collaboration through links, web guru Jeff Jarvis coined the phrase "cover what you do best and link to the rest," the notion that

> Instead of saying, "we should [cover] that [news]" (and replicating what is already there), [news organizations ought to] say, "what do we do best?" That is, "what is our unique value?" It means that when you sit down to see a story that others have worked on, you should ask, "can we do it better?" If not, then link. And devote your time to what you can do better. (Jarvis 2007)

Claims that newsrooms ought to collaborate more often in order to produce stories thus build on a variety of arguments. There is an economic logic to collaboration (in an era of diminished resources, you can save money by not repeating the work of others), a technological argument (hyperlinks and distributed work practices make collaboration easier than ever before), and a technocultural argument (the digital age is an era of intrinsic networked collaboration). One my of my goals in *Rebuilding the News* was to understand how interorganizational journalistic collaboration was functioning in practice, especially since so many technological and economic forces appeared to be making this collaboration increasingly likely and even desirable.

My findings were fairly stark. Not only were collaborations between news organizations in Philadelphia during the period of my most immersive fieldwork (2007–2010) rare, those that did occur were the result of a significant degree of formal institutional planning and lengthy bureaucratic negotiations. While this may not sound so surprising in 2015, at the time it was a finding that seemed to run directly contrary to the spontaneous, networked work practices then envisioned as the future of digital media production on the Web. Few news organizations even bothered linking to other journalistic enterprises, which seemed like the most trivial digital act imaginable They were more likely to institute formal collaborative guidelines and small, one-off projects with other companies following months of planning and organization.

Why was this the case? Based on my time in the field, the primary answer appeared to be that content management systems (CMSs) discouraged hyperlinking—the stumbling block, in other words, lay with the technical system. However I also wanted to probe the notion of linking and collaboration from a more historical vantage point. To that end, I undertook an analysis of the history of a 2007 campaign journalism project called "The Next Mayor" carried out by the public radio station WHYY and the *Philadelphia Daily News*, which involved studying various iterations of the project website and interviewing the principals involved in its planning and execution. I also examined the larger history of linking by a variety of "web-native" journalism projects in Philadelphia. The analysis of the history of linking practices in Philadelphia showed me that, with the exception of blogs, few journalism organizations did much linking (even if it was technologically possible). The history of "The Next Mayor" project," for its part, showed me that aggressive collaboration with other outlets not involved directly with the project was considered its own form of journalistic newswork, one that was nevertheless subordinate to other journalistic practices such as original reporting.

In other words, a form of genealogical analysis more sensitive to the history of journalistic work practices within different news organizations showed me that what I had originally thought was a purely technical problem—poorly designed newspaper CMSs—was actually the manifestation of far deeper cultural news values which could only be entirely understood by looking at the history of these values over time. Without being sensitive to the history of the Philadelphia news ecosystem, *Rebuilding the News* would have missed the most interesting parts of the collaboration story, and it would have misunderstood the manner in which an institutional aversion to collaboration told us a great deal about the evolution of news in the digital age. I think we see here a potential model for the kind of work I have in mind when I use the term genealogical ethnography. Values are posited (in this case, the value of digital collaboration), and these values are seen as embodied in specific organizational forms (collaborative cross-institutional teams) and material artifacts (hyperlinks). The deployment of these values are analyzed ethnographically, but also as particular constellations of work procedures, technological infrastructures, and passionately held beliefs that change over the course of time.

Journalism, Documents, and Objects of Evidence

Scene Four

It is midmorning at the coworking space WeWork in downtown Manhattan, and three young journalists are arguing over the definition of a newsworthy "event." They are spending the summer working at an experimental news project called Structured Stories, in which typical 800-word stories are dropped in favor of cataloging a series of one-sentence news "events" in a language that allows them to be creatively manipulated by human beings and read in bulk by computers. Key to the process of creating a structured story is determining the signal or initiating news "event" which ought to be placed in the story database. And so the question arises: is a New York City politician's press release the key initiating event for a particular story, or is the event the press conference they hold about the press release a few hours later? What is the epistemological status of a document, anyway, given that you cannot ask it questions? But it is easy to link to, which makes it useful for the database. ... The debate goes on for several more minutes.

I encountered the previous scene during a summer of fieldwork in New York City, fieldwork with the purpose of studying some of the major issues surrounding computational journalism in the twenty-first century. During the weeks I spent at Structured Stories, I was fascinated by the role played by material evidence in building the database of machine-readable stories; it was essential, the journalists and project advisors thought, to have a link in the stories to as much direct evidence of the events as they could provide. Given that the journalists were often not at the scene of the stories they were reporting, this would provide both digital transparency and would carry epistemological weight when assessing their truth claims. This use of documents, press releases, and online videos contrasted with some earlier data journalism projects I'd studied, in which the large-scale synthesis of documents was the most important use of these evidentiary forms. The contrast between these two uses of material evidence to justify truth claims, I thought, could give us interesting insights into the epistemological issues at play in twenty-first-century data and computational journalism, and in digital journalism more generally.

What kind of larger evidentiary contexts, in other words, did documents supply the journalists working at Structured Stories? Thinking about the question this way might once again point the way toward how we can carry out a form of ethnography that sits comfortably in quadrant four of the table above. We live in a moment in which the evidentiary standards of digital journalism, as well as the larger technical and narratival infrastructures which supply journalistic context, are changing rapidly. So how do data and computational journalists justify that they really know what they know, how have they justified these claims over the longer course of journalistic history, and how has the value of "journalistic truth telling" changed over time? Not only were documents incredibly important to the computational journalists I was studying at Structured Stories, but documentary evidence has played a major role in journalistic reporting procedures since the earliest days of the occupation. These historical linkages might thus make studying the different ways that documents and other forms of evidence have supplied journalistic context an intellectually fertile way to study the evolution of journalistic truth telling over time.

Conclusion: Ethnography, Genealogy, History

This chapter has advanced concept of the "genealogical ethnography," a methodological and theoretical perspective which attempts to combine the spatial, presentist, and reflexive methods of ethnographic research with a more humanistic and critical perspective that sees these lived-in, grounded work spaces as historically contingent and moving through time. As I hope the first section of this chapter has shown, this perspective is not new; rather, the historicity of newsroom culture and practices is more apparent in eras (like our own) of rapidly changing technological, economic, and political circumstances. Boczkowski, in particular, demonstrated the value of an ethnographic project sensitive to historical change in his 2004 *Digitizing the News*. My own goal in this chapter has been to push the notion of historical contingency past the study of occupational practices to the study of journalistic values and beliefs. In a sense, the goal is to fuse the focus of Gans with the contextual sensitivity of more recent journalism research.

Foucault famously defined the genealogical method as "gray, meticulous, and patiently documentary. It operates on a field of entangled and confused parchments, on documents that have been scratched over many times" (Foucault 1971, 76). I think we can fuse this documentarian impulse with an ethnographic one, particularly if we have an ability to see both human culture and technological artifacts as entangled in a flat sort of quasi-text. For our purposes, Foucault's most important comments about his method center around the argument that "a genealogy of values, morality, asceticism, and knowledge will never confuse itself with a quest for their 'origins,' will never neglect as inaccessible the vicissitudes of history. On the contrary, it will cultivate the details and accidents that accompany every beginning" (80). In essence, Foucault—like Nietzsche—is most interested in understanding the accidental, occasionally even haphazard, origin and evolution of knowledge systems and value claims.

What is the value added by this particular way of looking at journalistic work and journalistic values in the digital age? Why not just do either media history or newsroom ethnography? There is certainly nothing wrong with these tried and true research methods, and they obviously can stand alone, but I want to argue that, when it is possible, serious ethnographers will find a lot to value in incorporating a historical, genealogical perspective into their work. I want to conclude this chapter by discussing four of them. First, a historical approach to newsrooms has the salutary effect of "slowing down" the seemingly ever-accelerating present. Second, genealogy can help inoculate us against the tendency to find the origins of the present in the past or the past in the present. Third, combining genealogical and ethnographic approaches reminds us of the vital importance of media infrastructure. Fourth and finally, ethnography forces the occasionally unmoored tendencies of media theory and cultural critique into dialog with actually existing human beings embodied in real-life institutions.

I have written elsewhere (Anderson 2015) about the ways that research in the digital age can sometimes leave us breathless, always terrified that our object of analysis (the digital) will permanently escape our grasp. Like poor Achilles and his tortoise, digital technology seems just over the horizon, always out of reach. One way to compensate for this, I argue, is to extend our horizon of analysis backward in time. By reducing the present to a manageable size, we not only guard against the tendency to treat current digital trends as an epochal developments, unparalleled in human history, but we add a layer of context that can allow us to treat the journalistic present as simply one moment among many others—an important moment, to be sure, but not the entire story.

When journalists or public intellectuals bother to talk about media history in relationship to modern-day digital developments, their most common strategy is to either claim that there is nothing new under the sun ("short bursts of randomly assembled news in the pages of late nineteenth-century newspapers were the original Twitter") or that the digital era merely marks a return to longer, older forms of communication (for instance, arguments about the so-called "Gutenberg Parenthesis"). Foucault's writings on the genealogical method, however, specifically warn against these tendencies. "Genealogy does not pretend to go back in time to restore an unbroken continuity that operates beyond the dispersion of forgotten things; its duty is not to demonstrate that the past actively exists in the present, that it continues secretly to animate the present, having imposed a predetermine form on all vicissitudes" (81). Rather than using history to uncover eternal journalistic beliefs or epistemological commitments, research along genealogical lines is more likely to argue that the connections we see in our current era and the past are either the working out of particular power relations or the temporary alignment of fragile, provisional values.

Another way to think about journalism research of this kind is to see it as an ethnographic example of what Geoff Bowker and Susan Leigh Starr once labeled media infrastructure studies (Bowker and Starr 2002). This approach aligns closely with work in the science and technology studies and the history of science tradition, and I would further argue that infrastructure studies are inherently genealogical in spirit insofar as they seek to understand the background level systems and technological operations that shape the sociotechnical world. In a sentence that might have been written by Foucault, Peters adds a critical twist to the study of infrastructures; infrastructuralism is fascinated with "the basic, the boring, the mundane, and all the mischievous work done behind the scenes" (Peters 2015a, 33). However, the importance of infrastructuralism goes beyond the manner in which it draws our attention to the subtle organization of order and power, Peters argues. "Technical infrastructures are not limits on our humanity, but its conditions," he writes elsewhere (Peters 2015b). A genealogical sensitivity to the history of apparently mundane technologies and categorization systems can push our ethnographic research in exciting directions. It might serve as a bulwark against being overwhelmed by the newest, shiniest technology or social organization system to appear in the field at the moment in which we happen to be in it.

All of this looks like an excellent argument for media scholars to conduct critical historical research. What, then, is the point of ethnography? I want to conclude this chapter by reflecting on the ethnographic impulse, having subjected it (perhaps unfairly) to a series of trials and criticisms in the previous pages. We cannot hope for history alone to tell us everything we need to know about journalism and digital media work. Most fundamentally, immersion in the field forces us to run our exotic theories, our carefully constructed historical syntheses, and our lovely models of human and organizational behavior through the gauntlet of actual living, breathing, human beings. Research on digital media needs this grounding, I argue, and media theory perhaps needs it even more. More than other disciplines, the eclectic impulses of communication and media research allow it to often launch into the air in exotic flights of fancy. These flights of fancy make our discipline interesting (Havelock 1982; Innis 1950; Ong 1982), but they also run the risk of entirely unmooring it from the real world. Nothing grounds scholarship more than dealing, on a daily basis, with the petty arguments and bureaucratic minutiae that govern workspaces. Nothing reminds us of the gap between the literature review and the world than the need to ethnographically engage with the very human activities of eating, drinking, flirting, tacking Post-It Notes on the wall, convening meetings, picking up office supplies, engaging in video conference calls, and dealing (between calls) with the family back home. History is a drag on all forms of journalistic work, and human beings animate that work in a way that pushes it ever beyond the pull of history. Research on emerging forms of journalism, ideally, will grapple with both the drag and the pull, and will do so in a manner that sheds the most light on the nature of media production in the twenty-first century.

Note

1. I would like to thank Pablo Boczkowski, Bill Dutton, Elizabeth Hansen, and participants in the April 2015 "Remaking Digital News" conference at Northwestern University for their comments on earlier drafts of this paper.

Commentary: Reflections on Scholarship in the Study of Online News

William H. Dutton

The authors in this section identify criteria for judging scholarship in the field that resonate well with related fields, such as Internet studies, and with the development of journalism studies as a whole. This commentary draws on each chapter to identify criteria for scholarly research in this field.

New Challenges for a Maturing Field

Establishing a Two-Way Street

Boczkowski and Mitchelstein (chapter 1) see reason for concern over the propensity of the field to import rather than export ideas to other disciplines and fields—what they call a one-way street. Should this academic trade imbalance be taken as a symptom of a field that is not attracting researchers from outside its domain? This balance of trade in ideas has been a focus of concern for other developing fields. For example, this was a perceived problem for the fields of communication and information studies, more generally, such as in studies of the balance of citations between them (Rice and Crawford 1992). Such an imbalance could be a reflection of the degree to which researchers in a variety of fields are not looking at digital journalism for inspiration. Alternatively, it might also be a reflection of a relatively new and interdisciplinary field, rather than the place of journalism in the hierarchy of fields. Boczkowski and Mitchelstein call for the field to redress this imbalance in order to benefit fully from interaction with related disciplines and areas of study.

There is some evidence that this pattern might be changing already, such as with the growth of Internet studies, where citations to Internet research have begun to eclipse many other major fields (Peng et al. 2013). There is a value for digital journalism in finding ways to enhance the impact of its scholarship on other fields. A major challenge for all scholars is to influence the way people think about the worlds they study. It is a primary mark of accomplishment in academia, and this could be furthered by a greater focus on "building intellectual bridges" with related fields.

Description and Explanation

Rodney Benson (chapter 2) observes a prominence of description versus explanation in the field. He acknowledges that a focus on description is expected, given the novelty of digital journalism, which immediately raises questions about the nature and diffusion of innovative practices—from the arrival of blogging to video coverage on smartphones. However, the prevalence for description over explanation is by no means limited to digital journalism. Description is often easier than explanation in the social and behavioral sciences. So Benson is right to question what he calls a "new descriptivism," and argue that the field should seek to anchor research in more explanatory than descriptive questions.

At the same time, the call for theory and explanation reminds one of a provocative counterargument that I attribute to John Kenneth Galbraith, along the lines that we are more in need of facts than theories. From this perspective, economists and other social scientists are often agile at offering up explanations of events and processes, while too frequently lacking a solid basis on the actual facts of the case. Getting the facts, which is what is pursued in good descriptive research, can be a valuable aim in itself and even help generate inductive theory. This claim underscores the interdependence of explanation and description, and how much of what we know about the social sciences emerges from inductive as well as deductive theory.

The Value of Explanatory Research

Benson does not negate the importance of descriptive research, but asks for it to be accompanied by explanation. What counts as a good explanatory framework? Benson reviews important theoretical perspectives that might be applied to online journalism. His recommendations are posed as a useful set of questions that might help a scholar move from descriptive to explanatory research: What's at stake? Despite great complexity and heterogeneity, is there systematic variation across cases? And, why?

One can agree with this general plea for placing greater priority on explanation, but with an awareness of several potential warnings or risks. First, the push for explanation in the social sciences can sometimes drive research into overly narrow problems and questions—ones that are amendable to explanation. So it is important to also take Murray Davis's (1971, 309) sage advice to always ask oneself: Is this research "interesting" in that the theories "deny certain assumptions of their audience"? That is, for example, they challenge conventional wisdom about an important topic. This is what Steve Woolgar (1999) has called a position of "analytic scepticism."

Secondly, many descriptive studies are based on participant-observer case studies or related qualitative approaches that provide patterns and themes that might be of relevance to, and therefore build some validity or generalizability over time across cases even though generalization beyond the case is not their primary objective (Diesing 1971). The advantage of such cases is more often lodged in discovering the rich detail of behavior, what some anthropologists have called "thick description." Perhaps this reaches toward explanation in Benson's sense, but it does not necessarily yield a simple causal explanation of variations across cases. Clearly, many qualitative ethnographies, and participant-observer and historical case studies can yield pattern explanations of the object of study. Identifying the complex pattern of relationships uncovered provides an explanation of value to qualitative researchers, and it might be argued that some seemingly descriptive work is explanatory in this larger sense of the term, such as Manuel Castells' (1996) conception of a network society.

Finally, in searching for an explanation, we need to beware of falling into a journalistic mode. One editor of a popular news magazine told me that the editor's advice to journalists was to simplify and then exaggerate. Social science scholarship is not journalism, and in the pressure to explain, be read, and get cited, it is critical that scholarship avoid such mantras, even if this fails at times to identify the forest amid the trees.

The Value of a Historical Perspective

C. W. Anderson's (chapter 4) discussion of "technology without history" laments the ahistorical character of much research on technology in communication, and particularly in ethnographic research on the newsroom in the digital era. This issue is echoed by Boczkowski and Mitchelstein, and is in line with the experience of many other new fields around technology and society. For example, Internet studies has faced very similar issues (Haigh et al. 2015).

In digital journalism and Internet studies, this lack of a historical context is tied often to two misleading assumptions. One is that these technologies are so new that they have no histories to recount, but this idea seldom survives efforts to understand the origins of any innovation. The second is that it reflects a technological determinism that prioritizes a focus on the future. Most questions are about what digital journalism will bring—how it will shape the future of journalism, not about from where it came.

However, the histories of technologies can provide vantage points on the present and for anticipating the future.

The flip side of this problem is that historians of technologies often dismiss the futurism that surrounds new technologies. However, visions matter as they are one factor shaping current practice and future directions, and they need to be captured by historians. Ahistorical accounts of digital journalism are likely to diminish over time as the field becomes more significant. You can see this in Internet studies where, after more than a decade of research, histories of the Internet are only beginning to be taken seriously (Haigh et al. 2015).

In my own work on communication and technology, I found that practitioners in business and government share this ahistorical perspective. It was so prominent that I invented a label for it—"innovation amnesia" (Dutton 1995). One illustration is the videophone, which failed repeatedly, only to be reinvented again and again, and most often by people who never knew about the earlier failures. I actually thought this was very functional for the innovation process as there are indeed so many reasons why a technology might fail that might become less relevant over time. Lo and behold, the videophone eventually succeeded to a surprising degree with Skype and similar video communication services.

Bringing Newsroom Innovations and Ethnographies into a Historical Context

Anderson has a more focused concern with the degree that many contemporary ethnographies of the newsroom tend to be ahistorical. His approach to a remedy for what he calls a "historical amnesia" is gained through his reading of three strong examples of ethnographies of the news. Each of these studies is able to place these newsrooms in an historical context, but are varied in the degree to which they are focused on more enduring values and cultures of the newsroom versus more ephemeral organizational processes, which tend to be more subject to changing social and technological change.

One might expect ethnographic research to be more present oriented as ethnographers are most often anchored in being present and embedded in ongoing processes, motivations, and outcomes. History might be evoked in the present, but seldom studied as consciously as the present. As Anderson shows through his examples, this need not be the case, and he offers an approach which he calls "genealogical ethnography," which he describes as a "sensibility that sees newsrooms as both embedded within and subject to the passage of time." By approaching ethnography in this way, it might be possible to be more sensitive to the historical and social contexts that are shaping the present in ways that counter this ahistorical tendency.

Identifying Blind Spots

Victor Pickard's (chapter 3) paper employs a very effective form of understatement to critique research on digital journalism for failing to adopt a more critical perspective on technical change. He argues that major gaps in the scholarship in this field include valued traditions in the larger field of journalism, such as its contributions to democratic discourse. Rather than contribute to a growing number of studies that celebrate innovations in journalism, particularly technical advances, what Pickard calls "digital exuberance," scholars of the field should focus more on implications for journalism's normative foundations, on communication, media, and information policy issues, and also the changing economics of journalism. Moreover, he provides many links across these three "blind spots" where the democratic role of the news could be undermined by the focus on digital technical change that distracts scholarship from more fundamental normative, economic, and policy issues—a critical, political economy of the news.

The potentially unintended consequences of a focus on technological innovation has been a general problem with the study of technology and society, often leading scholars to concentrate on the means rather than the ends. In the 1980s, Leo Marx (1987) put it well, asking "Does improved technology mean progress?" This kind of question helps focus attention on normative issues, such as the ends that technology should be designed to serve, and how policy and economic issues are shaping their achievement. It is often the case that these ends lie behind many inquiries that are technically focused, such as whether social media might narrow or broaden access to the news. However, unless scholarship in digital journalism is continually reminded of the democratic roles the press is expected to play, there is indeed a danger of the normative questions being lost in the excitement of research on technical innovations.

Conclusion: Deciding What's Scholarship

In focusing on what does and does not pass various litmus tests for scholarly research, it might be useful to draw an analogy with Herbert Gans's (1979) seminal book, entitled *Deciding What's News*. Of course, Gans was focused primarily on the gatekeeping process in constructing the news, while these chapters are more concerned with the criteria of high-quality research—scholarship—on digital media. However, these chapters and the symposium that framed them also pose a gatekeeping question.

In the early years of online news and digital journalism, it was important to convey the nature of innovations to those in the mainstream of journalism studies and the educated public at large. It was useful to raise awareness of what it meant to get the news electronically and organize the newsroom in dramatically different ways. By 2016, this role is no longer sufficient to achieve scholarly recognition in this growing interdisciplinary field. The contributors to this section posit a number of key attributes of what they each regard as essential to scholarship in the field. The authors have each focused on different issues, but all, separately and together, are addressing perceived gaps that create challenges for digital journalism research.

As a means of summarizing and pulling together their major arguments, it is useful to put them into two major research traditions, which I will call disciplinary and interdisciplinary research (table 4a.1). A number of the contributions focus on what might be called key attributes of enduring traditions tied to disciplinary research across the social sciences. Put differently, interdisciplinary digital journalism research must address core demands of more mainstream research of the various disciplines. These attributes include the necessity to be methodologically sound, but also innovative methodologically in order to make serious advances in contributing to knowledge of the field.

Table 4a.1
Attributes of Interdisciplinary Scholarship in Digital Journalism

Tradition	Attributes	Explanation
Disciplinary	Methodology	Methodologically sound, enhanced by innovative approach
	Theory	Explanatory vs. "descriptivist": answering "why" questions
	Interesting	Challenges conventional assumptions
Interdisciplinary	Problem-Focused	Focused on addressing a problem central to journalism
	Multidisciplinary	Drawing from more than one disciplinary perspective in and beyond the social sciences
	Sociohistorical Context	Technology placed in a social and historical context, avoiding technologically deterministic perspectives
	Significance	Significant to the ends of journalism, such as its role in democratic governance

Disciplinary research is often focused on advancing a particular theoretical concept or framework in a field. While interdisciplinary research is more focused on problems, it is no less expected to bring theories to bear and not merely describe digital journalism.

Digital journalism research would benefit from Davis's (1971) plea to be "interesting" through research that challenges the assumptions underpinning the conventional wisdom of the field. While many sound research publications might not pass this "interesting" test, those that do can stand out in the field and have greater impact on subsequent research and practice.

However, a number of other attributes addressed in these chapters are more unique to interdisciplinary research, and some are particular to journalism studies (table 4a.1). A key attribute of interdisciplinary research, as has been mentioned, is a focus on problems. This might well be its primary distinction from disciplinary research, which is more focused on advancing theory. Researchers study various innovations, practices, or structures of digital journalism because they believe these are relevant to the ends of journalism, such as informing the public or obtaining high-quality information about developments and observations on the ground. If a theory is relevant to addressing these problems, then it should be used. If a theory emerges from thick descriptions of processes or innovations, such as a pattern explanation of a process, then that is valuable for informing future research.

Given a problem focus, it is seldom possible to address major research questions from one disciplinary perspective. A commitment to problem-oriented research often goes hand in hand with a need for more multidisciplinary, often team-based, research.

Digital journalism research is inherently technologically anchored—based on some assumption that the use of digital technologies will make a difference. Why study digital journalism if technology does not matter? However, that does not imply an acceptance of a technologically deterministic perspective. To the contrary, the contributions in this section argue for placing the study of technology into a broad social and historical context in order to understand the full range of social factors, including technical change, that are shaping outcomes. Early research in this area might have often been ahistorical as has been noted, but as the significance and centrality of digital journalism is increasingly recognized, study of its historical roots and trajectories, as well as its social shaping more generally, will become a higher priority.

Finally, the section raises the bar beyond problem-oriented research to argue for work that is significant for achieving the ends which digital journalism is designed to serve. Keeping the ends in mind, and not sinking into a fascination with problems that do not question the larger political economy of journalism is perhaps the most difficult challenge for the research community. What difference will innovations make to the quality of journalism, or the control of journalism by governmental, commercial, or populist forces? This is a question that Michael X. Delli Carpini returns to in his concluding commentary for this volume.

II Rethinking Key Concepts

5 Digital News as Forms of Knowledge: A New Chapter in the Sociology of Knowledge

Rasmus Kleis Nielsen

[News] does not so much inform as orient the public, giving each and all notice as to what is going on.
—Robert E. Park (1940, 677)

Knowledge about a thing is knowledge of its relations. Acquaintance with it is limitation to the bare impression which it makes.
—William James (1890, 259)

News is intimately related with knowledge. It is part of journalism's self-understanding that news helps people understand the world around them. Wanting to know more about the world is a key motivation for using news. The hypothesis that those who use news do in fact know more about the world around them has been a central focus of academic analysis of news. The idea that news leads to a more informed and knowledgeable citizenry and therefore a more well-functioning society and more robust democracy has been central to most normative theories of journalism. All these different self-understandings, personal motivations, academic hypothesis, and normative theories rest on a connection between news and knowledge.

But what kinds of knowledge might news be said to be? And how is news as knowledge changing as the social practices, organizational forms, and media technologies that create and constitute it change over time? Robert E. Park's 1940 essay "News as a Form of Knowledge: A Chapter in the Sociology of Knowledge" remains the seminal treatment of this question, a deeply *sociological* analysis focused not simply on the information contained in individual news items or on the impact that such information may have on media users, but on the significance of the wider category of "news" and the social, organizational, and technological factors that shape it. Park was, in a way, uniquely well positioned to write about this subject—before his pioneering academic work as one of the founders of the Chicago School of Sociology, Park had worked both as a newspaper journalist and as a publicist and researcher for the African American educator and political activist Booker T. Washington at the Tuskegee Institute. He was a man who, more than most, connected academic research, professional practice, and a lifelong concern for how knowledge is produced, recognized as such, and circulated, with how it orients, engages, and ties together communities in different ways.

The purpose of this chapter is to address the question of what kinds of knowledge today's digital news might be said to be and to offer a contemporary sequel to what Park called "a chapter in the sociology of knowledge" (which I take to be a field focused not on the epistemology of journalism—what and how journalists think they know what they think they know—or the validity of journalists' knowledge, but the actually-existing conditions under which different kinds of knowledge known as news arise, how they work, and how they change).[1] I am concerned with what changes in news content, the organization of news work, and the technologies involved in producing and disseminating news means for how we think about news as knowledge, and will discuss these more general issues on the basis specifically of past and present examples from the United States. In the first part of the chapter, I return to the United States in 1940 and Park's original analysis of news as *a* form of knowledge positioned between formal and systematic "knowledge of" and the more intuitive and unsystematic "acquaintance with" (a continuum sketched out by the pragmatist philosopher and psychologist William James). In the second part, I develop an analysis of digital news as *forms* of knowledge in the very different media environment of the United States today. In the third and final part, I discuss the wider implications of digital news as a new chapter in the sociology of knowledge.

Overall, my argument is that while Park could plausibly offer one broad ideal type for understanding news as a form of knowledge in 1940, such a unitary view is less useful today and we should therefore be increasingly wary of analysis or commentary that tries to make generalizations about news as such. News is a historical phenomenon, and its forms have changed over time just as it varies across space (Barnhurst and Mutz 1997). I suggest that much news today is still frequently characterized by many of the traits Park identified, but that our increasingly digital media environment offers far more diverse forms of news and also includes a growing amount of substantially different kinds of news closer to James's extremes of "acquaintance with" and "knowledge about." Today, we see simultaneously an increasing emphasis on presentist, minute-by-minute and second-by-second, breaking news *and* the growth of various forms of long-form journalism, explanatory journalism, and data journalism designed to overcome some of the perceived epistemological shortcomings of older forms of news—constituting new forms of news as knowledge that has greater staying power as content, but also because of certain affordances of digital media. Drawing on Park and his inspiration from James, I suggest we can think of digital news as involving at least three different ideal-typical forms of mediated, public knowledge today. First, we see the growing importance of forms of news-as-impressions, decontextualized snippets of information presented via headline services, news alerts, live tickers, and a variety of new digital intermediaries including search engines, social media, and messaging apps. Second, a recognizable descendant of the archetypical late-twentieth-century form of news remains important: news-as-items, published in principle as self-contained, discrete articles and news stories bundled together in a newspaper, a broadcast stream, on a website, or in an app. Third, at the opposite end of James's spectrum from acquaintance-with to knowledge-about, we see the rise of news-about-relations, combining elements of long-form "contextual" or "explanatory" forms of journalism well known from some twentieth-century newspapers, magazines, and current affairs programs (Fink and Schudson 2014) with new forms of data journalism, visualization, and interactivity enabled and empowered by digital technologies (Fink and Anderson 2015).

Digital news may be associated with the rise of news-as-impressions and a potential hollowing out of inherited forms of news-as-items—with more transient information for what Park in 1940 called a "specious present." Certainly many critics among journalists, academics, and other public figures complain about its "churnalistic" qualities. But digital news is far more than this and we should be suspicious of overarching generalizations about the nature of news today, which also involves a remarkable growth in news-about-relations, more oriented toward providing what James called knowledge-about, and news that today is more accessible, more timely, and more detailed and data driven that probably ever before. Recognizing the properties of digital news as *different* forms of knowledge—rather than *a* form of knowledge—will help us understand how journalistic self-understandings, popular conceptions of journalism, academic hypotheses about journalism, and normative theories of journalism might require rethinking as the basic connections between news and knowledge they all implicitly rely on change over time.

News as *a* Form of Knowledge

In his original 1940 essay on news as a form of knowledge, Park was interested in news specifically as a *mediated* and *publicly available* phenomenon. This is the kind of news we associate with journalism. One might draw the distinction between news in this narrower sense of information that is published or broadcast and on the public record versus news in the broader sense of new things, novelties, and tidings which can be largely interpersonal and often private. News in the first sense is deeply intertwined with a wide range of informal forms of social communication including rumor, gossip, and personal conversation—news makes people talk (Shibutani 1960). But it is distinct from it, because it is mediated differently, because of its public character, and because the category of "news" is *seen* as significant and distinct. Drawing on the sociological tradition of developing ideal types—conceptual constructs formed inductively by accentuating one or more salient characteristics of a given class of phenomena to enable categorization and analytical generalization—Park's essay aimed to identify the shared properties of the kind of news we associate specifically with journalism, to in turn be able to assess what kind of knowledge it might be said to represent.

As noted above, Park's analysis of news in this sense as a form of knowledge starts with the distinction between "acquaintance with" and "knowledge about" developed by the philosopher and psychologist William James (1890) in *The Principles of Psychology*. In James's work, "acquaintance with" is knowledge that is more informal, intuitive, and unsystematic, whereas "knowledge about" is relatively more formal, theoretical, and systematic (these are not a categorical distinction as much as a relative ones, a question of degrees). "Acquaintance with" comes with use and habit and is often based on first-hand experience, "knowledge about" comes with systematic investigation and is actively acquired, often in part from secondary sources. Here is James, who is worth quoting at length (1890, 221-259):

> There are two kinds of knowledge broadly and practically distinguishable: we may call them respectively knowledge of acquaintance and knowledge-about. Most languages express the distinction; thus, γνῶναι/εἴδομαι; noscere/scire; kennen/wissen; connaître/savoir. I am acquainted with many people and things, which I know very little about, except their presence in the places where I have met them. I know the color blue when I see it, and the flavor of a pear when I taste it; I know an inch when I move my finger through it; a second of time, when I feel it pass; an effort of attention when I make it; a difference between two things when I notice it; but about the inner nature of these facts or what makes them what they are, I can say nothing at

all. [...] But in general, the less we analyze a thing, and the fewer of its relations we perceive, the less we know about it and the more our familiarity with it is of the acquaintance-type. The two kinds of knowledge are, therefore, as the human mind practically exerts them, relative terms. That is, the same thought of a thing may be called knowledge-about it in comparison with a simpler thought, or acquaintance with it in comparison with a thought of it that is more articulate and explicit still. [...] What we are only acquainted with is only present to our minds; we have it, or the idea of it. But when we know about it, we do more than merely have it; we seem, as we think over its relations, to subject it to a sort of treatment and to operate upon it with our thought. [...] Knowledge about a thing is knowledge of its relations. Acquaintance with it is limitation to the bare impression which it makes.

For Park, as for James, the distinction does not imply a hierarchy, but reflects different ways of acquiring knowledge and forms of knowledge that play different roles in people's lives. Tacit knowledge and practical experience at the "acquaintance with" end of the spectrum is often a better way of know-how to do something or of knowing another person than "knowledge about" is, though more systematic forms of knowledge may be superior when it comes to, say, treating a serious disease or understanding the causes and likely consequences of a global financial crisis. Knowing that there are Russian troops in the Ukraine is about impressions, knowing why they are there or what it might mean is about relations.

Park's basic argument is that if one thinks of James's "acquaintance with" and "knowledge about" as a continuum "within which all kinds and sorts of knowledge find a place," then news has a location of its own between the two extremes (Park 1940, 675). It is worth recapitulating what he sees as its defining traits, since many remain instantly recognizable and relevant many years later (even though historical analysis suggests news in the US on the whole over time has grown longer, more analytical, and more abstract; see e.g., Barnhurst and Mutz 1997). News as a form of knowledge, according to Park, is more formal and systematic than "acquaintance with" and offers people knowledge of the world beyond personal experience because it is communicable and communicated in a way that tacit knowledge is not. But it is not the same as "knowledge about," because it remains focused on events, rather than processes, relations between events, or causes or meanings of events. In this sense, news is (and this is still a common observation) far better at the first four of journalism's famous "5 Ws"—it offers more on *what* happened, *who* did it, *when* it took place, and *where* it happened than on the fifth W, the *why*. (Let alone the *so what*?)

Throughout his essay, Park stresses that news is oriented toward events, and interested in the past and the future, in causality and teleology, only in so far as it recognizes such relations as throwing light on the actual and present. (Commentary and opinion is and arguably always has been a variation on this, perhaps less strong on informing people about events, and more oriented toward interpreting or asserting causality and teleology.) Making another point that is still commonplace today, Park argues that news can be said to exist only in what he calls a "specious present"—"present" because it is about the now (here today, gone tomorrow, first draft of history, etc.), "specious" because what qualifies as present news is determined not simply by actuality, let alone intrinsic importance, but by journalists' news considerations, organizational routines, publishers' commercial or other interests, and their various conceptions of what people are interested in. This combination leads news to focus on the authoritative, the exciting, the unusual and unexpected (though of course much of the actual present is rather more humble, mundane, usual, and expected). This focus on events and the present, in Park's view, makes news "transient and ephemeral," a constant flow of small, independent communications that can be easily and rapidly comprehended by people who by doing so can orient themselves in a wider world than that of their personal experience, but who still rely on other forms of knowledge (closer to the "acquaintance with" and "knowledge about" ends of the continuum) when it comes to navigating everyday life and work. News, in this sense, helps "orient" us in the world, especially the world beyond personal and professional experience, even if it does not necessarily cultivate acquaintance with or knowledge about things. Having read about Russian troops in the Ukraine is different from having seen them in your back yard or being privy to their plans or purpose.

Park's description of news as characterized by (a) an orientation toward events over processes, (b) little discussion of causality or teleology, (c) its transient and ephemeral nature is so recognizable and relevant today that it is easy to forget that his analysis is based on a very different media environment from ours. Briefly, the United States in 1940 was a heavily regionalized media system with little in terms of truly national media (let alone international media), and where news was mediated primarily via newspapers and radio (and to some extent magazines and newsreels). The most common media technologies of Park's time are material incarnations of his point about transient and ephemeral news, with daily newspapers often overtaken by events and radio broadcasts dissipating into thin air, here one moment, gone the next. The newspaper industry that survived the Great Depression was a vibrant one with high circulation (about 115 weekday copies per 100 households). Most of the country was covered by local papers, and large metropolitan areas sustained a range of competing morning and evening titles of different orientations and political persuasions—in Chicago including not only the *Tribune*, but also other dailies like the *Daily News*, the *Times*, and the *Sun*, the country's most prominent African American newspaper, the *Chicago Defender*, as well as a range of weeklies, community papers, suburban papers, and a number of titles in other languages. Radio, by 1940, had made it into about 80 percent of all households but was no longer the free-wheeling free-for-all of the 1920s. The 1930s had seen the transformation of what had been a large number of separate transmission towers, broadcasting independently and for a variety of different reasons, and vying for attention with many other, into a more consolidated and commercial business dominated by a few large networks linking a large number of stations across the country (Barnouw 1968).

This media environment was definitely a "mass" media environment in terms of large audiences of individual media users paying attention to the same print or broadcast content. But it is not the kind of more nationally oriented mass media environment characteristic of post-war radio and television broadcasting. Franklin D. Roosevelt's famous "fireside chats" in 1940 reached an estimated 25 percent of the radio audience, about half the audience share that major Presidential addresses would draw on television during the 1960s and 1970s (Craig 2000; Eshbaug-Soha and Peake 2011). Nor is it the kind of local and regional monopoly newspaper media environment that arose especially with the demise of most evening papers in the 1960s and 1970s. Most importantly for the purposes of this chapter, it is an environment devoid of the two most important news media platforms of today's environment—television and the Internet—and an environment in which media were a more confined part of most people's everyday life than is the case today, neither always-on nor ever-present the way television and the mobile Web increasingly are today.

Digital News as *Forms* of Knowledge

Even as Park's description of news as oriented toward events, as thin on discussions of causality (why does it happen) and teleology (what does it mean), and as transient and ephemeral may still strike us as capturing key aspects of much of today's news, our media environment is of course markedly different. The United States today still has a regionalized media structure with more than 16,000 radio stations, more than 2,000 television stations, and over 1,300 daily newspapers, as well as thousands of digital-only news sites of various sorts.[2] It is a media market where local and regional media are still an important part of most people's media repertoire, but there is also a range of truly national news media, most importantly television network news, cable television news channels, and the websites of a small number of high profile newspapers such as the *New York Times* and digital-only news media like the Huffington Post. (Where the US in the 1940s offered lots of local media and fewer national media, the situation in 2015 is increasingly lots of national media and fewer local media.) It is an increasingly digitized and digital media environment (though TV still accounts for most time spent with media), increasingly accessed via the Internet and navigated via branded websites but also search engines and social media.

This is a media environment, like Park's, with plenty of mass audiences, if by mass audience one means large numbers of individuals engaging with the same media content. But it is also, like Park's, not a media environment in which these mass audiences routinely approximate a *national* mass audience, the way they did in the 1970s and 1980s. President Barack Obama's 2015 State of the Union address drew more than 30 million viewers and a combined audience share of 20 percent (across 13 television channels)—not far from the audience Roosevelt could command for his fireside chats, but down considerably from the 67 million viewers and 44 percent audience share that President Bill Clinton drew for his first State of the Union address in January 1993.[3] Media events like the Super Bowl and the Olympics may still draw large national audiences, but everyday news media use is characterized by a combination of audience fragmentation and audience duplication rather than a few large audiences gathered around a few media outputs (Webster, 2014). It is a media environment where legacy media predating the rise of digital remain the most important producers of and sources of news for most people, but where their business models have been deeply disrupted by the migration of advertising to digital platforms and by increased competition for audiences' attention. It is an environment where the two most important media platforms for news are television and the Internet (Newman et al. 2015), both of course very different from the dominant news media of Park's time.

Scholars and commentators invested in a certain vision of what news may have been or ought to be have been highly critical of what television news and digital news offers today. Briefly summarized, many argue that increasing commercial pressures and intensified competition for audiences leads to more infotainment, more opinionated content, and to more sensationalism (e.g., Fenton 2010). James Fallows—journalist, writer, and former speech writer for President Jimmy Carter—summarized the overall assessment of digital news as "shallow, divisive, and unreliable" (Fallows 2011). (In the UK, the trend has been summarized as a move from journalism to "churnalism.") President Obama, for one, seem to share the assessment. When asked in 2010 by a student at George Washington University what most surprised him in his first years in office, the president responded:

> Well, where do I start? (Laughter.) You know, let me tell you something. On the one hand, I've been surprised by how the news cycle here in Washington is focused on what happens this minute as opposed to what needs to happen over the course of months, years. The 24-hour news cycle is just so lightning fast and the attention span I think is so short that sometimes it's difficult to keep everybody focused on the long term. The things that are really going to matter in terms of America's success 20 years from now when we look back are not the things that are being talked about on television on any given day or appear on the Internet on a blog on any given day.[4]

The charge from academics, commentators, and various public figures seems clear. News organizations may have been initially slow to reinvent the form of news and adapt to a digital environment, preferring to just transfer analogue content to digital platforms as "shovelware" (Barnhurst 2010). But now new forms of genuinely digital news are developing, and many critics are not impressed. If journalism has traditionally aspired to inform people by making the significant interesting and the interesting significant, the accusation is that today, in face of commercial and competitive pressures and enabled by technology, it is increasingly simply focusing on making the interesting as interesting as possible in the battle for attention, while leaving the significant aside. The charge in short is that news today is like the news that Park wrote about in 1940, only more so—more focused on events, even thinner on discussions of causality and teleology (even if offering probably even more commentary and analysis, some of it as opinionated and speculative as ever), and even more transient and ephemeral. At its epistemological best, such forms of news may provide a mediated, second-hand form of "acquaintance with" (this, that, and the other happened). At its worst, it may not make sense to think of it in terms of knowledge at all, but just a torrent of impressions—all noise, no signal (Gitlin 2001).

As made clear from the outset, I think the critics are partly right, but also that the charge is misleading if we mistake what may be the most voluminous forms of contemporary news for the wider, more varied, and abundant total output of news. Even recognizing the differences between newspapers, radio, newsreels, and magazines, Park could plausibly offer an analysis of news as *a* form of knowledge, positioned between "acquaintance with" and "knowledge of" in 1940. His ideal type never corresponded to all news, but it captured key aspects of most news, and the most important forms of news. But such a unitary view is less useful today, where we have certainly seen the intensification of some of the features Park outlined, developments that move some forms of breaking news and light news toward a version of the "acquaintance with" of William James's spectrum, but also, in parallel and arguably equally important, the development of forms of news that are closer to the "knowledge of" end of James' spectrum, while still being very much news—mediated, publicly available, and far more timely than other forms of "knowledge of" (like that offered by scientists). Digital news is live tweeting, live blogging, and live TV coverage of breaking news, sometimes of a character where one may not feel news is any form of knowledge at all. But digital news is also long-form journalism, in-depth, detailed, data-driven interactive features, and on-demand streaming or download of well-researched current affairs or documentary programs. As news grows even more diverse, we need more ideal types to capture the increasingly diverse real types of news that abound, and we should be skeptical of analysis and commentary that aims or claims to capture one general tendency common to all forms of news.

At the "acquaintance with" end of James's spectrum, we see the increasing importance of forms of news that are increasingly unsystematic reportings of what, who, when, and wheres, with less and less room for or attention paid to the whys and so whats, and where it is, as with first-hand experience, left to the media user to make sense of the stream of content, to transform news into some sort of knowledge. A Facebook posting, a 140-character tweet, a news alert on a smartphone or a smartwatch, or an image or a 57-second video shared on Instagram will give you an impression, and maybe some information, but not a lot about relations, causation, or teleology, no matter how otherwise self-consciously serious and substantial the news organization behind it is. In television, 24-hour news channels might be the clearest example of this, because they are not about not having the time or the space to cover relations (they are 24-hour channels, after all), but because they are built around a format where the imperative to go live and break the news in real time frequently leads to situations that seem to question the value of the whole enterprise, at least in terms of knowledge. Consider just a few from CNN, an organization arguably more committed to conventional forms of news journalism than some other US cable news channels: June 28, 2012, CNN splashed "Mandate struck down" on air, on their website, and on Twitter, with the subheading "High court finds measure unconstitutional," suggesting, erroneously, that the Supreme Court had ruled against the Affordable Care Act ("Obamacare"). In fact it had upheld it 5–4. April 18, 2013, while the manhunt was still going on, CNN wrongly reported that a suspect had been arrested after the Boston Marathon bombing. Both are high-profile events where cable television news audiences spike, and both are serious, high-profile, cases of getting it wrong.

Breaking news on television is not alone in providing a steady stream of news-like content. Constantly updated news feeds on websites as well as dedicated live blogs built around individual events like high-profile political speeches, product launches, or sporting events, as well as many streams of content distributed via third-party social networking sites ranging from Facebook and Twitter to Instagram and Snapchat, are other examples. Alf Hermida (2010) has coined the phrase "ambient news" for how microblogging platforms like Twitter can function as an "awareness system" that provides instant online dissemination of short fragments of information from a variety of sources. This is a subset of what media industry professionals see as a wider phenomenon, where digital information is, in the words of Eric Schonfeld, "increasingly being distributed and presented in real-time streams instead of dedicated Web pages."[5] It is Web organized, like breaking news on television, in a reverse-chronology based on nowness and a permanently unfinished flow of information. We see it with social networking sites like Facebook and especially Twitter, we see it with how news brands and others use these sites to distribute a stream of their own content (and on average have about 1 percent of users click on links, according to one estimate [Thompson 2015]), and we see it in digital news with initiatives ranging from Buzzfeed to the Times Wire (Madrigal 2013). This stream of content is the kind of phenomenon many academics, commentators, and public figures decry. It is still news, in the sense of being produced by journalists and laying claim to a category of "news" as distinct from the wider "news to me," it is mediated and publicly available, but it is also in many ways more unsystematic, closer to William James's notion of "acquaintance with" that the more distinct packages we associate with more traditional forms of newspaper and television bulletin news. It gives (mediated, second-hand) impressions, but it does not convey relations. And there is a whole lot more of it.

But it is not alone. This is a simple but important point.[6] There is also today far, far more mediated and publicly available news much closer to the "knowledge of" end of James's spectrum. While retrenchment in the news industry has led to an overall decline of specialized beat reporting, we have also over the last decades seen the rise of various forms of long-form journalism, explanatory journalism, and new forms of data journalism that all are clearly different from the more immediate and unsystematic news offered on 24-hour news channels and various digital streams and different from forms of news that Park wrote about. (Katherine Fink and Michael Schudson [2013] call this "the rise of contextual journalism.") Take one example, Dana Priest and William Arkin's investigation into "Top Secret America" for the *Washington Post*, extensive reporting on the growth in the national security and intelligence services in the aftermath of 9/11. The project took almost two years, involved 27 other contributors, and produced a four-part series of articles in the newspaper, a dedicated project website with additional stories, video, multiple interactive graphics, and searchable databases, a project blog, a Facebook page, a Twitter account, an hour-long *Frontline* documentary for TV broadcast by PBS (and a later sequel), as well as a book. Whatever this is, it is not acquaintance with, and it is not "churnalism." (It might have other shortcomings as a form of mediated, public knowledge, in not connecting to audiences and their everyday concerns, view of the world, etc.) Or consider the currently active *New York Times* visual guide to the Iraq-ISIS conflict, a continually updated and evolving digital offering that combines conventional reporting from the area (text and pictures) with multiple interactive maps, annotated satellite imagery of the Euphrates and the Tigris rivers, maps of US airstrikes, data visualizations of where ISIS fighters come from, historical maps of the area going back to the Ottoman empire, and much, much more. This is far more systematic reporting, still primarily oriented to the what, who, when, and wheres, but with far more attention paid to the whys and the so whats, and with forms of storytelling that go well beyond individual news items or the multipart series to embrace new forms of digital presentation. This is not impressionistic. It is about relations, much closer to James's view of "knowledge of," while still being mediated and publicly available. And it is far more timely than the "knowledge of" produced by NGOs, think tanks, and academics. The point here is not that this kind of in-depth reporting is intrinsically superior to the more ambient news provided by cable television and digital news feeds. It is not. Nor that it is perfect (data journalism, for example, is highly dependent on datasets that can be hard to verify, and well-resourced investigative reporters also make mistakes). The point is simply that it is different, that there is more of it, and that these different kinds of news coexist and that both are characteristics of news as different forms of knowledge today.

In a post sometimes invoked by industry professionals when talking about digital news, the American author Robin Sloan has suggested a distinction between the stream of impressions (which he calls the "flow") and the "stock" of more durable things that stand out.

> Flow is the feed. It's the posts and the tweets. It's the stream of daily and sub-daily updates that remind people that you exist.
>
> Stock is the durable stuff. It's the content you produce that's as interesting in two months (or two years) as it is today. It's what people discover via search. It's what spreads slowly but surely, building fans over time.[7]

In 1940, Park argued that news as a form of knowledge is positioned between the ends of William James's continuum, between intuitive, unsystematic, and first-hand "acquaintance with" and more formal, systematic, and often second-hand "knowledge of" as a form of knowledge that was mediated and publicly available and characterized by (1) an orientation toward events over processes, (2) little discussion of causality or teleology, and (3) a transient and ephemeral nature. Much of this is still true for much news. But the unitary view he offered is less and less useful when it comes to digital news, and a single ideal type is no longer adequate to capture a phenomenon that is growing more and more diverse and varied. (The intellectual utility of ideal types rest not on whether they correspond to an always messier reality but in whether they help us order and understand the most important aspects of that reality.) A part of news—by volume a large part—has, as many have argued, become more akin to James's "acquaintance with," as a steady stream of television news and digital news feeds flow freely and continuously and offer impressions, but offer little in terms of narrative structure or focus on relations. We might call this news-as-impressions. A part of it remains recognizably akin to what Park analyzed in the mid-twentieth century, even if some of the aspects he identified may have been intensified. This is news-as-items, coming in the form of discrete articles and stories. But simultaneously, the rise of combinations of long-form, explanatory, and data journalism exceeding the headline and the story suggests that some kinds of news increasingly offers knowledge that is closer to the more formal, systematic, and second-hand "knowledge of" James talked about, but is still recognizably news, based in part on on-the-ground reporting and bearing witness (vicarious mediated first-hand experience) and far more timely than, for example, the humanities and social sciences as other forms of "knowledge of" current affairs. We can call this news-about-relations. News still offers the first draft of history, but the draft comes in more and more different forms, ranging from instant impressions scribbled and shared in streams to more detailed, data-driven, and systematic forms of explanatory journalism that go well beyond what news offered in the twentieth century.

Part of this development—from news as a form of knowledge between acquaintance with and knowledge about in the twentieth century to digital news as more differentiated forms of knowledge including more acquaintance with but also more knowledge about—is about the professional innovation and journalistic ambition associated with pioneers like Jonah Peretti of Buzzfeed and Melissa Bell of Vox Media, as well as evolutionary thinking by established brands like CNN and the *New York Times*. It is being integrated into journalistic professional practices and news media's workflows as more and more organizations work not only with fast journalism but also slow journalism, and avoid producing content that is too long to be sharable, and too short to be in-depth. But it is also, at a more fundamental level, driven by the development of the larger technological systems within which news is produced and published. It is highlighted by Sloan's reference to "search" above, and it bears directly on the question of whether, and how, news is "transient and ephemeral." In terms of informational content, much digital news may well be transient. But in terms of how it is mediated and made publicly available, much of it is less and less ephemeral because of the combination of digital publishing and the development of search engines, most importantly Google. The digital humanities scholar Lev Manovich captured many of the key changes early on. In 2001, he highlighted five tendencies of "a culture undergoing computerization" (Manovich 2001, 27–48):

1. Numerical representation: new media objects exist as data.
2. Modularity: the different elements of new media exist independently.
3. Automation: new media objects can be created and modified automatically.
4. Variability: new media objects exist in multiple versions.
5. Transcoding: the logic of the computer influences how we understand and represent ourselves.

All of these developments can be observed in news, which takes the material form of digital data in the production process and increasingly at the point of use, that can be aggregated, shared, and remixed, and is increasingly seen through the cultural lens of the technologies that allow us to use news in new, digital, ways. In an earlier piece, Manovich (1999) highlighted how the Internet privileges databases over narratives, a prescient point given how search engines help us navigate the Web on the basis of link patterns and page ranks rather than semantic principles (in his analysis, the development represents a basic inversion of the relationship between syntagm and paradigm, from a situation in which we navigated by narrative and the database was implicit, to a situation in which the database paradigm is central and the syntagm [narrative] less central). In part because of the numerical representation of media content (as digital data) and its modularity (elements independence of each other) we can encounter news in new ways through new and increasingly important digital intermediaries like search engines and social media (Bell 2014; van Dijck 2013). No systematic data is publicly accessible, but conversations with digital editors suggest not only that search engines and social media account for larger and larger parts of overall Web traffic to news, but also that direct visits to the home page, traffic coming via search, and traffic coming via social, exhibit interestingly different temporal orientations, with home page traffic driven by news updates, search traffic by older pieces (and much of this by pieces that are weeks or even months old), and social media driven by recent but not breaking news pieces.

At the most basic level, the features of the rise of digital media that Manovich describes mean that even news content that is highly transient is no longer ephemeral in terms of our ability to access it. Search the Web for almost any topic, and you will find news items, sometimes years old, from highly visible news organizations all over the world. Click on links shared via social media on current affairs, and they may take you to historical context, not just breaking news. News media organizations themselves are increasingly aware of this and—in line with Manovich's point that the way in which digital media influences how we understand what we do and present it to the world—are trying to operate with more differentiated forms of distribution and content creation, marketing their archives (sometimes decades or even centuries back), prioritizing measures to increase the "shelf life" of new content, and using various tools and technologies to lead readers from one article (whether old or new) to other articles (whether old or new). This is a new way of encountering, experiencing, and exploring news, a transformation in how most of us can engage with news. This development is only in its early stages, as the technical limitations of search engines, database navigation, and the costs of storage and transmission (declining as they are) still prioritize textual content over images, audio, and video in terms of its "findability," but it points toward a future where much of what is published as news never needs to disappear (unless someone wants it to disappear, keeps it in proprietary databases, stops maintaining the site/app, or is forced to make it hard to find—as per the European Court of Justice's landmark "Right to Be Forgotten" ruling against Google in 2014).

A New Chapter in the Sociology of Knowledge?

Digital news, because of changes in the social practices, organizational forms, and media technologies that create and constitute news, represent a variety of different forms of knowledge today. Robert E. Park's analysis from 1940 still captures key features of much of what we encounter as news—it is oriented toward events over processes, rarely systematically deals with causality or teleology, and the content is often transient—but news is also growing more diverse and much of it is far from ephemeral.

Some forms of digital news—by volume probably most of digital news—may well be inching toward the "acquaintance with" end of William James's continuum of forms of knowledge, offering mediated impressions of things going on in the world beyond our immediate experience, but little in terms of placing those events in relation to wider processes in terms of why they are happening and what it means. Such news-flows are produced by digital media organizations like BuzzFeed but certainly also legacy media organizations like CNN (on air and online) and the *New York Times* (through the Times Wire, Twitter accounts, headline service on smartwatches, etc.) Many critics decry such forms of news, some for some good reasons.

But this development should not blind us to the parallel and simultaneous growth in forms of digital news that are far closer to the "knowledge-about" end of James's spectrum, forms of long-form, explanatory, data-enriched journalism that offers mediated, publicly available, forms of news very much concerned with questions of causality and teleology, with the relations between events, and that offers this in a far more accessible and timely fashion than other forms of "knowledge about" current affairs. Legacy media organizations like CNN and the *New York Times* invest significant resources in producing in-depth, detailed, and enduring investigations, but also sites like BuzzFeed have in recent years built investigative journalism teams in both the US and the UK.

These more diverse forms of digital news, which I have suggested we think of as news-as-impressions, news-as-items, and news-about-relations, are also in important ways less transient and ephemeral than Park's news. This is in part about the character of some of the content produced—the temporal orientation of news-as-impressions is often "now," of "news-as-items" the roughly last 24 hours we associate with newspapers and television evening news, but for news-as-relations, the orientation is often much longer, and breaks with an accelerated news cycle in favor of news meant to be more durable. It is also about technology, about content digitally published and left online by media organizations, and news that today can be accessed through a whole new range of database queries, search engines, and social media that gives us whole new ways of engaging with news, including news from yesterday or yesteryear. (Especially important for forms of news *meant* to be less transient and ephemeral.)

These changes in news as knowledge seems to merit a new chapter in the sociology of knowledge, and an appreciation of the many different forms of knowledge that news offer most people today. Of course, to understand the implications of this change, digital news as forms of knowledge has to be understood in relation to the different self-understandings, popular conceptions, academic hypotheses, and normative theories we hold of it. Very briefly, the evidence so far suggests that even as news grow more diverse and is accessed in new ways deeply embedded in wider everyday media practices (a glance at your phone at the bus stop, as part of a social media feed), most people still see the category of "news" as significant and distinct from the broader "news to me." But the ways in which people navigate the news on offer in an abundantly supplied, increasingly high-choice media environment underlines the increasing importance of understanding the changing relationship between news production (including journalistic self-understandings) and news use (popular conceptions, the social meaning of news, the motivations for using it, etc.) (Prior 2007; Neuman et al. 2012; Boczkowki and Mitchelstein 2010). As Alexis Madrigal (2013) has pointed out, the questions we face include how content producers position themselves between what people in the industry term "stock" and "flow," how different intermediaries facilitate stock and flow and offer us ways of dealing with each, and finally how we as media users combine stock and flow. (The supply side of news-as-impressions, news-as-items, and news-about-relations is important, but so is the demand side.) Leading digital news professionals are highly conscious of these differences and are experimenting with new ways of doing journalism that leverage the advantages of each form for different editorial (and commercial and other) purposes.

For academics concerned with the impact of news on what people know, the development of new forms of digital news challenges established theoretical and methodological designs in interesting ways (Bennett and Iyengar 2008). Consider three of the most important ways of studying the relationship between news and knowledge—surveys, experiments, and content analysis. Survey research relying on self-reported data on media exposure is good for many things, but not necessarily for capturing the impact of a six-second Vine news video or a detailed, interactive, piece of multimedia journalism. Experimental research under controlled settings can give us a far better understanding of the importance of certain variations in content and form of news, but increasingly faces problems of external validity and thus generalization if news use becomes a less discrete practice and news content a more fluid and diverse category. Content analysis, developed for analysis of news-as-items, also faces challenges when it comes to coding large volumes of very pithy news-as-impressions and more expansive forms of news-about-relations that go beyond the individual article or story, challenges that will become more pronounced as news organizations begin to use more responsive forms of design and potentially begin to automatically personalize text, images, and videos to different demographics on the basis of audience data.

It is not in itself surprising or interesting to point out that news has changed, also in terms of its relation with knowledge, since 1940. News is a historical phenomenon, always changing over time, just as it is a socially contextual phenomenon, varying across space. But let me highlight two key points that run through the chapter. First, Robert E. Park remains an insightful commentator on news and its relation to knowledge, its (changing) character as something that "does not so much inform as orient the public, giving each and all notice as to what is going on," and his 1940 analysis continues to capture certain key traits of what sets news as mediated, publicly available information apart from other forms of knowledge. Second, while Park's media environment too, of course, was complex and diverse, our media environment today is arguably even more so, and the idea of news as *a* form of knowledge should be replaced by the idea of news as *forms* of knowledge to acknowledge how much of it is pushing toward the extremes of "acquaintance with" (news-as-impressions) and "knowledge of" (news-about-relations) that Park placed it between, even as the middle, represented by twentieth-century forms of newspaper journalism and broadcast news bulletins (news-as-items), may well be eroding.

Notes

1. The epistemology of journalism has received some attention, including from Tuchman (1978) and Ettema and Glaser (1998). The validity of journalism is increasingly a subject of meta-journalistic coverage from, for example, fact-checking organizations (Graves 2016]).

2. Data from the FCC, the NAA, and the Project for Excellence in Journalism's 2014 State of the Media report.

3. http://www.nielsen.com/us/en/insights/news/2015/31-7-million-viewers-tune-in-to-watch-pres-obamas-state-of-the-union-adress.html (accessed March 10, 2015).

4. http://www.whitehouse.gov/the-press-office/2010/10/12/remarks-president-a-moving-america-forward-town-hall (accessed March 10, 2015).

5. http://techcrunch.com/2009/05/17/jump-into-the-stream/ (accessed March 16, 2015).

6. As Herbert J. Gans (1974) has noted, cultural critics concerned with "dumbing down" often overlook parallel processes of "smartening up."

7. http://snarkmarket.com/2010/4890 (accessed March 13, 2015).

6 On the Worlds of Journalism

Seth C. Lewis and Rodrigo Zamith

If only, as Michael Schudson (2013) has wished, we could get journalism to hold still for a moment, then we might assess what has happened to it, as an occupational field and paradigmatic form, during a period of seemingly unending and upending change. Then we might make sense of what has become of news—how it is produced, shared, used—in a world awash in digital media technologies. Such technologies enable networked arrangements that complicate conventional distinctions between production and consumption, professional and amateur, public and private, and so on. But while Schudson's desire is merely tongue-in-cheek, it speaks to a broader concern: the difficulty of understanding this thing called journalism at a time when technology has made a mess of what we thought we knew about a profession and practice intended to provide a first-draft accounting of public life.

In particular, technology has assumed an increasingly central role in every aspect of journalism (Anderson, Bell, and Shirky 2012; Lewis and Westlund 2015), much as in communication (Gillespie, Boczkowski, and Foot 2014) and media life (Deuze 2012) more broadly. Of course, technology is no more a single and stable "thing" than journalism is, but it is nonetheless apparent that a great deal of the recent change associated with journalism is technologically oriented: from ideas of "convergence" and then "digital first" becoming the norm in many newsrooms (Schlesinger and Doyle 2015), to discussions around training journalism students to write software code (Creech and Mendelson 2015), to grassroots movements to bring together journalists and computer programmers (Lewis and Usher 2014), to the growing role for news applications and interfaces (Ananny and Crawford 2015), to the adaption of news to suit the logic and flow of a social media environment (Belair-Gagnon 2015), to the heightened awareness of audience preferences via digital metrics (Petre 2015), and to the number of job advertisements looking for technologically savvy individuals who know how to harvest and analyze large-scale digital data, put algorithms and automation to work for journalism, and altogether bring a computational mindset and skillset to their work (Lewis and Usher 2013).

As technology becomes more salient, different actors once external to or on the margins of news organizations have moved closer to the center (Nielsen 2012; Westlund 2011). For example, the computer nerd once tasked with fixing email and shoveling content onto the website is now enrolled in coming up with new storytelling techniques, such as making news more interactive for users (Usher 2016). Some of these technologists are journalists turned coders while others are outside web developers and data scientists brought into newsrooms. Either way, these changes contribute to new contexts in which journalism is understood as a social system and applied as an occupational practice. In all, these shifts enable new kinds of symbolic interactions, which continually redefine the social meanings of various forms of news work. This, in turn, leads to the reinterpretation of journalism itself, and the boundaries associated with that domain (Carlson and Lewis 2015).

Our goal in this chapter is to consider what such developments mean for conceptualizing journalism and its interrelationship with technology. Rather than analyze a particular empirical case, this chapter makes a broad conceptual provocation about what we call the "worlds" of journalism. We argue that, to fully understand the nature of technological change in journalism, it is important to adopt a sociological lens that brings into focus the *collective* nature of journalism—its interconnected people, processes, and products—as well as the relative status, or *valuation*, afforded to certain actors and activities. Drawing on symbolic interactionism as a theoretical framework, and in particular Becker's (1982/2008) application of its ideas to the study of "art worlds," we call for considering journalism—and specifically ambient, data, and algorithmic journalism—as a series of distinct but intersecting "worlds." These worlds represent networks of social actors, labor activities, material infrastructures, and patterns of production that collectively enable and legitimize particular forms of journalism.[1] Put another way, particular and constantly changing configurations of actors, conventions, and cooperative activities permit and constrain particular forms of journalism, and confer upon those individuals, processes, and products a certain status that may not fully translate across the flexible and porous borders of those arrangements, or worlds.

Seeing journalism in light of worlds, we argue, helps accentuate at least three things: (1) the heterogeneity that exists among social actors (humans) and technological actants (machines) and their activities; (2) the development and negotiation of various conventions that give shape to certain creative works; and (3) the resulting arrangements that, while constantly in flux, lend distinctive value (and thus status) to certain people, practices, and products. Such valuations matter ultimately in shaping understandings of and expectations for journalism as a social enterprise that is increasingly technological in orientation.

Worlds as a Framework

Becker's (2008) ideas are in many ways rooted in symbolic interactionism, a major theoretical perspective in sociology that emphasizes the subjective meaning of human behavior and social processes. Symbolic interactionism assumes that people act toward things, including others, on the basis of meanings developed through social interaction, and that such meanings are continually reconstructed through an interpretive, sense-making process (Blumer 1986; Snow 2001). Drawing from this theoretical fountain, Becker articulates a way of thinking about artistic production, distribution, consumption, and legitimation that speaks to but can be applied beyond the arts, revealing a general schema for organizing the social world of actors, activities, and the conventions through which collective production is accomplished and meaning is assigned. Art worlds, according to Becker, refer to the cooperative networks of actors oriented around the creation and distribution of particular works that its constituents consider to be *art*.

Art worlds do not have clear and static boundaries, nor are they wholly distinct from one another, or from other parts of a society. An individual may belong to multiple art worlds simultaneously, yet perform different functions within their respective cooperative networks and, further, receive different degrees of acclaim—or no acclaim at all. In particular, the boundaries of a given art world and the valuation of particular individuals and their contributions also shift in response to the introduction of, among other things, new technologies, new ways of thinking, and the emergence of new audiences.

In Becker's view, art worlds should be understood as social systems, networks of people "whose cooperative activity, organized via their joint knowledge of conventional means of doing things, produces the kind of art works that art world is noted for" (2008, xxiv). As van Maanen (2009) has noted, collective (or cooperative) activity and conventions are the twin core concepts of Becker's analysis. However, to these two key concepts, we would add an implicit but no less important emphasis on the matter of reputation and status, or the relative legitimacy afforded to certain people, processes, and products in art worlds. Becker's final chapters take up "Change in Art Worlds" and "Reputation," and it is in and through such fluctuations and negotiations that status becomes a preeminent concern. For how change in art worlds occurs, and with what implications for the reputational character of art and artist, are of central importance for thinking about journalism in a time of turbulence.

Although Becker devotes great attention to the distinction between "art" and "craft" in his book, we find that discussion to be nonessential for the purpose of this chapter. We adopt Becker's proposition that art (or, more broadly, exceptional work) is defined not by the presence of certain aesthetic qualities but by whether members of a given world consider specific practices or products as being artistic in nature. That is, as Becker argues, rather than seeking out particular attributes, the analyst should "look for groups of people who cooperate to produce things that they, at least, call art" (2008, 35). While there is ample evidence that many journalists consider their work to be a kind of art (for examples, see Kerrane 1998; McNair 2005; Merrill 1993), what is most crucial to this chapter is the determination of what and who are considered to be ordinary and extraordinary—that is, what should be considered mundane support work and what should be considered something more, or what gets to be called "creative" and what is considered merely "technical." Thus, to not distract from this core aim, we focus here simply on "worlds" rather than "art worlds" while staying true to Becker's insights and using them to study changes in technologically oriented forms of journalism.

Key Concepts for Understanding Worlds

Central to Becker's analytic framework is the concept of *collective activity*, or the notion that art is the result of cooperation by multiple individuals. He contends that there are a series of activities that "must be carried out for any work of art to appear as it finally does" (Becker 2008, 2). Put differently, if certain activities were not executed, the work might occur in some other fashion, but it would not be the same work. Becker offers a provisional list of regular activities in the production of art, from the development of an idea, to the securement of supporting activities (e.g., copy editing), to winning the appreciation of an audience. This process culminates with the generation and maintenance of the rationale that those activities make sense and are worth doing. It is this final activity that yields the justification for why something is art, perhaps even good art, and explains its value to society. Ultimately, such a series of activities establishes cooperative links that are central to the production of notable work. Becker writes, "The artist thus works in the center of a network of cooperating people, all of whose work is essential to the final outcome. Wherever he depends on others, a cooperative link exists" (2008, 25).

In order to facilitate the requisite interdependence, *conventions* must be developed. Becker defines conventions as "earlier agreements now become customary" that "cover all the decisions that must be made with respect to works produced" (2008, 29), or "the ideas and understandings people hold in common through which they effect cooperative activity" (2008, 30). Conventions, which may be likened to norms, are important because they "dictate" the materials and abstractions to be used as well as the form in which those materials and abstractions will be combined, and they "regulate" the relations between the creator of notable work and his/her audience (Becker 2008, 29). Actors within a particular world, such as artists, may, and often do, break convention in order to stand apart or feel less constrained. However, in setting themselves apart, those actors run the risk of becoming marginalized, seeing the circulation of their works limited, or having the valorization of their talents decreased.

A final concept central to Becker's analysis is that of *reputation*. Reputations arise from consensus-building within the relevant world. That is, an individual's reputation is not something created by that individual, but rather by agreement among the various members of that world. Becker writes that, for works, makers, schools, genres, and media, reputations serve as "a shorthand for how good the individual work is as one of its kind, how gifted the artist is, whether or not a school is on a fruitful track, and whether genres and media are art at all" (2008, 362). Put differently, individuals' reputations are of import because they are central to the value accorded to them and their output.

Building on Becker: Shifting from Reputation to Status

The concept of reputation offers a useful starting point, but it is perhaps prudent to adopt the broader lens of *status*. Specifically, reputation is an important part of setting oneself apart from peers. However, the function that a person plays within a cooperative network is also important to determining that value. Within a given network, for example, the value accorded to an ordinary cellist and his abilities may exceed that of a renowned sound engineer. Status is therefore conferred as members of a given network classify certain forms of work as being more valuable than others—and, in turn, deem certain practitioners and their talents as more essential than others.

Although the production of notable work requires cooperative effort around shared conventions, making everyone and everything important, the concept of status reinforces that not all jobs and functions are created equal: members of a given world place different valuations on different forms of work and the actors associated with them.

Status, in turn, influences the all-important allocation and management of resources. This is true both of material resources (e.g., funding, equipment, and physical space) and social resources (e.g., delineating core and support personnel, and valuations of expertise). It determines who has access to what kind of resources, how such resources may be expended to produce and distribute "noteworthy works," and in what manner future resources are likely to be gained by pursuing a particular course. Ultimately, status gives shape not only to what the work looks like, but if it is to be considered exceptional at all (i.e., as art). Notably, and consistent with the core tenets of symbolic interactionism and Becker's application of art worlds, status is not singularly possessed but rather is a persistent negotiation among various parties to the production, exhibition, and reception of things that come to be viewed as exceptional.

Worlds of Technologically Oriented Journalism

Viewed through Becker's lens, journalism is comprised of distinct worlds forming around its various genres and practices. For example, sports journalism is a particular kind of journalism compared to investigative journalism, underscored by their separate professional associations—the Associated Press Sports Editors (APSE) in the case of sportswriters and Investigative Reporters and Editors (IRE) for investigative journalists. Similarly, these distinctions may be oriented around functions and practices that may be spread across different types of journalism, as evident in professional associations such as the Society for News Design (SND) for designers and the National Institute for Computer-Assisted Reporting (NICAR) for computer-assisted reporters.

While these distinct worlds surely share certain components and constituents—e.g., through networks such as the Society of Professional Journalists (SPJ), connecting journalists across domains—not only are their conventions distinct but also the reputation and status accorded to actors varies across them. For example, Nate Silver and his FiveThirtyEight website may command no special attention in a world oriented around narrative or literary journalism, even as they are held in very high regard in the data journalism domain. Similarly, while the inverted-pyramid style of storytelling may be conventional in the data journalism world, it is widely rejected in the narrative journalism world.

It is important to note here that such worlds are dynamic, changing continuously. They sometimes change gradually, through drift—minor shifts that do not require significant reorganization of cooperative structures and activities—or in a more substantive manner that requires participants to learn and do different things. Sometimes, they change abruptly and in a disruptive manner, a revolutionary process that demands major changes to the character of the works produced or the conventions employed. A change may be said to be revolutionary when "one or more important groups of participants find themselves displaced by the change, even though the rest remains much the same" (Becker 2008, 307). To be sure, Becker clarifies that not every pattern of cooperative activity needs to be changed for a revolutionary change to occur, and that for some members of a particular world, a given change may not be revolutionary at all.

In some cases, changes may be comprehensive enough to warrant the creation of an entirely new world—and as some worlds come into being, others dissolve. In particular, worlds may emerge with the development of new concepts and ways of thinking, as with the novel, which emerged partly as a result of the idea of "formal realism" as a mode of discourse in fiction (Watt 1957), and three-dimensional photography, which emerged as a result of the stereoscope (Becker 2008). Worlds may also emerge with the development of a new audience, with the artistic work itself remaining largely unchanged but new distributional arrangements allowing new markets to be tapped, as with the "'new' rock music of the 1960s [that] resembled what had preceded it" (Becker 2008, 313).

Technology Change in Journalism

One key driver in the emergence and the demise of worlds is technology, as innovations make new art products and distributional arrangements possible, though not necessarily inevitable (Becker 2008). As scholars have observed, journalism has been subjected to several fundamental changes in recent years—changes that are more revolutionary than gradual under Becker's conceptualization. For example, newspaper companies in many developed economies, long the largest employers of journalists, quickly transformed from cash cows to risky properties during the crisis of the mid-2000s, leading to massive job losses, a focus on restructuring, and growing pressure on newsworkers to do more with less (Picard 2011; Soloski 2013). Meanwhile, competition for audience attention in the media environment has intensified amid the growth of digital-native news sites, mobile and social media, and viral content providers, even as users themselves have gained greater opportunities for collectively and affectively shaping their news experience (Papacharissi 2015).

As Picard (2014) has noted, these and other changes—changes that have been greatly influenced by technological innovations—are fundamentally altering the nature of newswork, reshaping institutional logics, and leading to the emergence of new modes of news production. For example, Picard points to the rise of a service mode: news products are transformed into news services, content is streamed across platforms, and syndicated material, user-generated content, and linkages with other news providers all become increasingly central to a news organization's operation.

The Worlds of Ambient, Data, and Algorithmic Journalism

This shift to a service mode has led to, and is evidenced by, the growing importance of digital curation and aggregation skills. As ambient journalism, or a form of journalism that focuses on collecting and communicating news information drawn from streams of collective intelligence (Hermida 2010), has gained legitimacy, and as social media has become an increasingly important source of news (Anderson and Caumont 2014), a distinct world may be said to be developing. Within this world, the ability to efficiently sift through large volumes of information and quickly assess quality are demonstrations of skill, and the simple messages—sometimes no longer than 140 characters—that synthesize that deluge of information and break through the noise serve as exceptional works. These abilities are increasingly desired in job ads, as journalists are ever more expected to manage virtual communities, not only to encourage dialogue around stories and promote their circulation but also to unearth original information (Bakker 2014).

Another notable technologically supported development is the growth of data journalism (Fink and Anderson 2015; Howard 2014). As Coddington (2015, 343) puts it, "the goal of data journalism is to allow the public to analyze and draw understanding from data themselves, with the journalist's role being to access and present the data on the public's behalf." As such, data journalists seek to identify stories in data and/or tell stories through data. This process requires that the journalist not only be familiar with traditional skills of journalistic storytelling, but also have some familiarity with data structures and databases, statistics and statistical software, and, in many cases, design and visualization utilities. In response to these developments, data journalism courses have been introduced to the curricula in journalism schools (Splendore et al., 2015). These courses emphasize technical skills, such as how to scrape data, write SQL queries, and create visualizations using Google Fusion, Tableau, and related software.

These developments can be seen as part of a deepening dependence on digital technology in newswork broadly (Lewis and Westlund 2016). In addition to the proliferation of technologically *supported* practices, the field has also seen the emergence of technologically *oriented* developments such as algorithmic (or computational) journalism (Anderson 2013; Young and Hermida 2015). Algorithmic journalism involves "the application of computing and computational thinking to the activities of journalism including information gathering, organization and sense-making, communication and presentation, and dissemination and public response to news information" (Diakopoulos 2011, 1), and emphasizes the abstraction and automation of work (Coddington 2015; see also Stavelin 2014). Automatically rendering (natural) language from computational representations of information, algorithms have been used by start-ups to generate stories about sporting events and public financial disclosures (Dörr 2015; van Dalen 2012). Even mainstream news organizations such as *The Los Angeles Times* have used algorithms to automatically write blog posts about homicides in the area and populate a dynamic map (Young and Hermida 2015). Altogether, the infusion of increasingly technologically dependent forms of work (cf. Powers 2012) complicates traditional labor dynamics at the intersection of human and machine in newswork, leading to questions about how to conceptualize emerging relationships among social actors and technological actants in journalism (Lewis and Westlund 2015; Lewis and Westlund 2016.

Changes in Status

Ambient journalism, data journalism, and algorithmic journalism may all be viewed as distinct worlds—even as they overlap, as worlds do to varying degrees (see table 6.1). Specifically, they involve particular logics and skills, and their abilities and contributions receive distinct rewards. Prominent actors within each of these worlds are rewarded with greater status within the given world and are seen as being more valuable to their organizations. For example, Andy Carvin's numerous invitations to speak at conferences serve as an example of his increased status in the world of ambient journalism (see Hermida, Lewis, and Zamith 2014). Similarly, Paul Lewis from the *Guardian* and Ravi Somaiya from the *New York Times* gained prominence for their use of social media as a reporting tool during the London riots (Vis 2013). Furthermore, the emergence of prizes such as the Shorty Awards' Best Journalist in Social Media and the *Press Gazette*'s Social Media

Table 6.1
Examples of Emerging Technologically Oriented "Worlds" of Journalism

	Ambient Journalism	Data Journalism	Algorithmic Journalism
Collective Activity	Optimizing news for a social media environment	Making data more public, transparent, and interactive	Applying automation to expand the space of news production
Conventions	Using social media as key framework for gathering, verifying, and sharing news	Applying social scientific methods to identify patterns in sets of data	Employing computer code to automate traditional journalistic functions
Status	e.g., Andy Carvin's lauded Twitter-based coverage of the Arab Spring	e.g., Nate Silver's FiveThirtyEight as a prominent model in the field	e.g., The Associated Press relying on automated reporting to cover quarterly financial reports

Journalism Award, as well as new graduate degrees such as The City University of New York's MA in Social Journalism, help to delineate and legitimate a unique world.

In a similar vein, distinct news sites have emerged around data journalism, from subunits in large media organizations such as the *New York Times*'s The Upshot blog and the *Guardian*'s Data Blog, to websites such as FiveThirtyEight and Vox. In particular, Nate Silver, founder of FiveThirtyEight, has received extensive praise from organizations such as the Nieman Foundation for Journalism and the International Academic of Digital Arts and Sciences for his ability to effectively leverage, contextualize, and tell stories through data. Additionally, the *Columbia Journalism Review*, in partnership with the Tow Center for Digital Journalism, started in 2012 a dedicated column to "analyze, interrogate, and explore emerging work" in the area of data journalism (Codrea-Rado 2012). At NICAR's annual convention each spring, the leading lights of data journalism—in many cases journalists-turned-technologists, working for news applications teams at the likes of ProPublica, *The Chicago Tribune*, and NPR—are widely lauded as the next generation of computer-assisted reporters: artists in a world where data meets storytelling.

In the realm of algorithmic journalism, the case of the *New York Times*'s Ken Schwencke illustrates the growing valuation of journalists (and non-journalists) who are able to automate portions of their work. Schwencke, then at the *Los Angeles Times*, was celebrated for programming an algorithm that scanned information from the U.S. Geological Survey, identified newsworthy earthquakes, and then automatically wrote a headline and story, appended a map, and published it to the newspaper's blogging platform. This allowed Schwencke to beat his competitors as he slept, with a blurb that flowed much like a wire story. The Global Editors Network, a community of editors-in-chief, has featured sessions on algorithmic journalism in their recent annual conferences, including one talk entitled, "Robot Journalism: Don't Wait 'Til It's Too Late." Similarly, the 2015 meeting of SRCCON, "a conference for developers, interactive designers, and other people who love to code in and near newsrooms" (see srccon.org), listed multiple sessions on employing computational methods (e.g., machine learning) in journalism, selecting thought leaders in that world to headline those sessions. The Associated Press, in particular, has received a great deal of attention for employing an algorithm to write more than 12,000 articles a year using data from corporate earnings reports. Graduate programs in Computational Journalism have also emerged in prominent journalism schools, such as Syracuse University in the United States and Cardiff University in the United Kingdom. Proclamations that, within a few years, a computer program will win a Pulitzer Prize—the de facto mark of artistry in the field of journalism—ultimately point to the growing valuation of the ability to program algorithms to do journalistic work (Lohr 2011). More importantly, these developments point to the emergence of a world oriented around creating algorithms that can embody the ideals of journalism.

A common thread throughout these examples is that the mastery of technological actants is being increasingly viewed as valuable work within these worlds and within journalism more broadly, and those individuals who possess such skills are the beneficiaries of elevated status. A parallel to this development within conventional art worlds may be found in Becker's brief description of the evolution of sound mixing as an art form. Becker points to the example of the recording engineer and sound mixer, once largely viewed as technical support staff whose skill was measured by their ability to capture the sounds of a performance. However, as Becker notes, the introduction of high-fidelity recordings and multitrack recorders enabled those individuals to record different sound elements separately, manipulate them, and combine them in different ways. Soon, sound mixers were given prominent credit on record albums, and sound mixing itself began to be viewed as a distinct artistic activity requiring special talent. As Becker (2008, 18) puts it, "sound mixing, once a mere technical specialty, had become integral to the art process and recognized as such."

Within the realm of journalism, similar shifts may be found as individuals with technical abilities move from the periphery of news organizations to more central positions. For instance, as individuals like Aron Pilhofer shift from being "that nerd in the corner you'd call to help with a spreadsheet and maybe troubleshoot your email" (Pilhofer 2010, para. 4) to becoming executive editor for digital at the *Guardian*, such transitions reflect the recognition of their ability to effectively utilize technological actants. In particular, these abilities allow the newsworkers who possess them to put out news products—such as interactive, data-driven visualizations—that can be more easily differentiated in a crowded market, either because of their individual labor or their significant roles in larger teams. And, the demand for such technical skills is evident on job boards such as News Nerd Jobs, which states up front, "The news business needs people who can code in the public interest and build the digital news products of tomorrow. If you can code, there's a job for you." Whether such technical work qualifies as art or art-like may certainly be contested. The prominence and prestige—indeed, status—afforded to such individuals, however, is nevertheless apparent. Newswork may depend on cooperative activity around shared conventions, but the glamor associated with any particular role is always in flux—and, at the moment, it clearly favors the so-called nerds and ninjas, even those untrained in journalism, whose skills are so dearly coveted. Not surprisingly, then, resources have followed, as major news organizations engage in an arms race for top developer talent and build out, for example, teams of "data grinders and designers" (Phelps 2012, para. 1) focused on perfecting the art of exploring news through data visualization (Howard 2014).

Additionally, there is increasing recognition, driven largely by the emergence of discourse around big data as well as the growing availability of publicly accessible data (Lewis 2015), that individuals with technical know-how can lead a shift away from unrepresentative reporting that focuses on exceptional cases and toward more generalizable reporting that focuses on central tendencies. For example, as one journalist at the *Los Angeles Times* reported to Young and Hermida (2015, 390): "Mr. and Mrs. Outlier get covered really well in crime news. ... But as you know, that's an incredibly small fraction of the amount of crime that happens. ... But what data can bring us ... is to try to give some fuller sense of crime as a phenomenon in the city." The ability to tell comprehensive stories that reflect, with greater accuracy, the incidence and relative importance of newsworthy matters thus becomes recognized as valuable, if not artistic, labor.

Finally, technology has contributed to the commoditization of news stories, and thus reconfigured the status associated with writing a conventional news narrative. The isomorphic tendencies of news organizations have become heightened in a media environment that encourages minute-by-minute monitoring, leading to more and more homogenous coverage (Boczkowski 2010). One consequence of this homogeneity is the growing realization that news stories, in their current inverted-pyramid structure, are quite often redundant and less distinct from one another, as any Google search of a major current event will reveal. It's not so much that news routines have changed greatly—pack journalism has been around for many decades—but rather that the Web platform has more fully revealed the institutionalized nature of news production (Ryfe 2012). The upshot is that being able to write a conventional news story simply may not count for as much anymore; the status has declined as the differentiation has diminished. Not helping matters for traditional news writers is the rise of automated journalism, or "algorithmic processes that convert data into narrative news texts with limited to no human intervention beyond the initial programming choices" (Carlson 2015, 417). The drama playing out amid the growth of Narrative Science and other providers of robot-written news, as Carlson shows, reveals what happens when established conventions become disrupted: the relative need for (and thus status of) particular forms of human labor is called into question, leading to concerns about the authority of journalism and its normative role in a larger sense.

The Contributions of Worlds as a Framework

It is important to acknowledge that what we have hypothesized about ambient, data, and algorithmic worlds of journalism is precisely that: a series of *hypothesized* conceptions, developed from the literature as well as our own observations and fieldwork at various stages from 2011 to 2015. As such, these surmised worlds of journalism, among others that exist or are emerging, deserve investigation to clarify their actual boundaries, constituents, and implications. Even in their hypothesized form, however, these conceptions serve as provocations for a new way of thinking about the journalism–technology intersection. Namely, they orient attention to the shaping influence of (and influences shaped by) distinct but interlocking domains of collective activity, conventions, and status conferral. In this way, a "worlds" view brings at least three key considerations in the study of journalism into greater focus:

1. that journalistic products, including its exemplary works, are the result of the combined labor of a large set of social actors and technological actants that is more heterogeneous than typically is acknowledged in the literature;
2. that such cooperation is enabled by conventions, which both facilitate and constrain the creation of particular works; and
3. that the resulting arrangements are constantly in flux, with the valuation of particular actors, works, and forms of labor differing between worlds, even as they contribute to the general understanding of what we call journalism.

One of Becker's (2008) key contributions in *Art Worlds* is to puncture the myth of the artist toiling alone. As he shows, the artist benefits from a wide ensemble of social and material forms of support, including (and crucially so for this chapter) technicians and the technologies they manage in the service of objects deemed to be art. As Lewis and Westlund (2015) have proposed, inquiries into the processes guiding change in journalism should be at minimum conscious of the different social *actors*, technological *actants*, types of *audiences*, and work-practice *activities*—the Four A's—that are interconnected in the production of both ordinary and extraordinary journalistic products. For example, as Braun (2015) has documented, the creation and distribution of a single video on MSNBC.com involves the use of multiple technologies developed by a diverse set of actors that yields a particular set of affordances—many of which are unintentional and seemingly counterintuitive—for both news producers and consumers. Beyond process, even something as simple as an interactive crime map requires varied expertise. For example, Kirk (2012) points to eight hats of data visualization design: the initiator, the data scientist, the journalist, the computer scientist, the designer, the cognitive scientist, the communicator, and the project manager. Few individuals possess the expertise necessary to effectively wear all eight hats; rather, as Smit, de Haan, and Buijs (2014) note, collaboration among these actors—who generally have different backgrounds and priorities—is key to successful visualizations.

Journalism has always involved a diverse set of activities performed by myriad actors and actants. Nevertheless, technologically oriented worlds like ambient, data, and algorithmic journalism often involve actors, actants, and activities that have received limited attention in the literature, such as programmer-journalists creating JavaScript code to build an interactive graphic or using machine learning to extract key information from a large cache of documents stored in a relational database (for examples and discussion, see Usher 2016). Moreover, such worlds typically cater to particular audiences, whose appreciation can be readily measured through nonpurposive forms of feedback (e.g., article view counts derived from audience analytics; see Zamith 2015) and purposive forms of feedback (e.g., reader comments on articles; see Zamith and Lewis 2014). Such feedback can be used to justify particular logics that guide and distinguish those worlds.

In addition to highlighting the inherently collaborative nature of the production of artistic products, Becker (2008) emphasizes the role of conventions—agreements on how things should be done, which emerge from the interactions among the actors within a given world. While scholars have long called attention to the importance of routines in engaging in journalistic production (e.g., Tuchman 1978), a worlds perspective highlights the fact that conventions may not translate across worlds, or be decided by the same ensemble of actors and actants to fit the same set of activities. Indeed, as Lewis and Usher (2013; 2014) have indicated, many of the entrants into technologically oriented worlds follow logics that emphasize iteration and "tinkering," thereby promoting rapid development, fluidity, and experimentation rather than careful consideration toward a static, polished product. Similarly, those logics underscore the importance of leveraging collective intelligence by increasing and facilitating interaction. Crucially, as we emphasize below, it is those individuals—many of whom have little, if any, background in journalism—who are increasingly developing the systems and best practices, and in turn shaping the conventions, that guide technologically oriented worlds.

Finally, Becker (2008) points to the very fluid nature of such arrangements, and the discrepancies in the reputational cachet accorded to a given individual and product across segments of journalism. This, we argue, is key for understanding certain developments in journalism, especially when the observation is extended to account for the different roles involved in particular arrangements. Applying these insights to journalism and its increasing technological orientation, it could be argued that changes in media technologies (e.g., the rise of algorithms and automation, and the development of sophisticated Web frameworks) and the personnel connected with them (e.g., the need for technologists to maximize the utility of such technologies) may lead to new perspectives about what counts as a distinguished form of creativity and who counts as a distinguished creator.

Of course, this development need not be a displacement. Longstanding forms of news writing, such as literary journalism, have not suddenly gone out of style—and, within their respective worlds, such journalisms retain many of their same conventions and forms of reputational authority. However, as many news institutions reorganize themselves with an eye toward information technology, actors and activities once seen dismissively as "support" or "technical"—on the margins of journalism—increasingly move closer to the core enterprise of producing news. These moves strengthen the economic and symbolic resources that are available to such individuals, especially as the worlds of ambient, data, algorithmic, and related forms of technologically oriented journalism gain popularity and credibility. Such moves also simultaneously make a claim to other worlds of journalism about the relative esteem and aesthetic appreciation that ought to be afforded to technologically oriented actors and their "technical" work. Indeed, as Nielsen (2012, 975) points out, increasingly, "technologists do not simply execute decisions already made by journalists and managers. They play an active role, bringing not only technical know-how ... but also their own values and views on how [journalism and technology] ought to be done."

A worlds perspective therefore highlights that such determinations—of values and evaluations, of aesthetic acclaim and authority—are in constant negotiation and result from the interactions among the members of particular worlds. In technologically oriented worlds of journalism, this calls attention to the importance of studying both formal gatherings, such as online learning spaces designed to bring together journalists and technologists to develop open-source innovations for news (Lewis and Usher 2016), and their more informal counterparts, such as "meetups" among hacks (journalists) and hackers (coders) in many large cities around the world (Lewis and Usher 2014). Such interactions, facilitated online and offline, within and across institutional boundaries, help unite core members of worlds and, through such interactions, distinguish virtuosos, reconfigure conventions, and ultimately recognize exceptional work. However, such negotiation, it must be noted, also occurs beyond this immediate group, encompassing the symbolic interactions among actors ranging from the server administrators at content delivery networks to those who consume news. This larger negotiation is the continual struggle to define journalism, to shape the social boundaries around what counts as news and who counts as a journalist, as well as why such an occupation may be democratically useful.

Thus, valuations arising from worlds of journalism, while localized in their own right, matter by association. In the aggregate, in the network of distinct but interconnected worlds, meaning-making and shared interpretations established in one world both influence and are influenced by similar processes playing out in another. The broad character of this thing called journalism, we might say, is a bricolage of multiple worlds within it, each developing particular forms of collective activity, conventions, and status-giving that work in relation to (though not necessarily in harmony with) one another. In this way, to say that journalism is becoming technologically oriented is to recognize the rise and growth not simply of certain work tools and techniques, but indeed of *technologically oriented worlds that give symbolic meaning to the people, practices, and products underlying those developments*. As certain social actors, technological actants, and work activities attain greater prestige and position relative to others, the modification of existing worlds and emergence of new ones comes into view, providing an entry point for exploring what a technological orientation means for changing the nature of journalism: its taken-for-granted assumptions, institutional bearings, and normative purposes in society.

A "worlds" perspective thus offers journalism studies scholars a lens through which they can investigate and interpret shifting views on the creative nature and worth of particular actors, actants, and activities once viewed as being predominantly technical and supportive, while highlighting that journalism is comprised of complex networks of labors and laborers, guided by particular conventions, that produce and legitimize works. However, as we have illustrated, there is room to build upon Becker's insights. For example, while Becker emphasizes that all actors and activities are of equal importance to their worlds—he aptly contends that art simply would not be the same without the contribution of each component in the network—we have argued here that fluctuations in and across worlds are better understood by adopting a broader lens. Specifically, the view that we have outlined suggests that status, more than reputation, matters particularly for understanding how symbolic meanings are interpreted, translated into conventions and value judgments, and ultimately rendered into resource allocations. In effect, to understand worlds, art or otherwise, means unpacking not only their collective activities and conventions but also their forms of give-and-take around status and the positions (real or symbolic) of particular actors, actants, and activities within networks. That is, how worlds accord status to certain people, practices, and products ultimately reveals *who*, *how*, and *what* such worlds deem exceptional and worth emulating; that, in turn, shapes the fundamental orientation of worlds and their implications for interlocking aspects of social life.

Note

1. Becker (1982/2008) makes a point of articulating "art worlds" in terms of *collective* activity, which, as we discuss later, refers to cooperative networks of people and processes that bring about something deemed artistic. In drawing on his framework, we describe "worlds" of journalism using similar terms. However, we also recognize that *collective* can imply a degree of organized planning that may not be apparent in the more accidental forms of *collaboration* or *connection* that occur as various actors and activities come into contact with one another in and through emerging journalisms (such as the three discussed in this chapter: ambient, data, and algorithmic journalism). Such interactions, particularly at the boundaries of emergent and existing worlds, may be ad hoc and spontaneous. When repeated over time, however, they may mature into more formal conventions, akin to Bourdieu's notion of *habitus* (Benson and Neveu 2005). In a related sense, though one focused on the context of social movements and political organizing, Bennett and Segerberg (2012) have suggested that, in a digital media context, the familiar logic of *collective* action may be giving way to a logic of *connective* action as groups and individuals engage one another more loosely via social networks. Worlds of journalism, too, are characterized by connection among a range of social actors, facilitated by connective technologies like social media; nevertheless, such worlds also exhibit the kind of cooperative efforts around shared intentions that are reflected in Becker's use of the term *collective*.

7 The Whitespace Press: Designing Meaningful Absences into Networked News

Mike Ananny[1]

This chapter explores the idea of absence in the networked press. My aim is three-fold: to ground the concept of absence—generative whitespace and silence—in a normative model of the press as an institution charged with ensuring a public right to hear; to survey how other fields—from architecture and graphic design to musicology and urban planning—use absence to enrich their fields and challenge their assumptions; and finally, to show how journalism has historically neglected to use absence to help publics hear, and to suggest ways it might do so today.

I never mean to suggest that the press should be censored or unreasonably prevented from speaking with, for, and to anyone. Rather, at a time when the networked press is questioning its unique role and significance, whitespace is an opportunity—a way to prototype new forms of journalism that value listening as much as speaking, that help to ensure the public right to hear and self-govern.

Meaningful Absence as Democratic Self-Governance

In 1948, legal philosopher Alexander Meiklejohn argued that people can only consent to a democratic institution's control over them if that institution brings people "together in activities of communication and mutual understanding" that produce the wisdom needed to continue to let themselves be governed (Meiklejohn 1948, 17). Such institutions should not aim to produce consensus or agreement; rather, their focus should be on generating the "information or opinion or doubt or disbelief or criticism" critical to "planning for the general good" (Meiklejohn 1948, 26). They create a *system* of freedom of expression that ideally produces a never-ending conversation about the conditions under which people consent to government (Emerson 1970).

Meiklejohn and Emerson were founders of a small but influential school that sees democratic institutions like the press not only as vehicles to create and circulate speech—but as ways to foster the conditions under which people produce, regulate, circulate, and make sense of speech as part of the pursuit for public goods (Baker 1989; Christians, Glasser, McQuail, Nordenstreng, and White 2009; Fiss 1996; Scanlon 1972; Schauer 1998; Sunstein 1994). They saw this seemingly subtle distinction between the individual act of speaking and the public value of speech paralleled in the US Constitution's First Amendment which "does not forbid the abridging of speech … [but] does forbid the abridging of the *freedom of* speech" (Meiklejohn 1948, 19, emphasis added). Western cultures that tend to trust unfettered marketplaces of speech to prevent the "peculiar evil of silencing the expression of an opinion" and produce truth through its "collision with error" (Mill 1859/1974, 76) largely equate protecting *speech* with protecting an individual right to speak. The marketplace is assumed to do the rest. But this assumption dangerously ignores the fact that, to be meaningful and powerful, speech requires both speaking *and* listening.

Distinguishing between—and valuing simultaneously—the freedom *to* speak and *of* speech requires two moves. First, it means tracing how *systems* of expression make some speech more likely to be produced, circulated, and considered than others. Second, it means appreciating how the legal right to receive information (Kennedy 2005) and individual interests in seeking out information can be mistaken for the social and cultural process through which speech is heard, made actionable, and gains power. Someone's right to receive information and her interest in hearing information are always patterned by forces beyond her individual control. For *democratic* institutions interested in structures of self-governance, the "point of ultimate interest is not the words of the speakers, but the minds of the hearers" (Meiklejohn 1948, 25)—minds developed by enterprises that structure both freedom to speak and freedom of speech.

Commercial communication enterprises—focused on adience growth, advertising revenue, and competitive advantage—prize the production, dissemination, and interpretation of messages that achieve economic ends. They thrive or collapse in marketplaces of speech. But institutions concerned with *democratic* communication have a different, complementary goal. Ideally, their constituents not only make consumer choices, but also encounter people, ideas, and versions of themselves that markets may not reveal. These alternative selves need other people to speak and suggest other ways of being, but they *also* require environments that make it likely for people to encounter, interpret, empathize with, and take action because of that speech. To both know who you are and become someone else, people need communication systems that ensure freedom *to* speak and freedom *of* speech. This is the core challenge of hybrid institutions like the press, that have both commercial and public goals: to foster speaking *and* listening in environments governed by both economic and democratic logics (Baker 2002).

An emerging body of literature explores this challenge. As a political act with the power to change minds and spur action, democratic listening requires exposure: a social opportunity to listen *in* and hear speech (Lacey 2011, 6) and a legal "right to receive information" (Kennedy 2005). But it also requires an "ethical responsibility" to seek and listen *out* "for voices that are unfamiliar or uneasy on the ear" (Lacey 2011, 18). Listening is a willingness to consider someone else's version of reality (Eliasoph 1998)—an openness to joining an "intersubjective world … as you construct it for me" (Bickford 1996, 147). Listening also asks us not only to listen in to and listen out for, but to accept "the possibility that what we hear will require *change* from us" (149)—to accept that our freedom to both act *and* develop requires accepting the risk that speech will change how we think.

Individual autonomy is thus not only about *in*-dependence and *inter*-dependence, but also *intra*-dependence. It means looking within ourselves to see and nurture personal capacities that guide action. It is essential for people to cultivate "self-awareness, self-understanding, moral discrimination, and self-control" if they are to sustain "freedom in the sense of self-direction" (Taylor 1979, 179). Freedom means acknowledging a "possibility of incompleteness" (Bickford 1996, 151) that motivates you to listen out for voices unlike yours, a period of silence that becomes "shelter from" the expectation to speak (Brown 1998, 315)—a purposeful "stilling of the self" (Bickford 1996, 145) and "emptiness of awareness" (Lipari 2010, 348) that reflection and openness require. But, scholars caution, you should not become *so* overwhelmed by introspection or a "fear of being wrong" that you fall silent and stop others from hearing what *you* have to offer (Bickford 1996, 151). This complex theory of democratic listening promises collective decisions with greater legitimacy because people encounter ideas that marketplaces might not have surfaced, differ and disagree in ways that create novel compromises, and experiment with holding new ideas they never could have discovered on their own (Dobson 2012).

The press's democratic challenge is to create not only speech, but structures for listening. If "listening is to be understood as a political rather than a private phenomenon, then it must somehow *appear* in the world" (Bickford 1996, 153). How might the press enact its political mission by helping listening appear—by creating, making visible, and valuing the absences that accompany listening? Since silence and whitespace are rarely cultivated in journalism, it is worth tracing how other disciplines understand and create meaningful absence. How might we understand *their* whitespaces as inspiration for an institutional press that values a self-governing public's right to hear?

Whitespace as Meaningful Absence Across Domains

Japanese design philosopher Kenya Hara defines "emptiness" as an intentional absence that makes possible "a condition, or *kizen*, which will likely be filled with content in the future" (Hara 2009, 36). Distinct from "nothingness" that lacks energy, *kizen* lets "our imaginations roam free of any boundaries [on] every possible meaning" (ii). Such meanings, the related *wabi sabi* aesthetic shows, come from letting absences that *seem* to lack significance achieve power by complementing and contrasting against more visible, material elements (Juniper 2003).

Whitespace is thus often *intentional* and always *relational*: separating representations, people, ideas, or time periods into units that represent designers' intended meanings. In "controlling differences [and] retaining only those that are most essential" (Hara 2009, 19), designers use whitespace to *activate* audiences, focusing attention and forcing interpretations (Rand 2014, 48–73). Graphic designers, for example, use margins, kerning, and font sizing to create negative space that brings "serenity to the page"—a "silence [for readers] to better hear [a] message loud and clear" (Vignelli 2010, 92).

Though most commonly considered a print technique, whitespace appears in many media. The Sumerians invented notations for zero as placeholders for numeric absence and ways of anticipating future mathematical operations (Kaplan 1999). Musicians created the rest symbol to help conductors and performers coordinate silence among orchestra instruments, reducing the "spatial uncertainty" (Slobod 1981, 107) associated with of earlier notational systems that failed to synchronize the performance of silences.

But as, John Cage (1961) showed in his piece *4′33″*, even professional musicians synchronized through standardized notions cannot achieve silence. Convening an orchestra in a concert hall to play "nothing" actually reveals a different kind of omnipresent ambient music: whirring fans, coughing, shifting bodies, opened doors, and nervous laughter. In playfully attempting to achieve silence, Cage shows its impossibility (Kahn 1997) and reveals an entirely new "soundscape" (Schafer 1993) that could not otherwise have been heard. Indeed, much of the history of acoustics is the story of people inventing tools and practices to limit and direct sound in search of different types of silence. From Carlyle's construction of a soundproof room in the 1850s (Picker 2012) through the white noise of muzak (Jones and Schumacher 1992), noise-canceling headphones (Hagood 2011), the iPod (Bull 2012), an audio spotlight (Yoneyama, Fujimoto, Kawamo, and Sasabe 1983), and tinnitus-fighting remedies (Hagood 2012), people have tried to isolate themselves through and from sound. The entire field of "sound studies" (Keeling and Kun 2011) shows how acoustic absence means not simply inventing technologies to block sound waves, but struggling with the social, political, economic, and material forces that fill silences and create "sonic imagination" (Sterne 2012). Pursuing silence always means exerting power, discovering and resisting against forces trying to be heard.

Such struggles also occur in architecture and urban planning. As designers of physical environments connect and isolate elements, they create absence—anticipating how people might interpret and navigate a *lack* of materiality. For example, nomadic Basarwa, Navajos, and Euroamericans use rituals and habits to *conceptually* segment spaces. Instead of *materially* partitioning their homes they create absence through patterned movements that are largely incomprehensible to outsiders; people show their ignorance and nonmembership by failing to appropriately navigate materially invisible but culturally significant whitespace (Kent 1991). Similarly, in his meditation on Japanese architecture *In Praise of Shadows*, Tanizaki (1977) criticizes the "progressive Westerner" who ignores the value of shadows by going from "candle to oil lamp, oil lamp to gaslight, gaslight to electric light" on a "quest for a brighter light [that] never ceases." (31) The early twentieth century indeed brought dramatically new materials for making sound and light—but they were always complemented by equally novel ways of resisting and reshaping the material spaces made by others (Thompson 2002).

Cities have similarly struggled to make absence, forming "antinoise campaigns, antinoise conferences, antinoise exhibitions, and 'silence weeks'" (Bijsterveld 2008, 2) to cast silence as a *public good* that should be available to all, not only those rich enough to create private acoustic retreats (Dyos 1982). Such initiatives, though, were hampered by: industrial revolutions that prioritized economic growth over quiet; clashing definitions of desired versus unwanted noise; lawmakers' preference to let urban neighborhoods self-regulate; the dominance of visual over auditory senses; focuses on air and water quality over noise pollution; the mass production of personal sound-producing devices like radios and televisions; and the rise of avant-garde aesthetics that challenged centralized control through purposefully discordant sounds and acoustic disruptions (Bijsterveld 2008). The idea of a *public* soundscape has always struggled to gain traction. A nascent architectural movement, though, tries to find and rehabilitate urban absences they call "relingos": holes accidentally left when two or more buildings, freeways, or infrastructures collide incompletely, leaving "an ambiguous space ... that can be seized by the imagination ... used and reused for different purposes," rather than the "rational actors" who created the surrounding structures intended (Luiselli 2014, 74).

Similarly, absences can emerge from social interactions. Self-help and motivational books increasingly emphasize the power of simplicity (Maeda 2006), introversion (Cain 2012), decluttering (McKeown 2014), isolation (Maitland 2014), solitude (Johnson 2015), and disconnection (Harris 2014). These popular movements echo mental health best practices: silences engender empathy, facilitate reflection, or make time for clients and therapists to consider what to say next (Ladany, Hill, Thompson, and O'Brien 2004). Similarly, many artists describe the various significances of silence: choosing to "fall silent publicly" so that finished work can speak for itself (Kenny 2011, 87); having to explain years of seeming nonproductivity when they lacked inspiration or economic support to produce visible work (Olsen 1965); and the perennial struggle for enough visibility to secure patrons and economic stability, but not so much that they become beholden to any person or market (Star and Strauss 1999).

Indeed, silence is not uniformly desired: it is often evidence of power, pain, ethical transgression, or spiritual anxiety. Critical cultural studies of silence see it as elite governance through unspoken norms (Foucault 1994); political scientists see "spirals of silence" (Noelle-Neumann 1984) when people fail to speak because they sense that their opinion is in the minority (or soon will be); rhetoricians and feminist scholars see the silences of women, people of color, sexual orientation minorities, or others in positions of structural disadvantage as evidence of systematic oppression (Glenn 2004). Some holocaust survivors remain silent and refuse to record their memories because what they witnessed was too horrific to be recalled (Frosh and Pinchevski 2011); and some archives are silent on subjects because some people are been barred from collections or because archivists have failed to read "against the grain" of history to challenge systemic absences (Carter 2006). Finally, religious silence can be similarly complex. In some faiths, silence represents spiritual confidence and communal meditation;

in others, it reveals the pursuit that accompanies a *lack* of faith—quiet skepticism among those searching for evidence that will give their beliefs conviction (Lawson 2014).

This brief and incomplete sketch reveals diverse meanings of whitespace. It is not simply content that has failed to be produced or transmitted; rather, silence can be aspired to, made, endured, interpreted, and resisted through communication rituals (Carey 1989). These rituals may seem too diverse to be helpful but, together, they suggest a typology of purposeful absence that might guide the tracing and creation of *journalistic* whitespace—ways to make and signify absence that help the press achieve its democratic mission of ensuring a public right to hear.

Historical Meanings of Absence in the Press

The mainstream, institutional press has both created and resisted absence. The most obvious example is censorship: the "removal and replacement" of an author's text before publication, or the more subtle "dispersal and displacement" after publication, preventing an author from reaching her intended audience (Burt 1998, 17). Regardless of the mechanism, journalists have historically seen censorship as the primary force opposing their role as critical watchdogs who hold power accountable by publishing uncomfortable truths.

Journalists, though, sometimes self-censor (Anthonissen 2003). They might agree with the State that information hurts public interests—as Israeli journalists do when they determine the scope of investigative reporting collaboratively with government representatives (Christians et al. 2009, 206–211), and as US journalists did when they embedded with US military units during "Operation Iraqi Freedom" and largely avoided critiquing the military (Pfau et al. 2004). Publishers and senior editors might bow to state requests to avoid or delay publication, as the *New York Times*'s Bill Keller did when he agreed to Bush administration requests not to publish information on weapons of mass destruction or war-on-terror tactics (Boyd-Barrett 2004). Despite US Supreme Court decisions discouraging news organizations from self-censorship (D. A. Anderson 1975), many publishers continue to let fear of libel limit their reporting. In rare cases, an absence signals a news organization's protest against censorship—as was the case when inmate publishers of Vacaville Prison's *Vaca Valley Star* dramatically splashed the word "censored" across whitespace where four articles the warden "deemed excessively critical of the prison staff" were to have appeared (Truter 1984, 605).

In addition to censorship and self-censorship, organizational pressures can create journalistic absences. News organizations may engage in "scoop and shun" dynamics (Glasser and Gunther 2005, 386), seeing a story as "tainted" (Levy 1981, 24) because someone beat them to it first. They let their *lack* of coverage signal how they differ from their competitors. Absences also emerge from the routines and rituals of beat reporting (Molotch and Lester 1974; Tuchman 1978): stories tend to come from places where journalists and audiences *expect* news to come from (government offices, celebrities, stock markets), that commercial interests advertisers want to sponsor (automobiles, real estate, sports). Individual journalists typically have little freedom to ignore these forces and redirect reporting to other places that lack coverage (Murdock 1977). Even if they *were* able to broaden their beats, many locations—e.g., prisons, military bases, corporate offices—are hard to access, most countries lack foreign bureaus, and a reputation for too much independent exploration may damage a reporter's employability. The result is that many people, organizations, and locations are regularly and predictably absent from coverage.

Absences can also come from systematic underreporting of issues considered too deviant from public norms or newsroom cultures (Hallin 1986a). For example, most publications except the gay press ignored the HIV/AIDS crisis (Nelkin 1991) until powerful, agenda-setting news organizations covered it and signaled the professional acceptability of doing so (Rogers, Dearing, and Chang 1991). And based on evidence that suicide coverage correlates with suicide vulnerability among certain readers (Stack 2000), many news organizations either ban reporting on individual suicides or leave out many details as a matter of policy (Jamieson, Jamieson, and Romer 2003). Journalists also tend not to report on themselves—"silent bargains and silent routines" among staff make self-reporting rare (Turow 1994)—or people that their owners signal are beyond reproach (Chomsky 2006) or too powerful to cover (Castle and Dalby 2015).

Other types of absence emerge during reporting and broadcasting. Reporters may need time to confirm details, hone writing, cultivate sources, or secure their personal safety—unseen work that can seem like ignorance or idleness to outside audiences who only see final publications. Sources may speak on background, demand anonymity, limit attribution, or use pseudonyms that create an "absence of a source's identity" that makes it difficult to hold news organizations accountable (Carlson 2011, 2). And even though "all of us in radio live in fear of a second of dead air," NPR interviewer Terry Gross tells her editors to *keep* guests' pauses because she want audiences "to hear them think, to hear the *process* of them thinking." (*This American Life*, nd) Television news also uses purposefully empty "black hole" shots: even though there is "no activity occurring for a viewer to see," as long as such shots are "live" and "on-scene" images of "where events *had* occurred or *will* occur," broadcasters can hold viewers' attention (Casella 2012, 364). And, finally, stories often follow unexplained arcs that create the ultimate absence: they are inexplicably judged to be "done" or no longer relevant as the press falls silent and reporters and audiences move on to other topics.

Despite the dominant view of a free press as one that publishes whatever it wants and whenever it wants free of interference, absence is often a goal, result, or accident of journalism. Censorship, self-censorship, organizational routines, implicit collaborations, competition, access restrictions, taboos, and reporting styles all create journalistic whitespace. Some whitespace scaffolds the listening that the ideal democratic press enables while others represent pressures, habits, and rituals that limit the press's ability to help publics self-govern through hearing. Going forward, with these absences in mind, we might consider how silences appear in the relationships, practices, and materials that make up the *contemporary, networked press*—how absences in its networked relationships help or hinder a public right to hear.

Sources of Absence in the *Networked* Press

For the contemporary, networked press to be a listening institution, it must carve out silences with *public* value amid the forces governing online publishing. News now emerges not from news organizations alone but from a hybrid system of "technologies, genres, norms, behaviors, and organizational forms" (Chadwick 2013, 4) that, taken together, define a field of forces (Benson and Neveu 2005) within which human and non-human actors alike (Anderson and De Maeyer 2015; Boczkowski 2014) vie for the power to access information, create new knowledge, govern attention, and monetize audiences.

I highlight five sources of absence in the networked press: *professional norms, media avoidance, invisible audiences, organizational openings*, and *infrastructural holes*. While not an exhaustive typology to be sure, my aim is to sketch sites of absence in the networked press, to illustrate the origins, complexities, and opportunities of networked, journalistic whitespace.

Professional Norms

Many networked journalists engage in a "discipline of verification" (Kovach and Rosenstiel 2014, 98)—but such discipline creates absences that must compete with unceasing social media. Even in the context of high-profile breaking news, some news organizations stay silent. For example, immediately after the 2013 Boston Marathon bombing, the *New York Times* chose not to follow other news organizations in reporting "that an arrest had been made or that a suspect was in custody" (Sullivan 2013). *Breaking News* (2013) waited until at least two other news organizations confirmed information before it repeated specifics; WNYC's Brian Lehrer turned into a kind of meta-reporter creating and filling silence as he talked about what he would not report (*On The Media* 2013); and the European Journalism Center used the bombing coverage as a case study in how journalists and social media users alike should resist interpreting police scanner feeds or social media videos during crisis situations (Silverman 2014). Coverage of the bombing served as a kind of master class in journalistic self-restraint with news organizations that initially stayed quiet ultimately emerging as the most mature and reliable sources. Their lack of publishing was calculated caution—professional restraint amid noisy networks that were quick to broadcast rumor and ridicule silence (Ananny 2013a).

In contrast, during the Arab Spring uprisings, NPR's Andy Carvin (2013) openly shared information he could not verify, enlisting his followers as a verification network and calculating that the harms of staying silent outweighed the risks of circulating false information (Hermida, Lewis, and Zamith 2014). For Carvin, silence and caution lost out to transparency, speed, and trust of his network. Indeed, each new crisis seems to reveal new dynamics between professional journalists and social media (Brandtzaeg, Lüders, Spangenberg, Rath-Wiggins, and Følstad 2015) with individual reporters figuring out whether they have the time and professional standing required to endure the silences that come with verification. Silence seems to be a privilege and luxury available only to reporters who can resist their organizations' time pressures (Reich and Godler 2014) or who are part of an emerging "slow news" movement (Le Masurier 2014) that rewards tempered publishing. Finally, patterns seem to be emerging for deciding when news is "over." Few online stories are updated two hours after their publication (Saltzis 2012) and fewer still indicate that reporting continues—for example, KPCC appends some of its online stories with the phrase "this story will be updated." And the *New York Times* closes comment threads after a "discussion has run its course and there is nothing substantial to gain from having more comments on the article" (Sullivan 2012).

Silences are fertile ground for understanding journalist–audience relationships since the significances of such absences are unclear. They may represent: careful, unseen reporting by those privileged enough to slow time and resist social media; the lack of a trusted network capable of developing or verifying a story; a story's conclusion or shift to other topics; or the expectation that more information is coming.

Media Avoidance

Writing for the *Guardian*, Jesse Armstrong (2015) ended a month-long, self-imposed ban on news consumption observing that "what you read and watch is just a reflection of what you're interested in and who you are, and that's quite difficult to escape." Indeed, researchers are increasingly studying the purposeful non-use and avoidance of media and media technologies. Many television watchers actively avoid the news, agreeing with statements like "there is so much on TV that I seldom watch news" and "there is always a program that interests me more than the news" (Ksiazek, Malthouse, and Webster 2010). Such news avoidance is less common among Internet users but still prevalent among those interested in entertainment news, who tend to avoid current affairs reporting (Trilling and Schoenbach 2013). This finding agrees with the observation that whereas "journalists exhibit a strong preference for public-affairs news in the articles they consider most newsworthy, consumers lean toward non-public-affairs subjects in the stories they click most often" (Boczkowski and Mitchelstein 2013, 16).

There are similar patterns in the non-use of networked technologies—the tools and platforms that increasingly deliver news. Some people never adopt digital technologies (Oudshoorn and Pinch 2003; Selwyn 2003) or social media (Hargittai 2007) in the first place; others intentionally and conspicuously refuse them (Portwood-Stacer 2012); and others are caught in cycles of technological fasting, departure, and return (Brubaker, Ananny, and Crawford 2014; Lee and Katz 2014). People disconnect from technologies for a variety of reasons—for example, concerns about privacy, seeing social media as banal, fighting Internet addictions, trying to regain productivity (Baumer et al. 2013)—but the patterns reveal that people see "carving out quiet" (Plaut 2015, np) as an essential part of communication that entails avoiding media technologies.

Such non-use is a new challenge for news organizations. They may fail to reach audiences—and thus fail to ensure a public right to hear—not because of what *they* produce, but because the *platforms* they use to circulate news are caught up in complex patterns of avoidance, non-use, refusal, departure, and return.

Though Armstrong quickly returned to his regular news routines because his career depended on following the news, others avoid media and media technologies because disconnection is crucial to *their* identities. Some people cultivate "rational ignorance" (Downs 1957)—why educate yourself about politics if your vote is only one of many?—and an emerging field of agnotology finds that people resist knowledge in a variety of ways: a "native state" of unknowing (a blank slate), a "lost realm" of knowing (benign disinterest in correcting acknowledged ignorance), or a "deliberately engineered and strategic ploy" to stay ignorant (active resistance to learning) (Proctor 2008, 3). Instead of seeing a lack of knowledge or non-use as careless disengagement or a failure to reach an audience—Spanish researchers recently developed a system to push BuzzFeed content to mobile phone users when it detected that users were "bored" (Owen 2015)—the press might *engage* and actually support the various reasons people resist media and media technologies precisely because part of its mission is to help ensure the public right to hear that, sometimes, needs silence. Non-use, avoidance, and seeming nonparticipation may be evidence of cultivated ignorance or a press that is losing readers or revenue—but it may also be evidence of people creating space to listen and reflect upon what they have already received.

Invisible Audiences

Absences may also come in the form of audiences who exist but never appear. Most social media users leave few obvious traces (Hampton, Goulet, Marlow, and Rainie 2012) and are often still pejoratively referred to as "lurkers" (Nonnecke and Preece 2000) with secretive or asocial intentions. However, conceptualized instead as "listeners" who encourage "others to make public contributions" (Crawford 2009, 527), they are the imagined others who make social media platforms spaces for dialogue versus monologue. Though they rarely appear anywhere but in database logs, without them the Internet—and networked news—lacks the relationships that discourse requires. In the rare moments, when a media event is so captivating that it renders social media's vocal minority silent, we get a glimpse of what active receptive looks like—for example, Twitter fell largely silent during the shootouts that decided the 2014 Brazil vs. Chile World Cup match as those who usually tweet the most stopped to watch (Rios 2014).

Audiences may also be invisible because technological, social, and legal conditions fail to surface them. For example, Twitter news audiences are transient and fragmented in time (Lehmann, Castillo, Lalmas, and Zuckerman 2013); they exist briefly and are virtually impossible for news organizations to reconvene, update, or correct as a unified whole (unlike broadcast evening news or the morning newspaper). Online audiences may also seem much larger than they actually are: people routinely self-report consuming more news than they actually do (Prior 2009) and some issue-specific audiences dissolve into untraceable venues if they think the State is surveilling their online opinions or presence (Hampton et al. 2014; Rainie and Madden 2015). Finally, some news audiences run the risk of being more visible than others: in some jurisdictions, the identities of those who comment on online news stories are protected by journalistic shield laws while other jurisdictions guarantee no such anonymity (Bayard 2009).

These absences—listening, fragmenting, dissolving, dispersing, lying, relocating, anonymizing—show audiences navigating, and being buffeted by, political, social, technological, and legal forces that make them more or less visible—forces that might be supported or resisted by a networked press focused on ensuring a public right to hear.

Organizational Openings

Sometimes, online news organizations purposefully create absences and disconnect in order to achieve their missions. Subscription-based paywalls (Pickard and Williams 2014) sequester content in spaces that not all audiences can see but, sometimes, news organizations selectively breach these walls to strategically manipulate the scarcity that paywalls create (Ananny and Bighash 2016). For example, the *New York Times* now limits nonsubscribers to 10 (down from 20) articles per month on nytimes.com—other stories are effectively invisible beyond that limit—*except* if readers encounter articles through unspecified "search engines, blogs, and social media" (articles are accessible through such channels even if users have reached the monthly limit) (New York Times 2015) or are at Starbucks (allowing for 15 free articles per day) (Soper 2013) or if an event like Hurricane Sandy or Election Day warrants a temporarily dropped paywall and unrestricted reading (Beaujon and Moos 2013). The strategies of scarcity driving paywall designs reveal a networked press experimenting with how and why to sometimes make its stories *inaccessible* to audiences.

News organizations also control the visibility of their work through technological infrastructures. The *New York Times*, the *Guardian*, and NPR all use Application Programming Interfaces—software toolkits that selectively let outsiders access newsroom content through a system of licenses and keys—to both reveal and conceal data (Ananny 2013b). These strategic openings and closures leave clues about which information news organizations will share, and which is to stay hidden from public view. The *Guardian*'s "Open Newslist" was one of the boldest experiments in revealing not just data but practices. It briefly used a publicly accessible Google document to show reporters' topics, stories, and sources *while* they were working, *before* publication (Roberts 2011)—but the project is now largely defunct with such background work once again concealed. The *New York Times*'s "Times Insider" (2015) similarly offers behind-the-scenes accounts of how and why news is produced—but only to premier subscribers.

Finally, akin to the *Vaca Valley Star*'s use of whitespace to protest prison censorship, several online news organizations including Wired.com blacked out their websites to protest against the proposed 2012 SOPA/PIPA legislation (Wortham 2012). They simulated their *complete* absence to show what they thought the Internet would look like if the legislation passed.

From strategically designed paywalls to APIs, premier subscriptions, and protest blackouts, news organizations use absences to achieve organizational missions. Such absence can become a powerful way to understand how and why news organizations create whitespace that orients readers, generates revenue, and regulates access.

Infrastructural Holes

Other whitespaces result not from strategic, organizational intention, but from a confluence of intertwined and largely hidden forces that add up to absence. A particularly powerful type of silence emerges when news encounters social media personalization. Computational rules of platform algorithms can systematically include and exclude people or ideas (Gillespie 2014) that news organizations might *want* their audiences to consider—but they are effectively invisible because news algorithms fail to surface them (Ananny and Crawford 2014). News organizations may even purposefully place their data beyond the reach of search engines by using a robots.txt file to prevent crawlers from indexing their sites. "Big data" journalists are also susceptible to infrastructural absences. They risk creating stories that echo their data sets' exclusions if they fail to account for people whose "information is not regularly collected or analyzed, because they do not routinely engage in activities that big data is designed to capture" (Lerman 2013, 55). Such algorithmic exclusions are even creating *physical* whitespace, as the manufacturers of drones that news organizations are beginning to encode no-fly zones into their hardware, e.g., automatically landing a drone if it veers too close to the White House (Sydell 2015).

Without explanations for *why* such absences exist, people create folk theories (Mills 1959/2000) of whitespace—saying things like "I always assumed that I wasn't really that close to her" (Eslami et al. 2015) to explain someone's absence from their Facebook newsfeed, instead of attributing it to complex infrastructural intersections. Part of a whitespace press's responsibility ideally includes explaining *why* content fails to circulate on its partner platforms.

Social media absences can also be created because of how infrastructures organize time. Users might visit Twitter at a suboptimal time (12 p.m. and 6 p.m. if you want clicks, 5 p.m. if you want retweets [Bennett 2015]). News organizations might write tweets that decay too quickly—BBC tweets last longer than other news organizations' (Bhattacharya and Ram 2012)—but most news tweets are effectively invisible two hours after posting (Asur, Huberman, Szabo, and Wang 2011). The popularity of postings on Digg tended to stabilize quickly, but the same content is just as popular on YouTube about 30 days later (Szabo and Huberman 2010). Given how increasingly attuned news organizations are to metrics (Petre 2015; Tandoc 2014) and the influence they have on how editors conceptualize their audiences (C. W. Anderson 2011; Vu 2014), a *lack* of social media activity might reflect not only as disinterest or failed circulation, but intersecting platforms with nonsynchronized publication rhythms. Indeed, some online news never surfaces on social media at all, instead staying in privately circulated "dark social" domains like email and instant messaging (Madrigal 2012).

Finally, some content may be *published* but effectively absent from *public* discourse[2] because laws or policies explicitly prevent it from appearing on third-party, para-journalistic platforms. For example, though the content remains on their own websites, the BBC (McIntosh 2015), the *Guardian* (Ball 2014), and the *New York Times* (Cohen and Scott 2014) have all had stories removed from Google's European search index to comply with the European Union's "Right to be Forgotten" legislation. And Twitter blocked the account of *Independent* journalist Guy Adams after, as part of his critique of NBC's 2012 Olympics coverage, he tweeted the publicly available email address of a senior NBC executive, in violation of Twitter's privacy policy (Sherman and Womack 2012). Such absences—stories, links, reporters—are not the result of algorithmic accidents or platform idiosyncrasies but the private governance of infrastructures that news organizations rely upon but that are beyond their control.

Conclusion

Once you look for them, the networked press is rife with absences. Some are vestiges of early forms of journalism—professional traditions enduring in contemporary times—while others are newer, emerging from nascent technologies and relationships that create and circulate news online. Indeed, technological innovations have always suggested new tensions between speaking and listening. The director of the first German news radio service in the 1920s "warned that writing for listeners as if they were readers would be like 'trying to take a photograph with a violin'" (cited in Lacey 2011, 10). We are again in a time when traditional practices and expectations seem subtly mismatched with the contemporary moment. Simply creating and circulating *more* digital news without designing spaces for listening—prioritizing freedom to speak over freedom of speech—misapplies broadcast logics to networked spaces. It is like trying to take a photograph with a violin.

While a great deal of energy has gone into creating the conditions for voice and expression, we are only in the early stages of understanding what Razlogova (2011, 55) calls the "voice of the listener"—and what it might mean for online news organizations to design for networked listeners. Under what conditions might listening be seen as a full-fledged form of active, democratically necessary participation on par with more readily identifiable actions like speaking, writing, posting, tweeting, liking, and tagging? How might the press value a public right to *hear* just as much as its own right to speak? How might the scholars, practitioners, users, and builders of the networked press see the necessity of listening publics?

Scholars might start looking for absences in the networked press, asking when and why they are significant, what kind of power relationships create and sustain whitespace, and how they help or hinder a normative model of the press. Networked journalists, editors, and publishers might nurture professional norms that see silence as *intentional* and potentially *generative*, giving audiences the time and space to interpret preexisting speech instead of trying to tell everything to as many people as possible, right now. Audiences might see their desires to receive (tracked by surveillance analytics modeling their movements) as requests for the media to fill voids that could instead be generative silences; how might audiences resist the demands that content producers and advertisers manufacture, attribute to them, and then rush to meet? Perhaps most importantly, those building the networked press's technologies—digital media companies, software programmers, algorithm designers, and venture capitalists—might use radical, speculative, adversarial design techniques (DiSalvo 2012; Dunne and Raby 2013) to invent new forms of listening that help people thrive in, not merely endure, surfeits of online speech.

Many of the networked press's relationships to absence are about including or excluding people, beating competition, avoiding errors, struggling for visibility, manufacturing scarcity, overcoming information overload, and hiding from surveillance. Absence is largely seen as a chance to add new content, an algorithmic error, an oppressive silence, or a failure to move quickly—and *not* a necessary ingredient of democratic self-governance. To help this shift in perspective, designers and scholars of the networked press might create: criteria for trusting or challenging absences; institutions and infrastructures that create generative absences; and design languages for making and experimenting with absence.

My aim has been to sketch the democratic value, interdisciplinary richness, historical antecedents, and contemporary construction of journalistic absence. In the spirit of Tanizaki's (1977) celebration of *shadows* as intentional architectural elements with aesthetic and instrumental value, journalism might take up, debate, expand upon, and design for *whitespace* as a way to show how thoughtful silence and generative absence might ensure a public right to hear.

Notes

1. In addition to this volume's editors Pablo Boczkowski and C. W. Anderson, I would like to thank the following people for generously and thoughtfully reading early drafts of this essay, or for providing me with key background material: Ben Fry, Ted Glasser, James Grady, Wells Lawson, Zizi Papacharissi, Perry Parks, Ethan Plaut, Carol Strohecker, and Maya Williams.

2. Ongoing thanks to Theodore L. Glasser for this distinction.

Commentary: Remaking Events, Storytelling, and the News

Zizi Papacharissi

Technologies of storytelling express and connect. Well before Lippmann (1922) wrote about *the world outside and the picture in our heads*, technologies of storytelling enabled us to presence voices, events, and issues that would otherwise go unregistered in our collective vision and elude our collective memory. Storytelling facilitates knowing, and technologies of storytelling give form to different modalities of knowing. Journalism presents a particular approach to storytelling, defined by its own set of practices, protocols, and platforms. Like all approaches to storytelling, journalism evolves, and it is context-specific in response to varying historical, sociocultural, political, and economic realities. Modalities of storytelling evolve and the platforms and approaches to storytelling evolve with them, so as to move forward, and so as to remain relevant to contemporary context. Lippmann was writing about public opinion and a different crisis in journalism when he contrasted *the world outside* to *the pictures in our heads* so as to describe the importance of storytelling in preserving continuity between the two. By contrast, the crisis journalism has been facing for the past several years is a crisis of evolved platforms of storytelling and outdated practices of storytelling. It is a crisis of relevance, in that the resulting modalities of storytelling prevailing in journalism are no longer relevant. It is a crisis of context, in that the ensuing modalities of storytelling are not specific to contemporary sociocultural context. *Remaking the news* would require reimagining the form of journalism. The preceding chapters are ultimately about whether news can be remade, and if so, how. Their authors have specific recommendations about how the journalistic tradition of storytelling might be reinvented. I synthesize their approaches and offer my own thoughts on reimagining journalism and remaking what is news.

It is not accidental that I begin this critique by describing journalism as an approach to storytelling and not as a profession. Contemporary platforms invite us to think about journalism as a modality of storytelling that eliminates the distance between professional and amateur actors and actants of journalism. And it is not by accident that all authors of the preceding chapters make direct references to symbolic interactionism, as the inspiration for contemporary journalism and as its possible salvation. The emphasis on symbolic interactionism pays its respects to the early stages of our field, and also marks a full circle return to our roots. Symbolic interactionism was a progressive effort that placed communication at the center, as academics, scholars, and journalists pondered the world outside: industrialization, urbanization, the big questions of democracy and freedom, the role of press and media, the place of social progressivism and social welfare, the possibility for economic progressivism, regulation and antitrust mentalities, civil rights, feminism/women's rights, technology, the role of science, and a set of ethics to guide them all as they made sense of the changing world around them. We are faced with similar questions today, perhaps because we never came up with adequate answers to these questions in the first place, or perhaps because these are the questions of humanity, and we are destined to always ask them, answer them, and then revisit them again.

Much like the symbolic interactionists, Nielsen, Lewis and Zamith, and Ananny place communication at the center, albeit in different ways. Nielsen's chapter is ultimately about how modes of communication render news-as-impressions into news-as-knowledge, thus marking the difference between being "acquainted with an issue" and having "knowledge about that issue." It is a difference between short-form vs. long-form news, instantaneity in reporting vs. slow news, and chaos vs. curation. It is borne out of the desire to be the one telling the most comprehensive version of the story, even if that means being the last, not the first to tell it. Somewhere in that mess sits the contemporary citizen, monitorial in orientation (Schudson 1999), in an oversaturated information environment (Gergen 1991), immersed in active spectatorship (Kreiss 2015), overwhelmed. There are so many ways of knowing of things happening and so few ways to really know about them. For Lewis and Zamith, journalists can help untangle these worlds, and their worlds framework envisions a journalism that orients through intuitive, interpretive, and sense-making processes. Status, processes, and conventions are not static but are constantly in flux, in order for journalism to produce modes of storytelling that are not what Nielsen refers to as "churnalism," but are, instead, context-specific accounts about the world outside and how it connects to the pictures in our heads. Communication, again, is at the center, and for it to occupy that place, it must involve both speaking and listening. Storytelling must evolve out of the absence of storytelling; a true *whitespace*, in Ananny's words, that does not presume nor direct a particular form to a story.

All chapters then draw on the role of technology, with emphasis on how platforms transform the texture of knowledge (Nielsen), on the actors, actants, processes, boundaries, and modalities of storytelling (Lewis and Zamith), and on the privilege of autonomy in storytelling (Ananny). It is important to note here that technology is not perceived by any of us as an external force, capable of determining outcomes. Information and knowledge gathering mechanisms are human-made. Journalistic platforms are designed by humans and bear human assumptions, proclivities, and bias. Finally, organizing algorithms are constructed by humans, translating their own logic for curating information into code. Technology is still very much a force, but one rendered by humans, and as such capable of defining, restricting, and enabling our actions, in the same way many human-made things do, including our cars, our houses, our furniture, our books, our diplomas, our clothing, and other artifacts of our everyday lives.

Still, the unavoidable part to be played by technology in the storytelling project of journalism should not distract us from the objective of facilitating knowledge, of letting journalism attain a reflexive and historically sensitive form. And this would involve coming as close to whitespace as possible, out of which a pure and organic form of storytelling may emerge. Toward this point, all authors emphasize the role of actors in discerning acquaintance from knowledge (Nielsen), in engaging in disruption, hacking limiting architectures, and reorganizing boundaries (Lewis and Zamith), and in creating the circumstances for "stillness" out of which whitespace may emerge (Annany). My own position has always been that new(er) technologies lend themselves to all of these things, but that we, as human actors, reluctant to abandon the comfort of convention, seek to mold them into storytelling practices that they are simply not meant to serve. I find the nature of many of the new(er) technologies to be generative and corrective. For example, news was never meant to be a high-priced commodity. One cannot put a price on knowledge, of course, but the news of the day should be accessible to all at a low—if any—cost, and social media platforms correct the tendency to overcharge. Similarly, somewhere down the road to the professionalization of journalism, it became easier to trade in autonomy for status, power, and income. Finally, the storytelling conventions of broadcasting ushered in artificial divisions between those who broadcast and those who do not, concocted the advertising friendly abstraction of the audience, and made it easy for journalists to speak but difficult for them to listen. The technologies that seemingly have created the current so-called crisis in journalism are meant to correct all these tendencies. It is up to actors to interpret the place of technology, and put technology in that place. The point of these technologies was to deconstruct organizations that were never meant to be media conglomerates; to reintroduce heterarchy in existing hierarchies of framing and power; and to restore storytelling autonomy. News organizations are often perplexed by the money question in the new media paradigm of journalism. Forget the money question. The business of making the news lies somewhere between the nonprofit and the for-profit models, and ideally adopts a hybrid of the two. Emerging technologies are not built to sustain large profit margins for conglomerates, nor was journalism meant to pave the road to making millions. The technologies are equipped to pluralize and to sustain smaller, autonomous, and independent organizations that can still turn a profit, but of a different margin. The problem of journalism is not technological. Remaking the news is not a question of fixing the technology.

The question of remaking the news is both a sociocultural one and an economic one. Tackling it requires reinventing the journalistic habitus of norms and practices and reorganizing the economics of news production. It involves the sociology and the economy of news. The symbolic interactionists, whose work all three chapters go back to, and who laid the groundwork for the contemporary paradigm of journalism, never envisioned journalism as a multibillion dollar business nor a celebrity vehicle for those involved. Thus, remaking the news requires that news institutions adjust to the reality of the present economy and abandon the economies of scale mentality—or, at the very least, adjust the scale. It is difficult, but not impossible. Network stations accomplished this when they had to survive in a market saturated with cable channels. The music industry is in the middle of addressing similar issues. And it is only a matter of time before the movie industry confronts a similar set of problems. So that is that, and it presents the essence of remaking the news.

And so, the problem of remaking news is not just about technology. But, it is also about technology. Specifically, it is about finding a place for technology, instead of technology finding its own place. In order to do so, we need to understand what technologies of storytelling do for journalism, and in turn, what function journalism plays for storytelling societies.

Events and Stories

There are events, and there are stories that we tell about events. I have always understood journalism as the process of turning events into stories.

But whereas anybody can tell a story about an event, journalism is guided by news values, that is, the things that turn events into stories told by journalists (Hartley 2002). News values organize how stories about events are told, in ways that have been researched for decades by our field, and can be summarized as follows:

- News values prioritize stories about events that are recent, sudden, unambiguous, predictable, relevant, and close (to the relevant culture/class/location).
- Priority is given to stories about the economy, government politics, industry and business, foreign affairs and domestic affairs—either of conflict or human interest, disasters, and sport.
- Priority is given to elite nations (the US, the UK, Europe, etc.) and elite people.
- News values often involve appeals to dominant ideologies and discourses. What is cultural and/or historical will be presented as natural and consensual.
- News stories need to appeal to readers/viewers so they must be commonsensical, entertaining, and dramatic (like fiction), and visual (Hartley 2002, 166).

As I argued earlier, remaking the news is both a sociocultural and economic question. Remaking the news requires remaking news values. Without reorganizing news values, the problem of remaking the news remains. But there are ways in which technology remediates, amplifies, or introduces newer news values. Some of these are mentioned or alluded to by Ananny, Lewis and Zamith, and Nielsen, and are worth examining further.

First, social media platforms enable journalists, but also people committing acts of journalism, to observe, record, and report events in ways that instantly turn events into stories. While I am not suggesting that this is a feature unique to social media,[1] it is worth pointing out that these platforms amplify this ability in a way that further contributes to and cultivates a culture of instantaneity in news reporting (Grusin 2010; Papacharissi 2014a; Nielsen, chapter 5, this volume). This implies that a distinction between instantaneous and more thoughtful, curated news accounts must be drawn, highlighting the differences between fast and slow news (Gillmor 2009; Zuckerman2009); acknowledging the presence of many journalisms, or at the very least, many layers to journalism; and cultivating the literacies necessary for all, journalists and citizens alike, to process those.

Second, these platforms operate on a logic that introduces hybridity to both news values and news production (Chadwick 2013; Russell 2011). It affords journalists neither the time to process information, nor the privilege of being the first to report it. This change does not necessarily present insurmountable challenges, but it does require acknowledging that journalists are no longer the first ones or the only ones telling the story. The prevalent organizing logic of hybridity entails acknowledging the presence of many journalisms, or at the very least the many layers to journalism. These layers are frequently presented to publics in hybrid form, interweaved, not cleanly layered on top of one another, nor organized into fully formed stories that the public was previously accustomed to receiving (and journalists to delivering). So the job of the journalist is to untangle and curate these layers, and connect them back together, without sacrificing the integrity of the story or the perspectives of the *multiple* storytellers.

Third, as many—including Lewis and Zamith, Nielsen, and Ananny—point out, the new platforms of storytelling render journalism omnipresent, lending it an ambient, always-on presence (Hermida 2010). While not specific to social media, this property is augmented by such media, sustaining an environment of always-on listening and broadcasting. This implies the news is always on and always told, even when there is no new news to report. The latter frequently creates repetition and redundancy that only fuels the intensity with which news is reported without adding to the depth of the news coverage itself. This is especially harmful in crises, emergencies, or situations where access to media cannot be controlled or trusted. It is also particularly dangerous in circumstances involving the reporting of violent incidents, shootings, acts of terrorism, and events of a similar nature, where intensity can inadvertently contribute to the exposure and accidental glorification of behaviors that must be reported in a mindful manner. In the place of mindful reporting, we receive a mix of drama, opinion, news, and facts-in-the-making blended into one, to the point where it is impossible to discern one from the other. This brings me to my next point, concerning the form of storytelling each platform permits.

Mediality

Instantaneity, hybridity, and ambience play their part in reorganizing how stories are told on the newer platforms of journalism. But each of those platforms bears unique characteristics and lends the stories told a unique form, through a variety of affordances that include the algorithmic architecture that gives a platform its shape. And so we may think about the mediality each platform affords each event, as that event is turned into a story. For example, Lang and Lang (1953) famously studied the MacArthur

Day parade in Chicago, discovering that audiences watching the event from their TV sets had experienced a much more organized, personable, and warm event than the one experienced by the crowds participating in the parade in the streets. Similarly, we may think of the different realities experienced by people participating in the events themselves, as opposed to people following the events on their TV sets, or on Twitter, or via news sites, or on the radio, or even through a polymedia[2] environment rendered by a combination of all these. The texture of each story may give shape to a different event, depending on the form of the story and literacy sensibility of publics "listening in" to the developing event.

The remaining question then is, what form does the platform afford an event, as that event is turned into a story? What is the shape, the texture, and the feel of that story when it is told over Twitter, as opposed to, for example, Reddit, crossing the boundaries of which frequently affords a misogynistic glean? The form, which is the shape, the look, the feel of news is specific to historical and sociocultural context. It is also borne out of the mediality of storytelling contexts, and these are all questions for journalists, scholars, and citizens not necessarily to produce finite answers to, but to always be asking and thinking about.

Affective News and the Polity

Finally, instantaneity, ambience, and the organizing principles of hybridity, combined with the mediality of the new platforms of journalism frequently lend form to breed of news I have termed *affective*.[3] I define *affective news* as news collaboratively constructed out of subjective experience, opinion, and emotion, to the point where discerning one from the other is difficult, and doing so misses the point (Papacharissi 2014b). Characterized by premediation, affective news streams are driven by reports of events that are in the making, and thus frequently communicate a predisposition to frame the developing story, and in doing so, lay claim to latent forms of agency that are also affective and networked (Papacharissi and de Fatima Oliveira, 2013). Affective news is intense, phatic, and emotive. It blends broadcasting practices with the conversational conventions of interpersonal storytelling. It is filled with intensity and frequently prompted by repetition of the same news or opinionating about the news, especially with there is no new news to report. This becomes problematic when affect, or this form of intensity, is reported as the event, in the place of more in-depth analysis, new news, or, in Ananny's vernacular, absence; silence. Affective news does invite modalities of being acquainted with an issue, rather than knowing an issue, per Nielsen's analysis. But it is also a way for broadening journalism and lending it a "worlds" perspective, in the words of Lewis and Zamith, because it affords polyformity in news storytelling, and a way for citizens to feel their way into a story.

The fact that news is always on and always with us cannot but invite an affective connection, both in terms of how news is told and in terms of how news is absorbed (Beckett 2015). And while it is easy to exploit the emotional angle of news, the affectivity these newer platforms of news storytelling lend can actually help journalists finally reconcile divides between the subjective and the objective, the broadcaster and the audience, and emotion and reason that have longed defined and restricted the stories we can tell individually, collaboratively, and collectively. Emotion and reason, subjective and objective orientations, are symbiotic processes. Broadcasters and audiences are terms arbitrarily ascribed to actors that really are (and should be) coconversant and coevolving. The most masterful journalists, in their most memorable reporting, used technology to communicate a story that balanced emotion and reason and transcended broadcaster/audience boundaries.[4] Remaking the news is really about remaking the ways in which, societies turn events into stories. Technology is playing a part, but it has unfortunately been miscast. Technology is there to help us reimagine news values. That is the place of technology and that is the way to remaking news storytelling.

Notes

1. The 24-hour TV news cycle banked on this decades ago.

2. For the distinction between polymedia as opposed to multimedia, see Madianou and Miller (2012).

3. See Z. Papacharissi, "Toward New Journalism(s): Affective News, Hybridity, and Liminal Spaces," *Journalism Studies,* published online March 2014.

4. More thoughts on this, here: "The Return of Sentiment,"http://www.niemanlab.org/2012/12/the-return-of-sentiment.

III Interrogating Occupational Culture and Practice

8 Helping Newsrooms Work toward Their Democratic and Business Objectives

Natalie Jomini Stroud

Thought News represents a revolutionary new form of journalism. As a collaboration between journalists and academics, the new publication will offer a cutting-edge take on important issues facing southern Michigan and the country at large. By incorporating observations from academia, and from the field of philosophy in particular, articles published in *Thought News* will provide a fresh new perspective, distinct from the typical newspaper's focus.

The pitch has the ring of a submission to the Knight News Challenge (a grant competition sponsored by the Knight Foundation). Yet *Thought News* predates the News Challenge by over 100 years. The idea, announced in March of 1892, was born out of a collaboration between American philosopher John Dewey and former newspaper editor Franklin Ford. Ford was frustrated by the content of newspapers. He wanted to "replace the scattered facts reported by ordinary newspapers with an analysis of the deeper social trends which would give these facts genuine meaning and significance" (Westbrook 1991, 52). For Dewey, *Thought News* was a way to "spread his ideas about democracy beyond the classroom" (51) and to "make philosophy an instrument for social action" (52). Dewey's motivation, as Dewey described it, was not to try "to change the newspaper business by introducing philosophy into it," but rather to "transform philosophy somewhat by introducing a little newspaper business into it" (57). Although *Thought News* never saw the light of day, the issues motivating its proposed creation continue to resonate. What is the role of journalism in a democracy? How might academia engage the business of journalism?

In this chapter, I consider the intersections between academia and journalism and between journalism's business and democratic goals. I draw inspiration from John Dewey and from his fellow thinker Walter Lippmann. These intellectual contemporaries had much to say about the role of the news media and academia in society. Dewey, an academic philosopher, saw great potential in the press's abilities to educate citizens and to foster the development of democratic communities. Lippmann, a journalist by training, was more circumspect about the media's democratic contribution. Instead, he found a potential solution to democracy's ills in having scholars address social problems. Dewey's academic hope for journalism and Lippmann's journalistic hope for academia provide a valuable starting point for answering modern questions about the roles of journalism and the academy. I also incorporate insights from Paul Lazarsfeld, a pioneer of social scientific communication research who put some of the ideas emphasized by Dewey and Lippmann into practice. Lazarsfeld distinguished between scholars whose research was more practical, or administrative, and scholars who adopted a more critical and system-focused approach. His interest in the media and involvement in both academic and practical research raised questions about the role of academics in public life.

In the pages that follow, I begin by identifying potentially competing goals. Newsrooms can strive for democratic *or* business goals. Academics can be practical, administrative researchers *or* critical scholars. Although there is a utility to thinking about these sorts of tradeoffs, it also is limiting. To navigate our way around or through these tense waters, I use an approach from pragmatism, the philosophical tradition with which Dewey's work is associated—namely, scrutinizing dichotomies for their utility. Dewey, like most of the naturalistically inclined pragmatists, saw our world as a series of shaded continuums, and not as holding many either/or dichotomies. Most of the dichotomies we take as central are of our own making; many of these cause more problems as conceptual tools than they solve (S. Stroud 2011). When considering both the role of journalism and the role of academics, it is useful to think about continuums. Although there are tensions between democratic and business goals, and between critical scholarship and administrative research, there are also instances in which these goals are not in conflict. Reframing the debate by dissolving these limiting dichotomies is the purpose of this chapter.

The Media and Its Problems

The news media face a challenge: they are asked both to turn a profit and to inform a public. But what if informing the public is not profitable? Lippmann (1922/2004) saw the issue clearly early in the twentieth century: "We expect the newspaper to serve us with truth however unprofitable the truth may be" (174). Lippmann argued that newspapers had to reflect the interests of their audiences because strong circulation figures allowed newspapers to attract advertising dollars and remain afloat. Newspapers are caught in a bind, Lippmann wrote. The public expects them to function like schools, providing a public service. But newspapers have to function like businesses because they are not funded by taxpayer dollars. These business realities have democratic implications. As Dewey remarked (1927/1984), "For, it is argued, the mass of the reading public is not interested in learning and assimilating the results of accurate investigation. Unless these are read, they cannot seriously affect the thought and action of members of the public" (349). Thus the press face a choice: either print the material that the mass public wants or publish the hard news that enjoys only a limited audience. The dichotomy is no less relevant today. Taking a page from the pragmatism playbook, however, it is limiting to focus only on the conflict between business and democratic goals. Instead, democratic and business goals could be more comprehensively considered. As shown in figure 8.1, there are instances in which both goals are undermined. More useful, however, there are instances in which both goals can be advanced. In the paragraphs that follow, I explore the possibilities illuminated by each quadrant in figure 8.1.

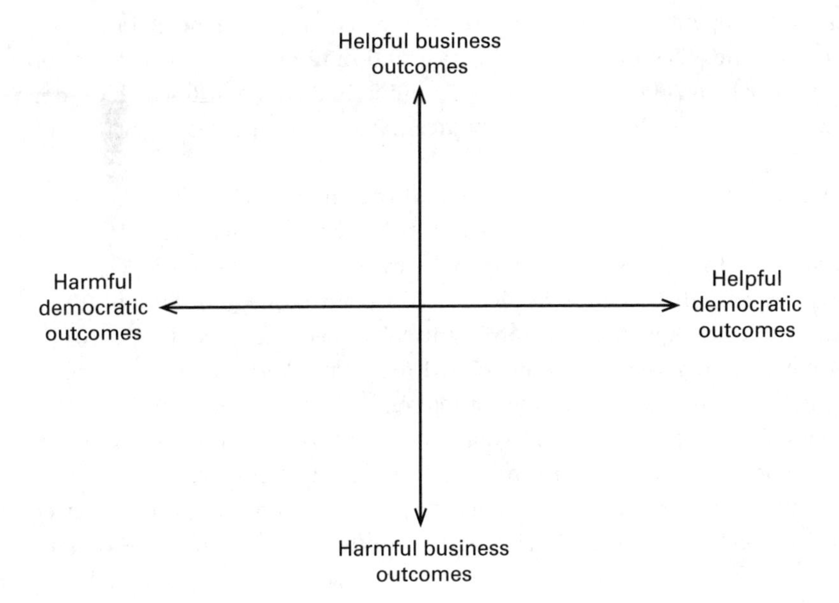

Figure 8.1
Business and democratic newsroom outcomes

The Depressing Quadrant: Harmful Business and Harmful Democratic Outcomes

The most worrisome quadrant of the grid is newsrooms facing both harmful business and democratic outcomes. Much research suggests that when the news media industry faces tougher economic times, democracy suffers.

Newsroom business models are showing signs of weakness as audiences increasingly move to digital forms of news. Traditional means of earning revenue no longer yield the same return they did even a decade ago. Advertising dollars, the main driver of news revenue, have declined. Print advertising revenue totaled $44.9 million in 2003. In 2012, revenues were $18.9 million (Pew Research Center 2014a). Newer forms of revenue have not made up for the losses. Online ad revenue has grown from $1.2 million in 2003 to just $3.4 million. Although local television news has fared somewhat better than newspapers, Pew Research (2014a) reports that between 2006 and 2011, their revenues dropped by more than 30 percent.

Modern information channels are a key reason that the business landscape has changed. Audiences are increasingly turning to digital news sources. Since digital advertising yields smaller returns, revenues decline. In addition, new digital competitors have cut into the news industry's bottom line. Classified advertising, a $19.6 million source of revenue for newspapers in 2000, declined to $4.7 million in 2012 (Pew Research Center 2014a). Other online classified sources, like Craigslist, now garner this traffic.

To deal with the business reality, news organizations have had to reevaluate their existing business models. To balance the accounting ledger, one strategy is to reduce costs. The Pew Research Center (2014a) reports that newspaper staffs declined by nearly 30 percent between 1989 and 2011. Expensive foreign affairs reporting also has been reduced (Enda 2011). Some news organizations have closed (e.g., the *Cincinnati Post*), or have switched to focusing on online content (e.g., the *Christian Science Monitor*). News sharing agreements also have increased (Potter and Matsa 2014), allowing news organizations to take advantage of economies of scale.

Another strategy is to explore new methods for earning revenue. Newer models include investing in side businesses that leverage the news organization's brand (e.g., the *Dallas Morning News*'s social media marketing firm DMNMedia), incorporating new services, such as crowdfunding platforms, into the news (e.g., Asahi Shimbun Media Lab), or hosting events (e.g., the *Texas Tribune*'s TribFest). According to an analysis from the Pew Research

Center (2014b), these newer revenue streams account for about 7 percent of the financial support for news. On digital platforms, news organizations also have been exploring paywalls, native advertising, and working with vendors who pay news organizations when site visitors interact with their content (e.g., Disqus, Outbrain).

Turmoil on the news business side has democratic implications. Democratic effects can be seen among citizens and in elections. Among citizens, civic engagement and voter turnout are affected by the vibrancy of the local media market. When local media cover U.S. House members less frequently, citizens know less about politics and are less politically engaged (Hayes and Lawless 2015; Snyder and Strömberg 2010). Hayes and Lawless (2015) coded over 6,000 newspaper stories about U.S. House races during the 2010 midterm campaign season. They overlaid these measures with survey data. Respondents in communities where the newspaper contained more stories about a House race were more likely to offer an evaluation of their representative's performance, to answer questions about where the candidates running for the House fell on the ideological spectrum, and to specify which House candidate they preferred. Although the analysis does not trace what happens when individual newspapers vary in how frequently they cover House candidates over time, the results show a correlation between voter knowledge and vote intention on the one hand and the amount of political coverage on the other. When local newspapers close, evidence suggests that civic engagement and voter turnout decline (Schulhofer-Wohl and Garrido 2013; Shaker 2014). Shaker (2014) tracked civic engagement in Seattle and Denver after the *Seattle Post-Intelligencer* and the *Rocky Mountain News* closed their operations. Declines in civic engagement in Seattle and Denver outstripped national trends. Further, the slump in civic engagement seen in Denver and Seattle was not replicated in eight other US cities where there were no major changes in the available local media. When the media aren't minding the statehouse, citizens aren't either.

A less-than-robust media also portends bad news for elections and political officials. House members are less active on behalf of their districts when they face less media coverage at home (Snyder and Strömberg 2010). Snyder and Strömberg reach this conclusion by analyzing how many congressional districts are included in a newspaper's circulation area. Newspapers overlapping with a single congressional district cover the district's House member more frequently than newspapers with readers spanning multiple congressional districts. The practical reason is clear; if newspapers catering to readers spanning multiple House districts covered any single representative, the coverage would be relevant only to the audience subset living in the district. The resources required to cover all representatives serving the newspapers' audience may not be worth the cost. House members from districts where the newspaper is able to focus on fewer representatives appear to be influenced by the heightened scrutiny. They more frequently appear as witnesses before committee hearings and place less ideologically extreme votes. Further, more federal money flows back to their districts. To the extent that the news media's collapsing business model reduces Congressional coverage, these results are disconcerting. If elected officials face less scrutiny from the press, they may slack in serving their districts. In addition to affecting the behavior of elected officials, media coverage also influences elections. Schulhofer-Wohl and Garrido (2013) examine the electoral consequences of the *Cincinnati Post*'s closure, concluding that "its absence appears to have made local elections less competitive along several dimensions: incumbent advantage, voter turnout, campaign spending, and the number of candidates for office" (78). As these studies show, elections are conducted and public officials act in less democratic ways in the absence of an attentive local media.

Research pertinent to this first quadrant demonstrates the connection between the health of the news industry and the health of democracy. Communities with a vibrant news media are more likely to have politically engaged citizens, competitive elections, and responsive elected officials. Flagging revenues for the news media mean more than a natural market correction, they have implications for the functioning of democracy.

The Off-Diagonal Quadrants

The two off-diagonal quadrants signal the trade-offs that newsrooms can face and represent the classic tension between democratic and business goals. On the one hand, a focus on business outcomes can lead a newsroom to shirk its democratic role, emphasizing entertainment content over hard news. On the other hand, a focus on hard news coverage could fail to find an audience that allows a news organization to stay in business. Evidence of these trade-offs is plentiful.

As a business, news organizations must pay careful attention to metrics that translate into revenues. For many news organizations, page views and unique visitors translate into digital advertising revenue. News organizations that can increase page views are rewarded financially. Tricks for increasing page views abound. It's easy to find a host of websites reporting ways to tweak a headline to increase click-through rates on social media. In larger newsrooms, some employees' jobs involve optimizing headlines to increase traffic referrals from search engines. The purpose of these activities is to learn, and then cater to, what generates the most clicks. The availability of these metrics can change how news organizations do their work. Vu (2014) conducted a survey of newspaper editors to learn about their use of digital metrics. Over 80 percent of editors whose jobs involved the online production of news said that they regularly attended to traffic metrics. Although not a strong relationship, editors recognized the payoff that comes from monitoring traffic: "The more editors feel that getting high readership brings economic benefits to the organizations, the more likely they are to make editorial changes based on online audience web analytics" (Vu, 1104). In this way, online traffic figures become a new gatekeeper, affecting what appears as news.

Financial pressures and the desire to increase page views push news organizations to think more about the "return on investment" (ROI). At a series of workshops hosted by the Engaging News Project at the University of Texas at Austin, digital news leaders repeatedly said that they wished they could compute the ROI. If they knew the staff time and resources it took to produce an article, digital teams could then compare these inputs to how well the article performed, such as how many clicks it garnered (Scacco, Curry, and Stroud 2015). This metric makes great business sense. News organizations could figure out which types of articles yield the greatest ROI and provide more of them. Yet how the ROI is calculated matters. An advertiser-driven page view metric should give us pause. If news organizations base their story selections on what content is expected to get the most clicks, a practice already in effect to at least some degree in newsrooms (see Anderson 2011), we may not like the outcome. Catering to audience preferences could have democratically meaningful consequences. When learning that an article about Afghanistan yields little traffic, editors would have to carefully consider whether to expend the resources for another Afghanistan article. As I show shortly, modern data demonstrate that audiences are drawn toward content that is not always in keeping with oft-expressed democratic norms. Before turning to an analysis of whether clicks-focused newsrooms are bad for democracy and whether democratically focused newsrooms are bad for clicks, it is incumbent upon us to identify normatively desirable democratic outcomes.

Although many democratic ideals could be identified, Dewey and Lippmann offer some guidance on the attributes of a "perfect citizen" in Lippmann's words, or a member of a "great community" using Dewey's language. Content that pushes people away from these ideals arguably has a negative democratic impact. Lippmann turns to textbooks to uncover what the "perfect citizen" might look like. Lippmann (1927/2011) concludes that "I do not see how anyone can escape the conclusion that man must have the appetite of an encyclopedist and infinite time ahead of him" (13). Despite his belief in its implausibility, Lippmann's read of the perfect citizen is one who is politically informed and then employs the information by placing votes on matters of public policy. Extending beyond an individual focus, Dewey (1927/1984) thinks about democracy in terms of the individual interacting with others. He conceives of the "democratic ideal" as follows:

> From the standpoint of the individual, it consists in having a responsible share according to capacity in forming and directing the activities of the groups to which one belongs and in participating according to need in the values which the groups sustain. From the standpoint of the groups, it demands liberation of the potentialities of members of a group in harmony with the interests and goods which are common. Since every individual is a member of many groups, this specification cannot be fulfilled except when different groups interact flexibly and fully in connection with other groups. (327–328)

Ideal democratic citizens, in Dewey's conception, recognize that they must work in concert with other individuals and groups toward common goals. This involves a level of trust in others and in democratic institutions and it requires groups working together to pursue shared objectives. Drawing inspiration from Dewey and Lippmann, we can analyze how the news media affect knowledge about public affairs, efforts to work toward shared goals, and trust in the democratic system. Further, we can consider how the content selected by an emphasis on business outcomes, such as page views, may affect these outcomes.

If news organizations follow audience desires, what sort of content would news organizations wanting to increase page views offer? And what are the democratic effects of content that caters to audience desires? Audience behavior with respect to election coverage, partisan and uncivil news content, and entertainment programming, shows exactly what is suggested by the off-diagonal quadrants: a clear contrast between the content that increases page views and the content that moves us closer to democratic ideals.

Campaign season represents a period when the media devote extensive attention to politics, but not all campaign coverage attracts equal public attention. In an early demonstration, Iyengar, Norpoth, and Hahn (2004) tracked where people went when they were asked to browse a CD containing information about the 2000 presidential campaign. The CD contained information on a host of topics, such as the candidates' background and issue stances. Audiences seemed drawn toward horserace coverage that described which candidate led in the polls and the candidates' strategies for electoral success. To the extent that the result replicates today, news organizations carefully monitoring digital traffic could conclude that they should boost horserace coverage during campaigns. The effects of this form of coverage, however, are potentially troubling. Cappella and Jamieson (1996) found that horserace coverage increased political cynicism, an outcome that seems to move us away from Dewey's notion of a great community. A focus on page views leaves this democratic consequence unmonitored; it may even risk increasing the use of publicity-gathering horserace coverage that hurts the democratic ideals we hope to instantiate through the institution of journalism.

Partisan and uncivil content also make more business than democratic sense. In interviews with news staff, several reported that partisan and polarizing news content can be good for an organization's bottom line (Scacco, Curry, and Stroud 2015). Inflammatory articles, news featuring polarizing political figures, and comment sections that generate fights among those with competing partisan viewpoints can dramatically increase page views, return visits, and time on site. In a tightly controlled experiment comparing civil and uncivil media coverage, participants in a study by Mutz and Reeves (2005) reported enjoying the uncivil version more. From a business perspective, uncivil and polarizing content is a good idea. From a democratic perspective, however, the case is not so clear. Like-minded partisan content can increase political polarization, whereby citizens increasingly favor their own political perspective over the opposition (Levendusky 2013; Stroud 2011). Uncivil comments left in comment sections also can polarize attitudes (Anderson et al. 2014) and negatively affect what people think about a news source (Lee 2012). In their research, Mutz and Reeves demonstrated that uncivil news content reduced trust in politicians and Congress relative to civil versions of the same content. Digital metrics showing a spike in page views for partisan and uncivil content, however, gives newsrooms a financial incentive to produce and promote this content.

Soft news and entertainment also attract audiences. Survey research, once a dominant way to understand readers' news interests, finds that the public has a strong interest in public affairs. Behavioral data from tracking where people go online, however, shows far less engagement with public affairs news content (Tewksbury 2003). The widespread availability of digital analytics within newsrooms only highlights this fact for newsrooms. Boczkowski and Mitchelstein (2013) compared the type of news most often read by site visitors to the type of news most often prioritized by editorial staff. Visitors looked to soft news, such as human interest stories, more than editors prioritized this content. There are democratic consequences when citizens abandon news in favor of more entertaining fare. When those with a preference for entertainment over public affairs news content have the ability to act on those preferences (such as when they have the breadth of content provided by the Internet), they do so. Those eschewing news have lower levels of political knowledge and turn out to vote less frequently when they are able to choose more entertaining content (Prior 2007). If preferences for human interest stories outweigh the desire for hard news, metrics-based newsrooms will provide more of the former and, if the overall availability of hard news content declines, even those with a preference for it may not be able to find it.

These potential democratic consequences accrue most clearly if many news organizations adopt a follow-the-page-view approach to news. To be fair, this is unlikely to occur on a mass scale. Segmentation will guarantee that some news organizations will offer in-depth, issues-based political reporting, whether because they have a civic commitment to it or because they can generate a financial return from offering the content. But in pressing economic times, news organizations are forced to figure out how to increase, or maintain, revenues. To the extent that the audience preferences documented here hold, unbiased political news has a much smaller chance of returning large traffic numbers compared to horserace, partisan, and entertaining content. By focusing on page views, there is evidence that the democratic impact of journalism would suffer. By ignoring page views, journalism may fail to make a profit. These disappointing trade-offs are valid and important—but focusing too much on the trade-offs ignores the opportunity made possible by the final quadrant to which I now turn.

The Fourth Quadrant: Helpful Democratic and Business Outcomes

Although the three quadrants reviewed thus far all point to undesirable consequences, the utility of drawing from the pragmatist approach is the recognition that there are some moments in which business and democratic goals do not conflict. Lippmann and Dewey famously disagreed in their prescription for solving the problems they saw facing democracy. Although Lippmann saw a limited role for the press, Dewey was far more optimistic about the media. In Dewey's (1922/1983) glowing review of Lippmann's 1922 book, *Public Opinion,* ("To read the book is an experience in illumination," wrote Dewey, 337), Dewey modestly offers a "suggestion"

(343). "Mr. Lippmann seems to surrender the case for the press too readily," Dewey writes, "To assume too easily that what the press is it must continue to be" (343). Instead, Dewey suggests, the news should be in the business of finding artful ways to present the news. He continues, "If the word 'sensational' can be used in a good sense, it may be said that a competent treatment of the news of the day, one based upon continuing research and organization, would be more sensational than present methods afford" (343). Figuring out how to make sensational news represents a promising opportunity to blend the business and democratic interests of news organizations.

Examples abound of newsrooms embarking upon precisely this challenge. The *Wall Street Journal*, for example, created a successful interactive Health Care Explorer to allow people to compare different healthcare plans. New entrants into the space also are exploring innovative ways of presenting the news. Vox introduced card stacks that provide background information about important issues—and yield strong digital metrics. The Skimm provides a daily email newsletter tailored for women with an impressive circulation. Foreshadowing the next section, academics also have provided some insight into how to present information in ways that help audiences to learn. Lee and Kim (2015), for example, show that infographics can result in news elaboration, particularly among those finding the issue in the infographic more important. To the extent that infographics are widely shared and accessed, they also could have a business benefit.

Today, there is considerable opportunity to take up Dewey's challenge. Digital analytics can reveal that an article about foreign affairs has low traffic. In all likelihood, it probably also didn't attract much readership when it appeared in the newspaper, either. Without digital analytics, though, we didn't know it. Now we do. The pessimistic take on what to do was presented in the earlier section—namely, stop reporting on foreign affairs. The optimistic take is Dewey's challenge: experiment with new ways of telling the story to find out what attracts and informs the audience. This is the hallmark of the underappreciated fourth quadrant.

News, Academia, and a Solution

In figuring out how to create sensational news, academia may be able to make a contribution. Indeed, a public role for academics is central to Lippmann's thinking about how to improve upon the democratic system. Lippmann proposed relieving citizens of the unrealistic burden of keeping abreast of public affairs and releasing the media from the impossible mission of providing the information necessary for the task. About the media, Lippmann (1922/2004) wrote that "If newspapers, then, are to be charged with the duty of translating the whole public life of mankind, so that every adult can arrive at an opinion on every moot topic, they fail, they are bound to fail, in any future one can conceive they will continue to fail" (228). The point is true in two senses: not only are newspapers incapable of covering this much material, but even if they did, the public may reward the tome with a dismissive yawn. Instead, Lippmann proposed that a bureau of social scientists should be commissioned and charged with conducting systematic and unbiased research on matters of public policy. The bureau also could transmit information to the press which could inform interested segments of the public.

Although Lippmann's idea is provocative, the role of academic research in public life is not without controversy. Forefather of empirical communication research Paul Lazarsfeld (1941) distinguished between two types of research: administrative research, which is conducted "in the service of some kind of administrative agency of public or private character" and critical research, which studies the "general role of our media of communication in the present social system" (8–9). Put more concretely, administrative researchers would empirically investigate which advertisements worked best to sell products, critical researchers would critique the American focus on consumerism. Just as journalists' democratic and business goals push in different directions, so too do the objectives of administrative and critical scholars. These schools of thought are treated as opposites—the former hueing more closely to Lippmann's vision and the latter consisting of academics detached from the spheres of business and government so as to have a better perspective from which to observe and critique these systems.

The pragmatic orientation of this chapter begs for a reconsideration of this divide. Although Lazarsfeld (1941), and later Rogers (1982), urged critical and administrative scholars to work together, I want to push the point further and argue that it is not productive for us to think of scholars as critical *or* administrative. As before, I present a grid consisting of one axis that spans from a critical to a noncritical approach and a second axis that spans an administrative to a nonadministrative orientation. I discuss each of the quadrants in turn, ending with a quadrant that has been underdeveloped: the critical-administrative scholar. Although each quadrant makes a valuable contribution, it is in this last quadrant where academics have an underutilized opportunity to affect the practice of journalism. (See figure 8.2.)

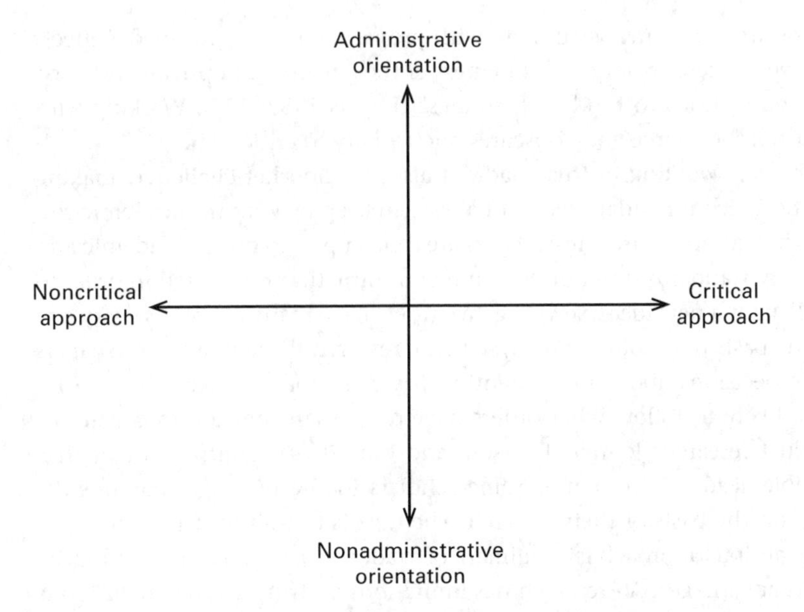

Figure 8.2
Critical and administrative academic orientations

Administrative Orientation, Not Critical Approach

Administrative scholars without a critical approach essentially operate as consultants, evaluating outcomes associated with business and government goals. Lazarsfeld, in the minds of some, epitomized the administrative researcher (see Rogers 1982). He received grants from publications like McFadden Publications and CBS Radio to do what amounted to market research. His famous coauthored book *Personal Influence,* although investigating how people came to decide for whom to vote, includes chapters on opinion leadership regarding fashion and movies (Katz and Lazarsfeld 1955/2006). The research answered a very practical question: what influences how people make decisions on both mundane and consequential matters?

The critiques of this approach are multiple. Some argue that administrative scholars "solve little problems, generally of a business character" (Lazarsfeld 1941, 8). Critical researchers charge that administrative work, like Lazarsfeld's, concentrates on smaller questions that support the status quo instead of asking questions about how an entire system or industry influences society, or how power is manifested. This mentality results, at least in part, because administrative researchers are beholden to the businesses and government entities funding their research. Researchers' "preoccupation with persuasion, with propaganda, and with media effects in general, stems from the willingness of the government, private industry, and foundations to fund such studies" (Rogers 1982, 127). Working with industry, the charge goes, obscures one's ability to criticize it.

Scholars working in this quadrant also face another challenge: making practical recommendations is not always in keeping with the academic culture, but administrative researchers are called upon to do so. Academic and industry standards differ on how much information is required in order to implement a new idea. News organizations make many decisions every day on the basis of absolutely no systematic research. When daily newspapers make decisions about implementing paywalls, for instance, they are far more likely to follow what others have done than they are to consult or conduct research (Jenner, Thorson, and Kim 2014). Contrast this with a possible academic response. Some scholars loathe making recommendations on the basis of their research. The urge is to wait for the next study. This cautious approach is a hallmark of academia, and it has much to speak for its advantages. All research has limitations and any given conclusion is stronger if the findings have replicated under many different circumstances. But much research goes unreplicated. Further, it's not clear how much replication would be required before one would be comfortable saying that the results have recommending force. One could always delay action and wait for more evidence, no matter what findings one presently has. There is a middle ground between hazardous brashness and paralyzing caution, of course. Scholars can figure out how to make carefully constructed recommendations that articulate the limitations of their research so that anyone deciding to act on the basis of the findings is informed. For members of the news industry accustomed to making decisions in the absence of any research, even some research is better than nothing. And by working with industry, academics can inform newsroom staff about our research results that typically hide behind the paywalls used by the publishers of scholarly journals. This does not, of course, solve the other critiques facing purely administrative researchers, but represents one way of countering this critique of administrative work.

The Nonadministrative Quadrants

Scholars working in the nonadministrative quadrants are not in the business of making recommendations that a newsroom could enact tomorrow. Critical scholars—those with a nonadministrative approach and a critical orientation—are interested in control, conflict, power, and social class structure. They examine the mass media as "agencies of social control, blocking pathways of radical social change" (Blumler 1980, 371).

Although the approach has much to offer in terms of illuminating the broader societal implications of the press, it has limitations as well. Critiques of this approach contend that it has limited practical application and a small probability of influencing the social systems it aims to critique. Identifying a patriarchal or consumeristic society is important, but it is not always clear what to do with the insight. Further, purely critical approaches have been critiqued methodologically because they presuppose the phenomenon they set out to uncover; as Lazarsfeld (1941) writes, "Although efforts are steadily being made to refine and corroborate [a "promotional culture"] theory it is taken for granted prior to any special study" (12).

Another scholarly approach takes neither an administrative nor a critical orientation. Scholarship in this vein may have a purely theoretical aim without a clear practical implication or an evaluative or critical tone. Content analyses of how the media frame an issue, for instance, could fall into this quadrant. The analysis may inform our understanding of the many frames used to cover an issue, but if the author makes no evaluation of which frames better serve the public or the press, it is of limited practical utility to organizations beyond the academy. It also doesn't advance a critical agenda by failing to challenge larger cultural patterns or power dynamics.

Administrative Approach, Critical Orientation

The final quadrant that is revealed by dissolving the dichotomy between critical and administrative scholarship represents an opportunity for academia and the news industry to work together without necessitating that academics compromise their critical edge. Collaboration between academia and industry has the potential to yield productive insights particularly as the news media industry faces challenges in living up to its business and democratic potential. Members of the academy are unburdened by the day-to-day pressures of the newsroom and the need to make a profit. Instead, academia is free to focus on the democratic aims of journalism and to identify and test practices that might advance the cause. Thus academia can push an agenda, an idea more in line with critical scholarship, while still making a practical contribution, as an administrative scholar would. In the following section, I give several examples of early steps in this direction.

Comment sections can be challenging for news organizations. From a business perspective, comment sections can be profitable. Some commenting platforms pay news sites each time someone interacts with advertising content. Further, commenters tend to be frequent and loyal site visitors, and comment sections give these audiences a space to engage with the news product and other commenters. But these business boons come with problems. Comment sections can be filled with uncivil and hateful comments that news organizations worry could undermine their journalism and hurt the news brand (Singer and Ashman2009). Several prominent outlets, like Reuters and *Popular Science*, have opted to remove their comment sections altogether. Others, like the *New York Times*, employ staff dedicated to moderating the comment sections and ensuring that only comments meeting editorial standards are posted. Viewed through a Deweyan lens, these could be spaces where citizens engage in conversations that serve as the basis for a great community.

Academia has much to add to understanding this space. Rich scholarly traditions have been dedicated to understanding deliberation (e.g., Delli Carpini, Cook, and Jacobs 2004) and this work provides insight into thinking about how these spaces can be more democratically edifying. This research already has begun, and I mention several promising projects here. Santana (2014), for instance, examined comments left in response to articles about immigration. Half of the comments appeared on news sites allowing anonymity and half appeared on news sites requiring site visitors to use their Facebook accounts or their real names to comment. When anonymity was permitted, 53 percent of the comments were uncivil. When anonymity was prohibited, 29 percent of the comments were uncivil. This study has limitations; commenters weren't randomly assigned to either a real name or anonymous comment section, which means that we can't be sure that real name platforms *cause* more civil discussion. But the research is a start. On the basis of these findings, newsrooms concerned that their brand is tainted by uncivil comments could explore platforms requiring commenters to use their real names. Even better, a newsroom could work with academics to randomize whether commenters are required to use their real names or not and we could learn about the causal effect of this practice.

As another example, Sukumaran, Vezich, McHugh, and Nass (2011) found that cues indicating that the comment section should contain thoughtful comments, such as a prompt in the comment box asking people to "Please try to make your contributions as constructive as possible," resulted in longer comments and more issue-relevant thoughts compared to comments left in a comment section not cueing thoughtful engagement (e.g., the prompt read "Have your say here!"). Because an experiment was used, causal conclusions can be drawn from this research. The limitation is whether the results will generalize an actual newsroom. But the idea is testable.

Working directly with news organizations to analyze their practices also is possible. Research that I coauthored with Josh Scacco, Ashley Muddiman, and Alex Curry (2015) involved working with a news organization to randomize newsroom involvement in the comments on their Facebook page. According to a randomized schedule, a popular political reporter, an anonymous newsroom staff member, or no one from the station conversed with commenters on political posts. The news staff modeled civil participation by asking questions, responding to genuine queries for more information, and complimenting strong comments. They did not directly engage those who were behaving in an uncivil manner. We analyzed whether the comments left differed depending on the station's involvement. Comments left when the reporter posted were more civil and were more likely to include evidence-based claims.

This in-depth look at comment sections provides one example of how academic research can provide industry-relevant insights. Democratic goals are advanced by envisioning ways of creating better spaces for citizen conversation. Business goals are advanced by limiting commenting behaviors that can hurt a news brand. It's also possible that these sorts of strategies could increase commenting overall, which could have a more direct financial return. By expanding academic work to directly consider business outcomes as well, we would increase the probability that our ideas would be implemented.

Analyzing comment sections is just one example. Work from the Tow Center for Digital Journalism, for instance, examines the practice of incorporating gaming into the news and play of newsrooms (Foxman 2015). Research from the Engaging News Project finds that online quizzes about news topics can increase time with news content and can help people learn—and learn better than sentences conveying the same information (Scacco, Muddiman, and Stroud in press). This research also suggests that quizzes can increase time on site, a direct business outcome. Pickard's chapter in this volume takes a different, but complementary, tack by advocating for nonmarket-based ways of funding journalism. He offers the view that the news is a democratic good and that academia can make a contribution by examining policy solutions.

Research in all four academic quadrants makes important scholarly contributions, yet the current crisis facing journalism begs for more attention to this fourth quadrant. Although work related to this quadrant has not been connected to Dewey and Lippmann explicitly, their perspectives can be easily imputed. The benefit of doing so is that it unites a research tradition around a common understanding of the media's important role in a democracy, the social scientific techniques that can help to bolster this role, and the creative and artistic means required to engage citizens.

Conclusion

Social science can be employed to test and identify news presentation strategies that increase desired forms of engagement. As Dewey hinted, these forms of presentation can be lackluster or artful (S. Stroud 2008). The promise of the latter, of course, was what he sought with his *Thought News* venture, and what he wanted in any successful reconfiguring of journalism. Thinking about artful journalism gives us a way out of a crippling dichotomy that forces us to choose to be either critical scholars or administrative researchers, or to advance business or democratic journalism goals. Figuring out artful ways of presenting public affairs content not only promises to aid the news industry's bottom line, it also coincides with the democratic mission of journalism. Dewey (1927/1984) seemed to have something like this in mind in *The Public and Its Problems* when he wrote about what it would take to turn a great society into a great *democratic* community:

> The highest and most difficult kind of inquiry and a subtle, delicate, vivid and responsive art of communication must take possession of the physical machinery of transmission and circulation and breathe life into it. ... Democracy will come into its own, for democracy is a name for a life of free and enriching communion. ... It will have its consummation when free social inquiry is indissolubly wedded to the art of full and moving communication. (350)

There is potential for academia to contribute to this mission. But we must be able to imagine new ways of doing journalism, as well as different ways of engaging in the craft of social scientific research. In answering such a call, scholars can help to identify news practices that promote democratic and business goals using measures and techniques that expand beyond page views.

I propose that there is no inherent opposition between industry and academia, at least with respect to research on how journalism can better engage its readers. The approach outlined in this chapter seeks to transcend this debate. Just as Lazarsfeld (1941) saw ways for administrative and critical scholars to work together, so too can academia and industry. To an industry focused on profits, academic scholars can bring their own critical agenda, namely one that focuses steadfastly on the democratic necessity of the news media. And by working directly with industry, scholars have a chance to implement change.

In fact, working with industry has the potential to broaden and enrich the sort of questions that academics ask. Academia often does the work of describing the social world and unpacking causal relationships among variables. This is a valuable role because without a systematic understanding of the problems we face, it's unlikely that we will stumble upon compelling solutions. Demonstrating that gerrymandering, geographic sorting, and partisan media contribute to political polarization, for example, provides policymakers and nonprofit organizations with insight on what they might do to minimize polarization. What Dewey and Lippmann suggest is that in addition to identifying problems, social scientists also can dream up and test creative solutions. Academia has the tools, and the time, to push an agenda that complements the democratic promise of journalism.

As history might have us believe, Lippmann and Dewey fought an epic intellectual battle. On the one hand, Lippmann bemoaned the overly burdensome intellectual demands that democracy asked of citizens and the news media. Rather than task the public with mastering the various nuances of public policy and the media with the charge of reporting the details, Lippmann thought that institutions made up of social scientists could provide policy makers with unbiased guidance on how to move society forward. On the other hand, Dewey thought that citizens and the media could be prominent engines of democracy and that creative and scientifically tested ways of engaging the public could fuel their interest in political affairs. By seeing life, and all of its interactions, as artful and democratic moments, a Great Community could emerge. These grossly discrepant visions of the world differed in their assessments of citizens' capabilities, journalism's potential, and the viability of democracy as a form of government. Or so the legend goes.

There was, and still is, a valuable use to which this narrative of conflict can be put in our disciplinary storytelling. But we must acknowledge the limitations of any dominant path of thinking, and be able to imagine different ways of seeing and solving society's problems. A hallmark of Dewey's pragmatist approach is to dissolve unnecessary dichotomies, especially when they portend more drawbacks than benefits as guiding concepts for our practical activities. It is ironic that history has retold the story of Dewey and Lippmann by relying so heavily on such a dichotomy. Rather than describing Dewey and Lippmann as having a collegial relationship and as concerned with many of the same issues, history has focused on their few moments of intellectual departure to cast their exchange as a great debate between adversaries. Some, like Jansen (2009), have made strides toward recasting this story. The present chapter wants to continue such efforts and to extend constructive retellings into the very practical domain of researching how journalism and academia operate. By building on the many commonalities between Dewey and Lippmann, we can see that there are harmonious ways to do the business of the news *and* careful scientific research evaluating democratic effects.

It is too bad that *Thought News* never came to pass. Its failure, however, is instructive. For Dewey, *Thought News* was "an overenthusiastic project which we had not the means nor the time to carry through" (Westbrook 1991, 57). For Ford, Dewey "had gotten cold feet when faced with the threat that radical journalism would pose to his academic career" (Westbrook 1991, 57). By leveraging the academic skillset and dissolving the dichotomy of academics as administrative or critical scholars, perhaps the academy can make a meaningful contribution to the practice of journalism.

This chapter attempts to seriously consider whether news organizations' business and democratic goals truly conflict and whether academia has a role to play in advancing these missions. Although it is true that newsrooms' business and democratic goals are sometimes at odds, this chapter argues that it needn't always be the case. An increasing number of scholars are approaching journalism research in a manner consistent with Dewey's thought, even if they do not explicitly recognize that their work continues this tradition. We need to do more of this, and explore how to do it better. And although academic and industry goals can clash, this chapter shows that there are valuable intersections. Lippmann's ideas of having social scientists play a role in policy development, and Dewey's insistence that social science could help to find artful ways of informing citizens, were not useless pipedreams. Creatively combined, they provide a path forward. This chapter unites these traditions with modern challenges facing newsrooms and academic engagement with the journalism industry. The result is a forward-looking take on how the news media can retain and enhance their important role in a democracy.

9 Journalism Ethics and Digital Audience Data

Matthew Hindman

[T]he supreme test of good journalism is the measure of its public service.
—Walter Williams, "The Journalist's Creed," 1914

What is morally required of journalists?

This is a question that goes to the core of journalism ethics and the boundaries of journalism as a field. But if the query is familiar, this chapter argues that the answer has changed. Digital media has profoundly altered the normative landscape of journalism and the duties that fall on reporters and editors.

Some shifts in the ethical calculus of journalism come from the industry's changing social role and its increasingly precarious financial position. Today more than half of respondents in Pew's long-running media survey list the Internet as a main source of news, up from just 10 percent 15 years ago (Pew 2013). The Internet has blurred once-clear notions of who is counted as a journalist (e.g., Friend and Singer 2007). Print newspapers, which had always produced most of the nation's news, have roughly half the weekday circulation they did a decade ago (Mutter 2014). Newspaper ad revenue has dropped by two thirds, and three out of ten newsroom jobs at newspapers have been lost.

Still, the most novel and important ethical challenges come not from new competition, or a blurring of professional boundaries, or even from the retrenchment of local newspapers. Rather, the ethical transformation of journalism comes from new forms of data about the audience. In order to maximize it social impact, *ethical journalistic practice now requires strenuous attention to data on audience behavior.*

Print journalists have for centuries been essentially ignorant of the audience their specific stories receive. Broadcast journalists have had more audience data, though typically only at the aggregate level and within demographic categories such as gender and age group. Now audience data is ubiquitous and incredibly detailed—user-level data that tracks audience members over time and across devices. Metrics now show precisely how far individual readers scroll, how many seconds of video they watch, even their patterns of mouse and finger movement. Importantly, descriptive data is accompanied by experimental data showing the effect of different headlines, or different ledes, or different site designs on the digital audience. It is no exaggeration to argue, as C. W. Anderson (2011b) has recently done, that "a fundamental transformation has occurred in journalists' understanding of their audiences" (see also Turow and Draper 2014).

This new understanding of the audience must be accompanied by a paradigm shift in journalistic responsibilities. Journalism ethics has long been largely *process driven*. As Berry (2012) notes, "The majority of [journalism ethics'] focus is on production rather than consumption." Ethical journalism has been defined by it verification of facts, its focus on issues of civic concern rather than mere entertainment, its fair treatment of sources and subjects, and its objectivity (or at least its transparency). Responsible journalists were those who checked all of these boxes, and who carefully weighed competing values.

This chapter argues that some journalists and editors—rightly, in the author's view—are shifting from *process-focused ethics* to *results-focused ethics*. Responsible journalism increasingly requires adherence to evidence-based practices likely to advance civic goals. Journalism has long justified itself in consequentialist language, but new audience data gives journalists vastly more information on the true consequences of their choices.

Shifting ethical responsibilities are, of course, not unique to the profession of journalism. In many professions, change in scientific understanding alters ethical obligations. When a major study found that estrogen therapy promoted rather than prevented breast cancer, for example, physicians altered treatment guidelines for millions of patients. Similarly, journalists have a growing obligation to learn about and follow practices that best inform the public. The irony is that, just as journalists have been declared a threatened species, their profession is finally acquiring certain features that long defined other professional groups.

This chapter starts with a brief discussion of journalism ethics, with a focus on how the long-standing goal of informing the public has taken on new meaning. It discusses improvements in audience measurement, the rise of new journalism organizations, and a new conception of what journalistic ethics requires. The chapter concludes by outlining research needed in order to map the new ethical landscape.

Journalism Ethics

Historically, the codification of journalism ethics has been closely tied to the professionalization of journalism as a field. In order to provide a more focused discussion this chapter concentrates just on the American case—although attempts to craft a truly global version of media ethics are an important subtheme of recent scholarship (Rao and Wasserman 2007; Ward 2010). In the American context, the liberal and objective models of journalism were dominant throughout the twentieth century (Ward 2009). Both emphasized the crucial role of an independent, watchdog press as a democratic institution.

Despite its claim to reflect objective or even "scientific" truth, many have argued that journalistic norms do not spring solely from philosophical first principles. The rise of "mixed" media and "new" media, beginning in the 1980s and accelerating in the 1990s and 2000s, unsettled the previous consensus. Communitarian, feminist, and critical theory scholarship have argued that journalism norms have long served some groups of citizens poorly. A still-growing body of research emphasizes that the availability of advertising revenue was crucial to development of an independent press (Starr 2004; Petrova 2011; Schiller 1981). Other work has portrayed appeals to journalism ethics as a rhetorical tactic focused on public persuasion (Ward 2005a) or a strategy for claiming professional jurisdiction (Waisbord 2013).

But despite contestation from within and criticism from without, there remain substantial areas of agreement in terms of core values and the appropriate approach to ethical reasoning. It is worth emphasizing several features that cut across different schools of journalistic ethics scholarship.

First, journalistic ethics commonly employs a balancing test as its central form of ethical reasoning. Journalistic ethics is applied ethics, not pure theory. It therefore strives—as numerous scholars assert—to offer useful guidance to daily practitioners.

But what exactly constitutes useful guidance? Conflicting demands and the inherent messiness of news work do not lend themselves to simple, rule-based decision-making. Journalism ethics involve decisions about "which values should be put in the balance and how much weight they should be given" (Klaidman 1987). Ward similarly argues for a "holistic approach" in which practitioners "stand back and take all of the major principles of media ethics into account and seek to balance them with respect to the facts of this case" (Ward 2011, 79; see also Black, Steele, and Barney 1999). The balancing test means that the normative calculus can change dramatically as different considerations are added to the scale, even if old obligations remain.

Second, most studies of journalism, and the beliefs of most practitioners, emphasize the primacy of public service. The opening lines of Walter Williams's "The Journalist's Creed" (1915) are among the most famous statements of this principle:

> I believe in the profession of journalism.
>
> I believe that the public journal is a public trust; that all connected with it are, to the full measure of responsibility, trustees for the public; that all acceptance of lesser service than the public service is a betrayal of this trust.

These sentiments still get overwhelming assent from journalists today—and indeed Williams's creed is set in bronze near the entrance to the National Press Club in Washington, DC. Even many critics of journalistic norms accept this premise, using the claim that journalists are failing the public as their central line of attack.

Third, there has long been latent tension between journalism's ostensible duty to the public, and the topics on which journalistic ethics has focused. Most discussions of journalistic ethics spend little time on how to engage the audience. This is a glaring omission for a profession that proclaims public service as its highest goal.

Consider the following list of ethical problems that are the focus of significant scholarship, adapted from Ward (2009):

—Verification and ensuring accuracy, especially under the pressures of immediacy
—Balancing access to sources versus journalistic independence
—How journalists present themselves to subjects and the public (for example, the ethics of undercover investigations)
—The ethics of graphic imagery (for example, the desire to accurately portray the reality of violence versus potential negative impact on the public)
—Source protection versus the interests of justice

Other journalism ethics books and articles highlight essentially the same set of problems (e.g., Gordon et al. 2012; Black, Steele, and Barney 1999).

This set of concerns centers on journalists' behavior in the process of preparing a story. If journalists are diligent about pursuing truth, if they are meticulous about verification, if they represent themselves accurately, if they protect their sources—etc., etc.—the resulting product will be ethical.

In contrast, attempts to measure the reception of the public are not just missing: they have often been explicitly discouraged. Scholars have spent a generation attacking the ratings-driven "if it bleeds it leads" mentality of much television news (Carter and Allen 2000; Kerbel 2000). A similar attitude has been applied to the Internet, with corresponding disdain for outlets like Gawker, the Huffington Post, and Buzzfeed. Kovach and Rosenstiel, authors of perhaps the most widely assigned journalism ethics textbook, note "newspeople's traditional resistance to using data to drive their editorial decisions":

> Even if the data were reliable, should news organizations that built their reputations on depth and quality in their legacy platforms build their online operations by chasing page views in the hope of selling more cheap banner ads? The metrics only repeated a familiar and predictable lesson: Wire stories about Justin Bieber would always blow away enterprise policy stories about Mississippi politician Haley Barbour (Kovach and Rosenstiel 2011, 251).

Nikki Usher found, as late as 2010, that similar attitudes were dominant in the *New York Times* newsroom, in which reporters were rewarded far more for a page one story than for a viral piece that received ten times more traffic (Usher 2014).

Each of these points highlights a broader contradiction. In philosophy more generally, several distinct (though overlapping) schools of ethical reasoning dominate (Plaisance 2013). One ancient school of ethics focuses on virtue, on nurturing proper character. Another focuses on duty, on the obligations that (for example) the role of journalist entails. Consequentialism constitutes a third school, one that judges acts based on their likely consequences.

Certainly the journalistic ethics literature is replete with mentions of virtue and duty. We read about "virtuous journalists" (Klaidman 1987), and the duties journalists owe to the audience and—especially—to the pursuit of truth. But this rhetoric notwithstanding, journalism ethics remains predominantly consequentialist in orientation. It rarely argues deductively from first principles. Instead, journalistic ethics mostly makes *empirical claims about what will happen if norms are not followed.* If journalists behave badly, the logic goes, there will be harms to individuals, to the stature of news organizations, and to the very process of democratic accountability. Balancing tests are a characteristic outgrowth of this kind of consequentialist reasoning.

Journalism's implicit commitment to consequentialism governs the Web's impact on journalistic ethics. Consequentialist philosophy judges actions both based on their intentions, and on their reasonably foreseeable consequences. It is the second part of the equation—the *reasonably foreseeable* part—that demands a wholesale reworking of media ethics. Much more of a news story's audience and impact is reasonably foreseeable today than in the recent past. When the actual readership of print journalism was largely unmeasurable, at least at the level of individual stories, journalists could become habituated to ignoring audience behavior. In the balancing tests that journalists are supposed to perform, actual consumption habits have been left off the scale.

This ignorance of how citizens consume the news is no longer ethically defensible. For journalists to fulfill the role they claim for themselves, they must attract a wide audience. Journalists must now make a good-faith effort to anticipate the consequences of their professional choices. This requires both new organizational patterns, and new responsibilities for individual journalists.

Three Archetypes

Journalism ethics addresses both the "micro" level—the actions of individual journalists—and the "macro" level, the collective behavior of journalistic organizations. On both levels, weighing audience behavior can alter seemingly settled ethical issues. We will consider both levels in turn, but will begin with a discussion of the institutional implications.

Consider a very rough taxonomy in which news organizations can be divided into different groups based on both (1) their *commitment to journalistic ethics* and (2) their *use of audience metrics*. These are ideal types, and no news organization fits perfectly into these categories. Most news organizations are hybrids or bundles, combining a wide variety of different content types made with different levels of attention to journalistic ethics. Still, classifying news organizations based on archetypes can be useful in thinking through the ethical issues raised by rich audience data.

The Traditional Ideal

We will term the first group of news organizations the *traditional ideal*. These organizations focus on hard news, especially on topics important to democratic self-government. They assign their resources based on elite judgments about which stories are most important, not which stories are likely to get the largest audience. One might think here of highly stylized versions of the *New York Times* or the *Washington Post*, or golden-era *60*

Minutes, or in-depth journalism on public broadcasting stations (both television and radio).

But these sorts of news organizations are not just found in a mythologized past. Today ProPublica and the Center for Public Integrity are perhaps the two news organizations closest to this ideal. Both are primarily digital outlets that focus on in-depth, long-form, investigative reporting. Both are nonprofit. Both focus overwhelmingly on the type of public accountability coverage that wins accolades from other journalists. Indeed, in recent years both organizations have taken home a raft of journalistic awards, including a Pulitzer each.

Traditional journalism norms and institutional incentives privilege in-depth stories over smaller, incremental ones. CPI and ProPublica both commit a remarkable portion of their journalistic resources to difficult investigations. For example, Chris Hamby's Pulitzer-winning story for CPI on benefits for black lung disease required a year of full-time reporting (Center for Public Integrity 2014).

The traditional ideal is also characterized by the tone of its writing. Ethical journalism connotes an editorial perspective that is scrupulously fair. In practice, this pursuit often lends hard news stories an unrelentingly serious tone.

ProPublica and the Center for Public Integrity, of course, have produced consistently excellent journalism. Yet despite a steady stream of accolades, these two examples suggest that there are important limits to this model. Most significantly: neither news organization has been able to build a broad or deep audience for its work on its own site. The author's conversations with sources familiar with the CPI's internal metrics confirm that readership numbers are below metro dailies with similarly sized reporting staffs.

This failure to build a large, independent audience has led ProPublica and the Center for Public Integrity to partner with newspapers or broadcast news outlets for their highest-profile stories. These partnerships are essential if this journalism is to reach the wide audience it deserves. At the same time, these partnerships can be difficult to manage, and in some cases contentious. CPI's and ProPublica's difficulty in generating readership on their own sites is consistent with a body of research on digital audiences (Hindman 2014). Sites require a steady stream of constantly updated content to sustain audience growth—not staff members who contribute just a few stories a year.

Commercial Metrics

The second archetype stands in stark contrast to the first. Let us call this group the *commercial metrics* outlets. This breed of news outlet uses digital metrics and instrumentation to maximize attention and revenue. And while these organizations still want to get the facts right, they are unburdened by many other traditional tenets of media ethics.

Commercial metrics outlets are the digitization of the tabloid ethos. Perhaps the purest example of this archetype was the Gawker network of sites. Gawker was founded to publish "the stories that journalists talk about with one another in private but never write" (Bercovici 2014)—an explicit rejection of traditional journalism ethics and norms. BuzzFeed and the Huffington Post, especially in their earliest years, were also exemplars of this approach. Both still have elements of this outlook embedded in their organizational DNA, even as they have expanded their production of genuine hard news. The commercial metrics crowd has often engaged in enthusiastic embrace of sensationalism. Sites like the Huffington Post, for example, will often pair a screamer headline with otherwise ordinary news reporting.

Perhaps the defining characteristic of the commercial metrics archetype is a commitment to instrumentation and testing to maximize audience. This goes far beyond site design and site features. The central elements of news stories—headlines, ledes, story topics, and artwork—are all routinely A/B tested. Some sites even use predictive models to do story assignment. AOL's BlogSmith system, for example, uses search volume and other metrics to identify "breaking" stories, "seasonal" high-traffic topics, or "evergreen" stories that are perennially popular (Johnson 2011).

Whereas traditional news organizations are staffed largely with reporters, metrics-driven organizations instead hire "writers" and editors. Commercial metrics news organizations are agnostic about the type of content they feature. Kitten slideshows, celebrity quizzes, and even content that verges on the mildly pornographic are all fair game. Indeed, Gawker assigned staff to "link whoring duty" on a rotating basis, in which they were told to scour the Web for "[whatever] they feel would garner the most traffic," from "dancing cat videos and Burger King bathroom fights [to] any other post they feel will add those precious, precious new eyeballs" (Daulerio 2012).

Of course, many of these commercial metrics organizations also engage in wholesale content aggregation. There are many instances of metrics-focused news organizations rewriting stories from other outlets, and in the process generating far more traffic than the original story. Despite handwringing from many traditional journalists, metrics-driven organizations have enjoyed remarkable success over the last decade. Several have become top digital news outlets, and they continue to take audience share from traditional news organizations.

At the same time, the fate of Gawker suggests that extreme versions of this strategy carry novel risks. Silicon Valley billionaire Peter Thiel began a secret campaign to destroy Gawker after being outed as gay by the publication in 2007. Thiel covertly bankrolled Hulk Hogan's sex tape lawsuit, which led to a $140 million judgment against Gawker, the bankruptcy of Gawker Media, and the shutdown of the Gawker.com website. While Gawker was perhaps the most gleefully transgressive online news outlet, many of the techniques it pioneered remain standard practice at digital news organizations of all stripes.

Ethical Metrics

This leads to our third archetype of news organization—what we will term *ethical metrics* organizations. These outlets' goal is not (just) to produce important stories on topics of civic import, nor do they attempt to maximize revenue. Rather, their aim is to *maximize the audience for civically valuable content.* The goal of informing the public is unchanged, but the journalism practices adopted can be significantly different.

Even more so than the other two types, there are no pure examples of this third archetype. But strong elements of this viewpoint can be seen in new digital start-ups such as a Vox and Medium, which blend an explicit commitment to informing the public with criticism of long-standing journalistic norms. Practices at the *Atlantic*, or recent shifts at the *Washington Post*, can be seen as having a similar rationale. Upworthy—a site that is not itself a news organization, but that is similarly mission driven and that disseminates a lot of public affairs content—is a prominent fellow traveler. More broadly, ethical metrics has a lot in common with the so-called explanatory journalism movement, which includes both new sites such as Vox and FiveThirtyEight as well as verticals run by established news brands (such the *New York Times*'s The Upshot, Slate's The Explainer, or Bloomberg's QuickTake).

The ethical metrics approach still requires elite decision-making about which stories qualify as important. A large dose of traditional news judgment remains essential. But once the judgment has been made about which stories are most civically valuable, testing and metrics can be used aggressively to maximize attention.

A crucial prerequisite of the ethical metrics approach is the emergence of richer and more robust usage metrics. For the first two decades of the Web, the three dominant metrics for online usage were *page views* (or more informally "hits"), *visits* (one or more page views with no more than 30 minutes between clicks) and *unique visitors* (which attempt to tally the number of people who visited at least once over some time period, usually monthly).

These metrics are insufficient in several ways. Their development was never driven by theoretical concerns; rather, they were a byproduct of early technical limits on user tracking and the commercial desire to sell advertising. These metrics come with significant measurement error—especially unique visitors, which can overstate the true audience by a factor of four or more (e.g., Graves, Kelly, and Gluck 2010). Worse, optimizing for these metrics rewards bad journalistic behavior. So-called clickbait headlines generate page views based on false pretenses, offering an often-misleading view of the story. From a democratic perspective, traffic metrics such as page views are only very loosely connected to the kinds of engagement that the democratic citizenship ideally requires.

Yet in recent years metrics have improved dramatically. Several of the news organizations that exemplify the ethical metrics approach, such as Medium and Upworthy, have also been at the forefront of creating new measures of online attention. New audience measurement suites, some developed by digital news organizations themselves, offer a level of fidelity that previously required users to install monitoring software on their computer. These measurement packages now give a much more detailed picture of who visits a site, for how long, in what depth, and how consistently. Newsroom subscriptions to services like Chartbeat similarly offer more detailed and accurate data about user behavior, even across sites and devices.

This technology has created a shift in emphasis from pageviews to better, multisignal measures of time on site. As Upworthy argued in support of its "attention minutes" metric, the headlines and stories that generate the most clicks are usually *not* the ones that generate the most total user attention. Optimizing for engaged time rather than clicks leads to a dramatically different set of stories, frames, and headlines. Even sites like BuzzFeed have explicitly repudiated clickbait tactics on the grounds that they are self-defeating. As BuzzFeed editor Ben Smith (2012) declared, "clickbait stopped working around 2009."

Old Values, New Tensions

This new data exposes powerful latent tensions between different elements of journalism values. To begin with, consider something basic: how long should a news article be?

Nothing in the canon of journalistic ethics declares a specific word limit for ethical journalism. For certain types of stories—such as breaking news—short articles have long been expected. Indeed, the inverted-pyramid style that emerged in the late nineteenth century conveys the basics even to those readers who do not finish the full story (Mindich 2000).

Still, professional norms strongly privilege lengthy, comprehensive articles over shorter ones. Lengthy stories allow journalist to detail their sources and evidence, and they give greater opportunity for "both" sides to be heard. Almost without exception, the stories that receive awards and accolades are lengthy, detailed features.

However, audience data challenges the pride of place that longer stories have enjoyed. Few readers who click on an article read the whole thing. Data from Chartbeat shows that a majority of readers do not go past the halfway point in a typical new story, and fewer than 10 percent scroll down to the end (Manjoo 2013).

Online data was cited this spring when on the same day—apparently coincidentally—the AP and Reuters both sent staff memos declaring that normal news stories were not to exceed 500 words (Wemple 2014a, 2014b). The AP's memo used metrics-driven logic to justify the shift on several ethical grounds. It argued that longer stories did not better serve readers: "Our digital customers know readers do not have the attention span for most long stories and are in fact turned off when they are too long" (Wemple 2014a). The AP also invoked an ethic of stewardship, suggesting that longer stories were wasteful: they took up more of reporters' time and consumed more editorial resources, without increasing readership—in fact, length may have *decreased* readership.

Similar dynamics can be seen in fights over so-called "hamster wheel" journalism. Digital news has been accompanied by increased pressures for immediacy, and demands to produce more content more quickly (Usher 2014). In a widely cited essay, Dean Starkman decried what he termed the "hamster wheel":

> The Hamster Wheel isn't speed; it's motion for motion's sake. The Hamster Wheel is volume without thought. It is news panic, a lack of discipline, an inability to say no. It is copy produced to meet arbitrary productivity metrics. (Starkman 2010)

Certainly Starkman and others are right that this behavior challenges journalistic norms. Yet ethically, the lower quality of these individual stories is not the whole picture. Many lines of evidence show that a necessary, though not sufficient, condition for audience growth is a steady stream of constantly updated content. Building traffic to a static website is an impossibility. Indeed, better measurement of traffic helped convince initially reluctant news organizations to publish more short, frequently updated content.

One of the pioneers of this strategy was the *Atlantic* (née the *Atlantic Monthly*), which faced the problem of attracting readers to content that only changed a dozen times a year. The core of the *Atlantic*'s plan was to hire a large stable of already-prominent bloggers, therefore ensuring a constant flow of new content. This shift toward more short, fresh content has even affected sites focused on longer material. In February 2015, Medium announced a series of changes to promote shorter but more frequent pieces of journalism, after initially branding itself as a home for in-depth content (Williams 2015). Even the Center for Public Integrity has recently changed gears, increasing the number of short articles on its website.

Not only do traditional journalism norms favor articles that are long, they strongly favor articles that are *new*. Yet this assumption, too, is potentially problematic from an ethical perspective. Vox founder and editor Ezra Klein argues that a journalist's desire to impress editors and peers often gets in the way of service to the public:

> Your editor will often want something "new." That is to say, they will want something that they, a highly educated hyper-consumer of news products, hasn't seen before. But your readers don't necessarily want the stories your editors haven't read. They want the stories that explain their world to them. Those stories are often absurdly basic, and they might feel like repeats of past stories: What's in this bill? Why do we care about inflation? What does the Fed do? (Klein 2015)

In keeping with this philosophy, Vox cofounder Matthew Yglesias reported on an effort to update and republish 88 older pieces on topics still in the news, reportedly reaching 500,000 readers in the process. Yglesias noted that "What was interesting—though not completely unexpected—was that no one even seemed to notice that we were flooding the site with previously published content. … No one seemed gripped by a sense of déjà vu" (Yglesias 2015).

Individual Journalists

As the examples above suggest, much of the adaption to this new ethical landscape has to occur at the organizational level. Individual journalists on their own cannot buy a Chartbeat subscription or build a multivariate testing framework. Indeed, with some important exceptions, journalism ethics as a field too often focuses on reporters instead of institutions, emphasizing individual virtue over organizational policies and culture.

Yet the metrics revolution certainly alters responsibilities for individual reporters as well. The first obligation is simply to understand their audience.

Most digital journalists can tell you exactly how many page views their past dozen stories received. Descriptive data is a crucial first step, especially when detailed audience data has been unavailable. Yet even detailed description, on its own, is not enough for the balancing tests we ask of journalists. Consequentialist reasoning requires explicit consideration of counterfactuals.

News organizations are increasingly hiring specialized staff to conduct and analyze experiments. While this is important, journalists make implicit counterfactual-based judgments dozens of times in the process of preparing their stories. Judgments about which story frame is most compelling, or which new piece of information is most intriguing, depend on a journalist's ability to anticipate audience reaction.

Understanding the behavior of one's own audience thus cannot be outsourced or delegated, because it is so deeply woven into the task of reporting at every level. Journalists have an ethical responsibility to continually test the reception of their stories.

New tactics may improve journalists' audience predictions. Recent research on forecasting in other contexts suggests that professionals' predictions can improve dramatically with training, practice, and teamwork (Mellers et al. 2014). Forcing reporters to make formal, probabilistic forecasts is indispensible for learning these skills. Journalists might be asked to forecast how many attention minutes one lede receives over another, the difference in traffic between two different headlines, the difference in readership between a longer story and a shortened version. Improved forecasting ability better equips journalists to balance their competing obligations.

Second, journalists need to think more expansively about journalist ethics. Many issues once seen as secondary are actually crucial to journalism's civic goals. In particular, more attention needs to be paid to style and presentation as an *ethical* issue.

Style has not been totally ignored in discussion of journalism ethics. Walter Williams and Walter Lippmann, among other early journalism luminaries, emphasized the importance of clarity in journalism, and the dire consequences of a sensational journalistic style. More recently Kovach and

Rosenstiel (2011) have admonished journalists to "make the significant relevant and compelling," making this the seventh of their ten commandments of ethical journalism.

Beyond the importance of clarity and sobriety in reporting, though, style is often considered ethically irrelevant. User data shows that this is far from the case. A large body of industry data shows that a host of factors—among them headline choice, article framing, lede choice, overall article topic, and personalized content recommendations—strongly influence readership numbers (Hindman 2014). Each of the topics above has strong ethical implications.

For example, consider the task of writing headlines. It is a truism that reporters do not worry about their headlines because they usually do not write them. Ethically, though, there is a strong case to be made that journalists should spend *less* time on reporting and editing, and *more* time writing better headlines. Some will view this suggestion as heretical, an endorsement of style over substance. From a consequentialist standpoint, though, big increases in readership will usually outweigh small losses in precision, nuance, or balance.

Of course, none of this eliminates the need for factual accuracy. Journalists cannot inform the public, after all, if they get the facts wrong. But *being boring can be nearly as damaging to the goal of an informed public as being wrong.*

New Research

This chapter has emphasized that we can't understand the ethical dimensions of journalism without better understanding how news is actually consumed. Yet as the previous section implies, the most crucial issue for new research involves the nature of media preferences—their structure, their distribution among different members of the public, their variation over time, and ultimately their malleability.

As far back as the 1920s, academic research has shown a significant gap between the news preferences of the public and what editors deem to be "important" based on journalistic ethics and norms (Gallup 1928). One of the best recent books on this topic, by Pablo Boczkowski and Eugenia Mitchelstein (2014), found a significant "news gap" between front page stories chosen by editors and lists of "most read" stories. Their work shows that the news gap is large on average—but also that the gap varies substantially across the electoral calendar, and across different news organizations.

As important as such work is, there is much more to be done. As James Webster (2014) outlines in his recent book *The Marketplace of Attention*, preferences for both news and for other types of content remain the subject of substantial uncertainty, despite decades of academic and industry study. Survey research, for decades the primary academic tool for studying preferences, produces massive and systemic measurement error that is correlated with our main variables of interest (Prior 2009). Directly measuring media usage largely solves this problem.

To understand just how basic our (mis)understanding of preferences is, it is helpful to deconstruct news preferences into four different components:

First, across all citizens, some stories are more interesting than others.

Second, some citizens are more interested in news than other citizens, no matter what the story is about.

Third, there is an interaction between the characteristics of a particular news story and the interests of a particular citizen. Some citizens like style and celebrity news; others may prefer sports or politics.

Fourth, some parts of news preferences are essentially random noise that cannot be modeled.

We simply do not know how much of users' news preferences fall into each of these four boxes. The very largest digital news sources have mountains of descriptive data on users' revealed news preferences. Even better, many use extensive A/B testing and randomization to draw inferences about users' preferences. Firms like Google know far more about news preferences than anyone in the academy, and likely more than anyone in the newspaper industry.

A better understanding of users' baseline preferences is a start. Even more important, though, is the extent to which news outlets can *influence* these preferences. For the first time news organizations have the ability to systematically examine the features that build audience, with data so fine grained that researchers can separate the decision to *start* reading from the decision to *stop*. Journalists can see precisely which features make readers click, which paragraphs lose audience, and exactly when users click away from a news video.

Understanding the character of news preferences is the key to several ethical questions. One concerns the extent to which news consumption is a zero sum game. The ethics of kitten slideshows and celebrity quizzes hinge on whether this content is consumed *in addition to* hard news, or *instead of* hard news. Do those who come for lighter fare stay for coverage of the Middle East and the mayor's race? Or will more reporting on civic affairs topics simply push readers away?

Understanding preferences is also crucial to understanding whether a taste for hard news can be cultivated. It has long been observed that the media preferences of the public are a powerful constraint on democratic politics (e.g., Entman 1989). Yet greater familiarity with a particular story may, in some circumstances, increase the appetite for follow up coverage (Neuman 1990). Journalists would have a responsibility to adopt practices that enlarge the hard news audience and improve the store of civic knowledge—if only we could figure out what those practices might be. As yet, we know too little about preference cultivation to say how it might be accomplished. This question demands field experimental data over a time horizon of weeks or months.

Another urgent topic for future work is journalism's ability to reach those with lower incomes, less educational attainment, or simply less interest in the news. Print news readership has long been disproportionately a habit of the educated. Broadcast television news was especially important because it increased the political knowledge and participation of less educated citizens (Prior 2007). But if journalism is to be judged by its service to the public, that obligation must apply equitably, even to those with lower levels of formal education or political interest (Elliot and Ozar 2009).

Conclusion

Journalists have long declared that their paramount goal is informing the public. Actually measuring the public's attention, though, has been viewed as a vice, a capitulation to commercial pressure or even an implicit endorsement of tabloid journalism. The crudeness of audience measurement in print and broadcasting helped preserve this latent tension. Even with the shift to digital news, metrics such as page views and unique visitors remained highly inexact measures of citizen engagement.

Yet times have changed. In the past few years audience measurement has become far more detailed and accurate. There has been much scholarship and industry discussion about how better tracking of attention is altering newswork and newsroom economics. The ethical implications of better audience data, though, have been underappreciated.

This chapter argues that journalists *now have a positive obligation to use these new audience measurement tools.* If a doctor refuses to use a new and more accurate diagnostic test, or if a lawyer fails to check relevant case law on LEXIS-NEXIS, we correctly fault them for ethical lapses. Journalists now have similar obligations to use new measurement technologies. It is not enough to diligently follow the traditional reporting process, or to rely on (often false) folk wisdom about audience behavior. Journalists need to test whether their own headlines are gathering and keeping an audience, and to see which parts of their stories lose readership. They must measure for themselves how different reporting and stylistic choices affect their ability to inform the public.

This revolution in measurement opens up important new research questions for journalism scholars. Many once-mysterious aspects of audience behavior can now be directly observed. While normatively focused empirical work has begun to explore this new terrain, much remains to be done. If journalists are to weigh competing considerations against each other, they need help in understanding what, precisely, the trade-offs of their choices are.

This new knowledge and these new tools also need to be incorporated into the classroom. Formal education in journalisms ethics has been overwhelmingly qualitative. This approach needs to be broadened, and students taught how integrate audience analytics into principled decision-making. Basic statistical numeracy is now a key prerequisite of ethical journalistic practice.

Given the tectonic shifts shaking the profession of journalism, it might be reassuring to think that the bedrock principles of journalistic ethics have remained unchanged. Paradoxically, though, it is steadfastness to the public service ideal that demands that journalists alter their behavior. Old values plus new metrics equals new responsibilities for journalists. Ethical journalism is now more complicated. At the same time, journalists can know more than ever before about whether they are meeting their civic goals.

10 Reinventing Journalism as an Entrepreneurial Enterprise

Jane B. Singer

As digital forms of journalism have inexorably grown to prominence over the past two decades, journalism studies scholars have deployed many innovative strategies in seeking to understand the dramatic transformations within legacy newsrooms.

But within these legacy newsrooms is no longer the only place we need to look.

The term "entrepreneurial journalism" (Briggs 2012) has been stretched to encompass almost any innovation involving information delivery. However, in contrast to "intrapreneurialism," the creative initiatives that push existing organizations in new directions (Baruah and Ward 2015; Boyles 2016), it pertains particularly to stand-alone enterprises dissociated from legacy media yet with some sort of journalistic mission. Journalism produced outside the traditional newsroom is growing rapidly in quantity and quality, from investigative journalism consortia to hyperlocal websites to niche offerings of all sorts.

For news organizations and their employees, the launch of new journalistic enterprises extends challenges over jurisdictional turf that became evident more than a decade ago with the rise of independent bloggers (Lowrey 2006)—prototypical information entrepreneurs. This chapter considers three overlapping issues that are especially disruptive—to journalists, news organizations, or both—and ripe for scholarly investigation in a context that explicitly engages entrepreneurial enterprises rather than the legacy ones that dominate digital-age scholarship to date. The issues involve normative boundaries, economic imperatives, and journalistic roles.

Normative Boundaries

In some ways, entrepreneurial journalism can be seen as among the latest in a series of disruptions to twentieth-century practices, from the emergence of the Web in the mid-1990s through the rise of multimedia, blogs, "user-generated content," social media, and more. Journalists' first response to each of these digital-era changes has been characterized largely by resistance on normative grounds. Repeatedly over the past two decades, journalists have evoked professional ethics as identity markers ("Who I am") and boundary markers ("Who I am not") in an effort to assert their own continuing centrality in a rapidly changing information universe. Eventually, that contestation has given way to accommodation, as journalists folded the once-new thing into their own work (Singer 2015).

But while the normative challenges raised by these earlier changes have largely been ones of degree—the extent to which creation of multimedia content takes time away from a reporter's pursuit of the full story (Singer 2006a), for instance, or to which social media affect verification processes (Hermida 2012)—entrepreneurial journalism raises issues that go to the core of journalists' self-perception. These issues relate particularly to autonomy or professional independence, encompassing such interrelated matters as objectivity, conflicts of interest, and interactions with audiences and advertisers.

Western journalists, particularly in the United States, have fiercely articulated a need for near-absolute autonomy in their pursuit of information they deem to be in the public interest. Over a period stretching back 100 years and more, they have positioned independence as a requisite condition to their ability to carry out that job in the face of the multilevel influences under which they work (Shoemaker and Reese 2014). Although historically rooted in the First Amendment safeguard against government infringement on press freedom, the norm has been widely invoked by US journalists defending the need for autonomy from two other influences more central to a consideration of entrepreneurial journalism: commercial entities and audiences.

The former have drawn more attention. Advertisers' real or potential influence on newsroom output and the decisions made in creating it, along with the fear that audiences may not distinguish between commercial and editorial content, has been of significant concern for decades (Eckman and Lindlof 2003; Soley and Craig 1992; Upshaw, Chernov, and Koranda 2007) and remains salient today, reflected in controversy over such contemporary permutations as native advertising (Carlson 2015; Coddington 2015; Howe and Teufel 2014). Relationships with other revenue sources also have drawn critical attention, from the intricacies of sponsorship arrangements (Foreman 2015) to the interconnections between philanthropic organizations and the nonprofit journalism they back (Ward 2014).

Historically, the interaction between journalists and news audiences has drawn less attention, perhaps because there was relatively little of it. Virtually all a journalist's relationships were with sources and colleagues; the newsroom walls and switchboard (not to mention the security guard in the lobby) created nearly impenetrable boundaries around the workspace. As the effective squashing of fledgling "civic journalism" initiatives in the 1990s suggested (McDevitt 2003), journalists rather liked maintaining a safe distance from any actual reader or viewer. Many were discomfited a decade later when the rise of user comments, followed by the explosion in social media, eradicated such distance (Hermida and Thurman 2008; Lewis 2012; Gulyas 2013). Increasing reliance on web analytics—instantly available, excruciatingly detailed usage data—also has put audience reading patterns front and center in newsroom decisions about what deserves good play or even coverage at all (Anderson 2011; Tandoc and Thomas 2015).

Clearly, journalistic autonomy from the influence of advertisers and audiences has served an important purpose in providing the freedom to report and write "without fear or favor," in the famous 1896 phrasing *New York Times* publisher Adolph Ochs. But the nature of boundaries is to fence certain people (journalists, say) in, as well as to keep others out. As discussed further below, maintaining autonomy from audiences has meant journalists know little about them—their interests, their media habits, what angers them, what they appreciate. For a journalist turned entrepreneur, however, such ignorance is untenable. No business can survive long without in-depth and nuanced understanding of the people it hopes to serve.

Advertisers or other financial sources also are key constituents. Open and frequent communication with them is vital for content creation, marketing, and revenue generation—three interconnected areas integral to any start-up but separately ring-fenced by traditional practitioner ethics. In a traditional media world, maintaining autonomy from commercial entities became a rationale for journalists to remain disconnected from, even ignorant about, the economics of the business that employed them, a stance not uncommonly intertwined with general concerns about the negative impact of a profit motive on the quality of journalism (Beam 2006).

But for an entrepreneurial journalist whose enterprise relies in whole or in part on an ad-based revenue model, as most do (Sirkkunen and Cook 2012), pursuing and securing advertisers is at least as important as pursuing and securing audiences. Some journalistic start-ups are large enough for that job to be delegated to someone not directly involved in content production, but many are not, at least in their crucial early stages. Many also rely on crowdfunding, which seeks money from audiences for often quite specific purposes (Hunter 2015).

Moreover, regardless of the organizational structure or funding model, audiences and revenue sources typically are interchangeable in a start-up environment, particularly for niche sites covering a narrow geographical or topical area. Relationships can become very tangled very fast; conflicts of interest for journalists who double as fundraisers have been labeled "the looming ethical problems of an entrepreneurial age" (Ward 2009, para. 19).

Key Issues for Scholars: Normative Boundaries

The transformation of journalism in the digital era has opened up many opportunities to examine, critique, and apply existing normative principles and theories (McBride and Rosenstiel 2014; Zion and Craig 2015). But most such explorations continue to posit the newsroom as the primary base from which journalism operates: The emphasis has been on how changes associated with the transition to a digital environment have affected the people performing traditional journalistic tasks within traditional news organizations.

How entrepreneurial journalists are navigating the dramatically different waters described above, particularly in light of their historically passionate demand for radical autonomy, begs for exploration. Involvement in a start-up almost invariably means a shrinking of distance between the journalist and a range of actors who once were the concerns of someone else in the news organization, with a concomitant challenge to the normatively grounded professional boundaries that journalists have drawn around themselves.

Of particular interest are intertwined issues related to separation of "church and state," or commercial and editorial activities; distance from audiences, which is being compressed across the journalistic spectrum but perhaps especially in small start-up environments, as discussed further below; and objectivity, particularly the ability to fairly cover people and issues despite a vested interest in them. Do the principles that have undergirded journalism within the newsroom for a century and more hold when journalism is practiced in non-newsroom locations? Do entrepreneurial journalists reify existing norms, modify them, or adopt new ones? To take just one example: If, as some believe, transparency is replacing (or should replace) objectivity as a core journalistic norm (Hellmueller, Vos, and Poepsel 2013; Karlsson 2011), what can enterprises that rely on crowdfunding for their revenue and thus are explicitly answerable to their backers tell us about how transparency might be operationalized?

Options for the development and application of expanded theoretical frameworks also merit consideration. An obvious one is a normative approach that builds on our extensive understanding of both idealized journalistic behavior and real-world influences on actual behavior (Shoemaker and Reese 2014). Recent advances in our exploration of journalistic boundary work (Carlson and Lewis 2015; Lewis 2012) are likely to be invaluable in these considerations. Useful as well would be deeper interrogation of the intended purposes behind normative prescriptions and proscriptions. To what extent do they serve the professed goal of safeguarding independent investigation of important issues in the public interest, in contrast to serving the rather less noble goal of enabling journalists to ward off uncomfortable questions about their activities and even their claim to a special place in society?

Economic Imperatives

Entrepreneurialism gave birth to American journalism, from the colonial printers to the nineteenth-century publishers who turned fledgling news initiatives into empires. But twentieth-century expansion followed a different story line. Socially, the paradigmatic "power of the press" has been a collective force, wielded by institutions capable of conveying information to large numbers of people and therefore commanding the attention of those whose attention mattered. Gaining and retaining that power required securing reliable and steadily rising profits. Indeed, profit levels became truly remarkable, particularly in the latter part of the century as dozens of news-producing enterprises consolidated under corporate ownership—mostly in publicly held companies with boards answerable to their stockholders (Bagdikian 2000).

But such organizational structures were more likely to encourage conservative practices than risk taking. When the Internet emerged as a force to be reckoned with, many of the executives who had led the media for decades were unable to respond quickly or creatively (Anderson 2013). Few had an entrepreneurial mindset—build fast, fail, learn, repeat until you get it right (Babineaux and Krumboltz 2013; Polgreen 2014) – and most additionally were bound by corporate mandates against knowingly risking failure. In the recent past, then, "entrepreneurship of any sort is not a concept that has been closely identified with the media industry" (Compaine and Hoag 2012, 30). Yet the ongoing viability of the news media in a digital age arguably rests on multifaceted innovation (Pavlik 2013).

Many factors made the US media so extraordinarily prominent and profitable, including demographic trends, favorable regulatory structures, and technological innovation in content production and distribution. But what stands out in retrospect is the near-monopoly that media institutions collectively held on the supply of something for which there was a high demand: information. That monopoly obviously no longer exists in an Internet age (Chyi 2009), and the resulting challenges to the media have been described as systemic, structural, and likely irreversible (Communications Management 2011). Indeed, their business models rest on "technical, economic, political and information environments that no longer exist" (Picard 2011, 8). The exponential increase in the number of entities providing something identifiable as "news"—from individual observers with a video-streaming app on their phones, to substantive businesses with hundreds of employees—has eradicated that highly profitable lock on the information market.

Legacy media outlets whose power rests on the attention of large numbers of readers, viewers, and listeners are losing them in a digital environment that invites entry and facilitates experimentation (Briggs 2012). Larger newcomers are challenging legacy outlets for the mass, general-interest audience at home and abroad, through rapid expansion into underserved markets such as India, Nigeria, and the Arab world. A few large start-ups and countless smaller ones also are luring niche audiences, from hyperlocal enterprises seeking to fill the coverage gap left by a steep decline in staffing and thus reporting at local legacy outlets (Anderson 2013; Enda, Matsa, and Boyles 2014) to topically focused sites covering every special interest imaginable. Such entrepreneurial endeavors can leave legacy outlets bleeding from a thousand pinpricks, each a tiny hole left by a user who is spending time elsewhere in the online universe.

Nor do the audiences of tomorrow offer much hope of regeneration. Millennial consumption of legacy products is minimal and likely to decline further as mobile devices continue their predicted march toward news dominance (Fidler 2013; Newman 2015; Westlund 2013). Legacy outlets risk falling behind nimble, digital-only competitors—with their significantly smaller investment in infrastructure, significantly less embedded production routines, and significantly more nuanced insights about their audiences—in developing an appealing mobile product. The parallel trend toward personalization of online information (Newman and Levy 2014), which enables individuals to modify the content, delivery, and arrangement of messages to their own "explicitly registered and/or implicitly determined preferences" (Thurman and Schifferes 2012, 776), adds to the challenge for traditional outlets.

Of course, start-ups face significant challenges of their own, not least their need to compete with entrenched, widely recognized, and relatively well-resourced companies. Start-ups inherently operate in highly uncertain circumstances, dependent on fickle users and investor whims; indeed, the elusiveness of sustainability for new media enterprises suggests that "survival in itself must be recognised as a form of success" (Bruno and Nielsen 2012, 102). But two related points stand out. One is that those that do succeed, even temporarily, are inherently disruptive to their legacy brethren. They are disruptive in terms of the information needs they fill, the audience attention they attract, and the money that flows to them one way or another. They are disruptive, as well, in the human resources they command: the reporters, editors, designers, and other journalists who provide them with creative energy and talent, plus the programmers, marketing experts, and support staff who fill additional roles vital to any media enterprise. The people who might be helping move traditional media in new directions are helping someone else instead.

The second overall point is that amid the challenges, entrepreneurial enterprises offer a valuable lesson for legacy media: Diversification of revenue models is all but vital for sustainability (Kurpius, Metzgar, and Rowley 2010; Sirkkunen and Cook 2012). In a twentieth-century world, media organizations thrived by maximizing profits from two key sources, in varying combinations: audiences and advertisers. Most carried a version of that model into the digital era. Indeed, the all-but-universal early decision to give away content for free rested on the notion that the value of news audiences is primarily their ability to attract advertisers; hence the bigger the audience, the better.

In hindsight, charging nothing for something that is very expensive to produce in order to artificially inflate the audience may seem foolish. But that is only because in hindsight, we know more about what works online, as well as what does not work and what works differently. Legacy outlets that now are erecting "paywalls"—charging users a fee to read any, a few, or a lot of their stories (Pickard and Williams 2014)—may in fact have learned precisely the wrong lesson; essentially, paywalls are the old model in new clothes that are especially ill-fitting in a digital environment where so much is free. Moreover, a paywall model does not represent a diversification of revenue sources beyond audiences and advertisers. News start-ups offer insights into more creative solutions, from event hosting to crowdfunding to consultant services.

Key Issues for Scholars: Economic Imperatives

With notable exceptions, few journalism studies scholars are well grounded in either economic theory or in the narrower field of management studies that forms the context for much of the entrepreneurial scholarship to date. Historically, literature about economically driven impact on the news industry has been either descriptive (Bagdikian 2000; Compaine and Gomery 2000) or critical (McChesney 2015), and has focused on legacy outlets; little attention has been paid to the economics of media start-ups. That said, Robert Picard (2002; 2005; 2012; Naldi and Picard 2012) has long been a leader in applying the tenets of capitalism across the news industry, which has not generally fared well in such an analysis, and recently, a small but growing number of scholars have extended his work within the digital realm. Several key studies have been published under the auspices of research institutes housed at leading universities such as Oxford University and Columbia University, respectively including Levy and Nielsen's (2010) Reuters Institute volume analyzing the changing business of journalism and the implications for democracy, and work for the Tow Center on the business of digital journalism (Grueskin, Seave, and Graves 2011).

Interest also is growing in exploring alternative revenue models for media enterprises; some of it is focused on entrepreneurial initiatives (Sirkkunen and Cook 2012), while other work has encompassed innovation within existing organizations (Boyles 2016; Kaye and Quinn 2010; Nel 2010). More theoretically grounded considerations of new revenue models such as crowdfunding (Carvajal, García-Avilés and González 2012), native advertising (Carlson 2015), and nonprofit journalism (Nee 2013) also are emerging. More broadly, there have been calls for a change in emphasis from a production to a consumption economy, one that acknowledges ongoing shifts in the nature of information supply and demand (Colapinto and Porlezza 2012).

Also relevant is the rich body of work from the field of management studies, where scholars have been studying entrepreneurialism at least since Peter Drucker published *Innovation and Entrepreneurship* in 1985. Innumerable books, journals, academic conferences, and business school programs are now devoted to the topic (Kuratko 2005), creating a sizable body of knowledge. But within journalism studies, although some work is beginning to incorporate concepts related to fostering innovation and entrepreneurship (Hollifield et al. 2015), the focus has again been on legacy media and on the management of changing but still institutionally housed newsrooms (for examples, see Ekdale et al. 2015; Gade 2004; Sylvie and Gade 2009). For the most part, application of management principles to entrepreneurial journalism continues to rest on anecdotal evidence (Briggs 2012) and trade press coverage, much of it vaguely positive (Vos and Singer 2016) but virtually none of it theoretically conceptualized or empirically scrutinized.

Economic theory—from basic supply and demand concerns (Naldi and Picard 2012) to innovative applications of rational choice theory (Fengler and Russ-Mohl 2008) or of concepts related to organizational ecology (Lowrey 2012)—and management theory, including concepts tailored to managing uncertainty, change, and creativity (Küng 2011; Lowrey and Gade 2011), would likely be fruitful in understanding entrepreneurial journalism initiatives. Additional insights may come through drawing on Bourdieu's (2005) connection of journalism with notions about economic and cultural capital (Vos and Singer 2016).

In general, the journalistic enterprise has been described as "a contested practice embedded in larger political, economic, and cultural struggles" (Carlson 2009, 273), and perhaps nowhere is that struggle clearer than in entrepreneurial initiatives. Indeed, entrepreneurial journalism is an exemplar of the interplay between economic resources and other social entities and structures.

Journalistic Roles

Countless academic studies and trade press reports have documented the impact of change on news practitioners who have had to modify their storytelling tools, skill sets, work practices, and relationships with audiences in a digital age. Here, too, virtually all the work has focused on journalists employed in traditional newsrooms. Tens of thousands no longer are or indeed never were, as journalistic work has become precarious and piecemeal. Many frankly find it less than enjoyable (Willnat and Weaver 2014), and hitching their wagon to a journalism start-up—or launching their own media enterprise—has become an increasingly viable and appealing alternative, particularly for journalists motivated by a desire for independence or self-realization (Block and Landgraf 2016).

For legacy journalists who leave a traditional newsroom to join or create their own news enterprise, perhaps the most jarring change they encounter is the dramatic narrowing of the distance between themselves and two key (and overlapping) constituencies: those who consume their product and those who finance it. That change was suggested above and is explored in more detail here because it is central to the shifting roles evidenced in entrepreneurial journalism.

As indicated in the discussion of normative boundaries, the traditional relationship between journalists and their readers, viewers, or listeners is a one-directional one: It consists primarily of journalists producing and disseminating information in the public interest, a term defined broadly, often vaguely, and almost entirely by journalists themselves. Despite the overarching goal of service to society, journalists only rarely have seen the need to engage directly with audiences to fulfill that role or even to identify what it should entail. The result has been a growing gap between audience interests and legacy media content (Boczkowski and Mitchelstein 2013)—a gap that seems unlikely to close as long as news consumers are seen as a "mass" audience, a faceless aggregation presumed to share a concern with matters judged (again, mostly by the journalist) to be of civic importance or general interest. To journalists working at news outlets larger than the tiniest community-based ones, readers, viewers, or listeners generally remain an undifferentiated and amorphous blob.

Entrepreneurs, in contrast, need a great deal of concrete and detailed knowledge about their audience. A vague conception of undifferentiated people in need of something broadly defined as in their own interest is not nearly good enough. No new business gets off the ground without extensive research that yields a precise understanding of audience desires, interests, habits, and willingness to pay, not least because financial backers—from investors to advertisers to donors, and many start-ups rely on all three (Sirkkunen and Cook 2012)—will inevitably demand such information. And the best way to obtain it is through personal communication. The entrepreneurial journalist has no newsroom walls, no marketing or circulation departments, to act as an audience interface—or buffer.

Because the people interested in a topic are apt to already know a bit about it, start-ups that rarely are flush with the resources needed to create and sustain a steady supply of content often solicit contributions as well as suggestions from readers. Journalists who become entrepreneurs unavoidably need to find, engage with, and nurture "outsiders" who can contribute reliably, cogently, credibly, and regularly. Meeting these needs demands relationship skills well beyond what journalists may have developed with traditional sources, whose only role was to provide information that was then routed through, and vetted by, the newsroom. Legacy journalists have few partners in their core content production tasks and role.

Entrepreneurial journalists must work collaboratively to survive. The few that succeed do so in no small part through personal connections, interactions, and information sharing, all of which remain difficult for larger outlets that historically have distanced themselves from their audience (Rosen 2014).

Even more challenging, as indicated in the earlier discussion of economic imperatives, that collaboration must also involve funding sources. Traditional journalists certainly have an investment in their careers, as well as a general interest in seeing an employer prosper, but those are relatively attenuated concerns. And while legacy journalists are well aware that the content they produce is for sale, their involvement in the transaction is quite deliberately nil. Figuring out where the money to support the journalism comes from and how to get more of it are roles filled by the publisher or owner or board of directors, not the journalist.

The learning curve for journalists starting their own business is therefore steep and sharp. An entrepreneur needs to think hard about every cost: how big it is, what value it adds, how essential it is to success—and then whether to support it and how. The scant research to date on entrepreneurial journalism suggests that a common mistake made by founders of news start-ups has been to put too much money into what they know best and love most: the journalism. Many have hired sizable news staffs before securing their financial underpinnings or attending to other aspects of a successful enterprise, such as creating effective marketing channels and forging key partnerships (Osterwalder and Pigneur 2010). Naldi and Picard (2012) labeled such misplaced priorities "formational myopia," or unrealistic expectations about the demand for, and the economic value of, journalistic work. Journalists without practical business skills or experience not infrequently make dubious fiscal and management decisions in negotiating the transition from employee to owner of a news operation (Bruno and Nielsen 2013).

These new sorts of relationships also raise broader questions about the extent to which entrepreneurial journalism veers away from the conception of journalism as a public service that enables an informed electorate to make sound choices and toward something more explicitly rooted in marketing concerns. Hanitzsch (2007), for example, distinguished the conceptualization of journalism as independent watchdog from a market orientation, associated with giving audiences what they want to know "at the expense of what they should know" (375). Market-driven journalism (McManus 1997) addresses audiences as clients and consumers rather than citizens, and the digital environment that hosts most entrepreneurial efforts has long been portrayed as an arena where journalistic practices are notably vulnerable to market influence (Cohen 2002).

Yet entrepreneurial journalists, despite the enormous potential for role conflict, still have to maintain what in business school jargon is termed their "value proposition" (Osterwalder and Pigneur 2010, 22). Here, too, a start-up environment changes the game. The value of individual journalists within a news organization rests on their professional expertise, defined through contributions to that overall mission of informing the public. Again, of crucial importance (notorious newsroom scandals notwithstanding) are the normative principles exercised in maintaining credibility and in fostering trust that, in theory, attracts and retains audiences. Notably, that individual value is connected to an institutionally held value that incorporates brand reputation, reach, and long-term patterns of stakeholder interactions dating back decades if not centuries. At both the individual and institutional levels, then, the value proposition for legacy journalism rests on delivery, over time, of an appealing information product to audiences, in turn creating an appealing proposition to advertisers seeking to reach them—a value deeply challenged in a networked digital environment.

Many of the information start-ups that have enjoyed big success thus far are less about serious reporting and fulfillment of civic roles than they are about edginess and trend-riding and visuals and speed (and, not atypically, kittens). Creation and delivery of credible information is surely valuable to democracy. But democracy is a big concept, while audiences consist of individuals, seeking individual gratification—and, thanks to the accessibility, openness, and unrivalled diversity of the digital world, finding it far more easily and cheaply than at any time in history. Personalization, and personal connections, can seem more gratifying than the "spinach" of news that's good for society writ large. Journalists both individually and institutionally have little experience in—and often little appetite for—relating to audiences at that level.

As journalists have sought to turn what they know and love into a going business enterprise, then, many have applied traditional views of what news is or should be, often along with misperceptions of the market for it. Most really do believe in the power of a free press and in the value of a well-informed public to civic and community life. And they believe that they know—know best—how to provide that value. Yet life as an entrepreneur raises many intertwined questions familiar from other fundamental shocks to journalism as those over age 20 knew it. What is my role as a journalist? What value do I offer—to whom, how, and how much?

What is my place in society, and how can I secure it? What hats can I not live without? What new ones do I need, and how do I get them to fit without chafing? Which relationships are the ones that matter? How can I nurture them and safeguard them from corruption in various guises? If success isn't leading the newscast, and maybe not even serving that nebulous thing called democratic society ... then what exactly is it, and how do I attain it?

Key Issues for Scholars: Journalistic Roles

Clearly, entrepreneurial journalists do more, and different, things than a traditional journalist does. Some of the resulting pressures, including normative ones, are comparable but need to be considered in a dramatically changed context. Others are new, requiring the journalist to adopt uncomfortable and ill-fitting roles. How are we to understand these things in relation to the place of the practitioner and the practice in democratic society?

Scholars have recognized that the occupational field is "losing its traditional bearings and casting its practitioners in a new entrepreneurial ideal of being free agents" (Deuze and Marjoribanks 2009, 558). But although good work is under way, the publication of theory-driven empirical work to explore the meanings of these and associated changes has been scanty. In addition to the economics work of Picard and his colleagues already highlighted, probably the most extensive look at entrepreneurial journalists published so far has been a descriptive study entailing exploratory interviews with 30 US media entrepreneurs—defined as founders of an independent content business with a clear revenue model—by Compaine and Hoag (2012). Their assessment was generally positive; they noted an environment hospitable to media start-ups, as "technology and economics have conspired to undercut many of the barriers that had existed to would-be media entrepreneurs" (43).

Perhaps the best approach is to continue, and to expand beyond the traditional newsroom, the ongoing scholarly exploration of cultural changes in the practice of journalism (Deuze 2005; Lewis 2012; Witschge 2013). Those relate to the norms and the economic pressures already described but go deeper to consider fundamental issues at the core of questions about who a journalist is and what such a person does. They relate as well to examinations of journalists' "rediscovery" of the audience (Loosen and Schmidt 2012) and of new forms of interaction with people as participants in the cocreation of information (Heinrich 2013; Robinson 2011; Singer et al. 2011).

How do these new or renewed or reconfigured relationships affect journalists' view of their own role in the information environment? How do they affect the process not only of making news but also of thinking about what news is? Is the entrepreneurial journalist still a gatekeeper? An agenda setter? Where does agency lie, and how is it shared around the relevant information network? For example, a number of journalism scholars (Couldry 2008; Domingo, Masip, and Meijer 2015; Primo and Zago 2015) have highlighted the utility of actor network theory in understanding twenty-first-century journalism. The theory treats objects, including technological ones, as actors within social networks. Given the centrality of digital platforms to most entrepreneurial journalism enterprises, as well as the interrelationships among many kinds of actors, the theory could be productively applied to our understanding of start-up culture.

One additional place outside the traditional newsroom to look for exploration of cultural change stemming from the move toward entrepreneurialism deserves mention: the classroom, where crucial early socialization of future journalists occurs (Mensing 2010). Scholars already have begun to examine the disconnect between industry disruption and journalism education, which does not typically offer the skills, knowledge, and qualities needed to take advantage of entrepreneurial options (Baines and Kennedy 2010). Several studies have documented efforts to bring entrepreneurial concepts into the curriculum. A small-scale study by Ferrier (2013), for example, found that instructors cited changes in the nature of media industry work—short-term contracts, self-employment, temporary group work on specific projects—among their prime motivations for creating classes that introduced journalism students to business concepts and market research; they sought to "empower students with the knowledge and skill sets to create their own jobs" (229). In the UK, university workshops have encouraged students to recognize synergies between the seemingly disparate fields of business and creative industries, and to "consider themselves not only within a framework of business and entrepreneurship but also as creative, imaginative individuals with a unique contribution to make to a sector that is in need of rejuvenation" (Hunter and Nel 2011, 15).

Yet the sledding is tough. Surveys in the United States (Blom and Davenport 2012) and Europe (Drok 2013) found that at least as of a few years ago, journalism educators and administrators placed entrepreneurial journalism classes relatively low on their list of curricular priorities; respondents in the European study said they preferred to emphasize doing journalistic work "without interference from market forces or the public" (Drok 2013, 156).

More work is needed not only to document what exists but also to bring concepts into the classroom, methodically measure and assess their utility and impact, and share the lessons learned.

In considering where scholars might devote their energies in seeking to understand entrepreneurial journalism, this chapter has focused mostly on concepts and theories with little attention paid to methodology. The reason is that existing methods widely used by journalism studies scholars should serve us well in these explorations. Qualitative methods are likely to be especially useful. Although more quantitative approaches present challenges related to definitional and sampling issues, qualitative ones are ideally suited to the diversity of contemporary journalism and can facilitate a focus on crucial questions about how and why journalists are reconstructing themselves as entrepreneurs.

As suggested already, there is a significant need and opportunity for more research within entrepreneurial journalism settings. Ethnographic work is ideally suited to obtaining a rich understanding of how, exactly, journalists are dealing with the issues outlined above. Other methodological options likely to prove fruitful include discourse analysis—how do journalists talk about the many issues surrounding entrepreneurialism?—and the more quantifiably oriented approach of network analysis, which can yield insights into the interactions of entrepreneurs with their multiple stakeholders, as well as the support structures they form when they no longer have the bonds provided by newsroom work. In-depth interviews also would be invaluable in probing topics that may be sensitive. For instance, we know little about what lessons entrepreneurial journalists might be taking away from failure. What do they believe went wrong, why, and what was learned from the experience? What comes next for journalists whose attempts at entrepreneurialism have proved unsustainable?

In summary, this chapter has highlighted three areas of particular interest in the exploration of a growing industry phenomenon, entrepreneurial journalism. In all three of these areas—the norms, financial underpinnings, and societal role of journalism and journalists in a digital age—work to date has been overwhelmingly about journalism in a place recognizable as a newsroom, housed within a news organization whose shape and structure have been familiar for a century and more. But unlike the previous transformations of the digital era, which can be seen as changes contained and accommodated more or less adroitly within a traditional work environment, entrepreneurial journalism invites us to reconceptualize how and where news work is done and news is made. To date, we understand very little indeed about this subject. This chapter has suggested places where we might begin.

However we go about it, the need to more fully understand entrepreneurial journalism as an inherently disruptive response to industry disruption is only going to grow. As mobile and wearable technologies, along with drones and virtual reality and a host of affordances we cannot yet even envision, open up fresh new opportunities for entrepreneurial initiatives, more journalists will seize the chance to prosper outside the newsroom. They will be shaping the ways our society understands itself, and we must do our best to understand them in turn.

Commentary: Blurring Boundaries

W. Russell Neuman

The three chapters in this section have a common theme—the blurring of traditional boundaries. One central theme of Stroud's chapter is the divide between the culture of the academy and the culture of professional journalism. Hindman addresses the way in which the new media environment breaks down the traditional boundary between the journalists and their audiences. Singer steps back to review the fundamental economics of the journalistic enterprise and questions the celebrated divide between advertiser and journalist, sometimes dubbed the appropriately separate domains of church and state. I'll review each of these themes and explore how each may as well be contributing to further weakening of conceptual and structural boundaries in the public sphere.

Let's start with Stroud's provocative challenge. If you think about it, the role of the journalist and the scholar is structurally quite similar. The mission of each is to seek out new knowledge using the traditional tools of their respective trades and to broadly promulgate this newly discovered information to the benefit of the broader community. Academics who stray from their traditional audience of fellow academics and students and write for a broader public, however, are infrequently honored for the efforts. The tradition of public intellectuals is weak, especially in the United States. Academics tend to view journalists as inveterate simplifiers and popularizers. Journalists are prone to view academics as jargon-bound obscurantists. Can these two professional cultures be bridged? Should they be?

Stroud's argument is that they can and they should. I find her case compelling. To examine this argument, let's go back to the original comparison of the role of the professor and of the journalist. The professor has an audience of students, the journalist an audience of news consumers. Each professional feels an intriguing tension between what their audience impulsively wants to know and what in their professional judgment their audience should know. The students are a bit more of a captive audience so the professor traditionally feels free to dictate the syllabus while the journalist, in a more open environment of public attention, needs to add a spoonful of sugar to the complexities of the policy debate to which the public really ought to attend.

In the final analysis the challenge in each is to make the complexities of the real world accessible to an audience distracted by day-to-day personal concerns. Drawing on Dewey, Stroud calls this "artful journalism." Social science, she insists, can be employed to test and identify news presentation strategies that lead to engagement. Traditionally journalists have been resistant to the idea of submitting their journalistic judgment to systematic empirical assessment, but given the economic pressures on journalism in the digital era, there is the potential for a new openness to such a joint enterprise between researchers and journalists, a boundary well worth bridging. The key to collaborative success will be moving beyond simply assessing what news consumers want to influencing those very interests and concerns.

Some years ago, with my colleagues Marion Just and Ann Crigler, we undertook a study of the journalistic process that was published as *Common Knowledge* in 1992. A key element of our research design was to interview citizens and probe their thinking and recall of a specific set of magazines, television, and newspaper news stories. Further, we were largely successful in interviewing the journalists who created these stories. Given this design, we were drawn to ask the journalists what feedback on the stories they had received from viewers and readers, and, accordingly, how their audience had reacted to the piece. The overwhelming response from our journalists was that they sought feedback from colleagues and editors but none from readers other than the comment of an occasional neighbor or airplane seatmate. In fact they seemed to think our question a bit strange.

Hindman points out that the culture as well as the technology of journalism is undergoing dramatic change. Journalists routinely include their email at the end of an article. The most read and most commented upon articles of the day are prominently listed for journalists and news consumers alike to ponder. Citizen blogging and journalistic blogging intermingle online. It is the nature of the two-way character of online interaction in contrast to the inherently one-way publishing and broadcasting of the industrial era. But the prospect of two-way interactivity requires more than technical facilitation, it requires a change in cultural expectations. I suspect present-day journalists would not find our query about feedback to be strange at all.

Finally, Singer raises an even more basic question about institutional boundaries. Advertisers make independent professional journalism possible. The culture of journalism makes it clear that while the financial largess is appreciated, the interests of the advertisers must be carefully and energetically insulated from the journalistic process itself. Will the technical and attendant economic restructuring of journalism challenge this traditional independence?

Singer's emphasis is clearly expressed in her choice of a central organizing concept—entrepreneurial journalism. She sees it as a matter of existential necessity and opportunity. While some observers seek out new income streams for legacy news organizations and ways to save traditional journalism, Singer wants to find new and multiple definitions of what news is. More than the boundary between advertiser and journalist, she wants to explore the boundaries of news and non-news. In fact she frequently simply refers to information and information rather than news. Further, she explores the idea of crowdsourcing as an economic engine to support the enterprise. In such a case, the consumer becomes, in effect, the direct financial sponsor of the news process rather than indirectly via advertisers, further blurring what was traditionally an easily identified church and state boundary. She points out that the invention of what are now taken-for-granted news institutions were once the risky inventions of investors and entrepreneurs exploring the potential of steam-driven roll printing presses to make possible the nineteenth-century's famous penny press.

Do these perspectives raise further questions about the blurring of structural boundaries? Yes, indeed.

Take, for example, perhaps the most traditionally honored global boundary, that of the nation-state. Broadcast and print-based news in the industrial era was bound by state boundaries. The news process was bound by national legal structures. The state, more frequently than not, had the will and the capacity to censor within its boundaries. In a world of an instantaneous. real-time, two-way network with increasingly near universal access, the will of the state may continue to be strong, but the capacity to enforce censorship much less so. Further, international satellite broadcasting in the model of Al Jazeera and CNN International reinforces pressures toward the globalization of news beyond national boundaries.

Stroud, Hindman, and Singer, each in somewhat different ways, address the traditional distinction between hard and soft news. The most simple and perhaps familiar distinction is one that posits hard news as most central to the democratic process and softer human interest journalism as a distraction, perhaps a necessary one given audience preferences. But in

Singer's spirit of entrepreneurial invention, what about exploring new ways to humanize hard news themes and draw on natural, empathetic audience concerns in blurring the hard news–soft news divide. Newspapers traditionally utilize banner headings to identify sections on world news and on food and wine, and separate these sections in the way the broadsheet is printed and folded in sections. Such traditions have been exported over to online journalism, but perhaps other models should be explored.

Another boundary that merits attention is the service area of the journalistic enterprise. Newspaper and television news, and their supporting advertisers, are well aware of and responsive to the definition of the metropolitan statistical area of coverage. It is how the boundaries of what is local news are determined. But what about hyperlocal news—a mixture of professional and citizen journalism defined by neighborhoods and geographic communities. The investment of AOL in hyperlocal information in its Patch project to redefine local news and news boundaries failed to establish itself as a sustainable economic enterprise, but one experiment may not turn out to be the final word.

Many of us have stereotypic images of news institutions and of journalists from media lore if not from direct experience. One thinks of, for example, *Citizen Kane, All the President's Men, Good Night and Good Luck, Broadcast News*. What is common to this imagery is a singular notion of the role of the professional journalist. Journalists work for established news organizations. The difference between news and non-news is unambiguous. Who is a journalist and who is not is crystal clear.

The analysis at hand suggests a different imagery—as suggested by this volume's title—a process of remaking the nature of news. It appears that whether we are pleased or disappointed by it, we live in an era of blurred boundaries. Recounting the golden age of dominating newspaper and network news and calling for its return offers little promise. Perhaps a more promising course is to invent anew and follow Jane Singer's call for an entrepreneurial spirit. We should anticipate a pluralism of institutions, of journalistic roles, and of the very definition of news.

IV Foregrounding Underexamined Themes

11 Check Out This Blog: Researching Power and Privilege in Emergent Journalistic Authorities

Sue Robinson

Currently in media studies we have a rich, well-theorized body of scholarship on journalistic production and its evolution in the digital era. One gap within this literature is the lack of sustained conversation about how societal power dynamics—especially regarding privilege born from race and class—play into who assumes authority to speak in public information spaces such as news sites. There are few examinations of how reporters or bloggers struggle with issues of privilege when producing public content. There is scant research into how digital news production spaces—those populated by journalists but also those by citizen contributors—are reinforcing/undermining the status quo under the guise of authoritative storytelling. The task of press theorists is not simply to document how digital technologies offer a chance for new roles and voices. It is also to examine critically who gets to speak in the public news realm, in which spaces sanctioned as journalism, and to what end? The key research question central to this work is: as traditional journalistic authority wanes in a world of globalized interactivity, what enquiries should media scholars who study content production pose as they document the transition into a new regime, in order to expose privilege in the jockeying for legitimacy?

When examining phenomena concerning journalistic authority, the very concept *authority* must be scrutinized, particularly insofar it is evolving in response to digital changes. In this chapter, I combine an institutionalist perspective that considers authority to be born of voluntary adherence to institutional norms with Matt Carlson's model of authority (forthcoming). Carlson's model assumes an interactive, relational concept that varies according to content producers and the expectations of their audience. However, I suggest that as people attain new, public spaces, network with new kinds of audiences, and build capital through new digital production, how authority manifests depends on the nature of those spaces and networks as they reflect, augment, or alter offline circumstances. Thus, as we theorize for a networked, interactive, global world, we must account for the offline power dynamics that endure (and yet also evolve) for online production. We must also take into account the new performative, discursive rituals developing in digital spaces that are altering these dynamics.

This chapter disentangles five interrelated concepts—*authority* as it is related to *power, privilege* as it informs *networks,* and *values* that become shared as a result of these production dynamics. Journalistic authority is examined as a part of the structural, institutionalized power relationship among sources, journalists, and audiences. As part of an interactive, web-based infrastructure, digital networks alleviate offline hierarchies that breed exclusive, authoritative, behavior in information worlds. This essay examines previous findings of online scholars in related fields. It looks at actual production practices of journalists and citizens and the changing nature of authority in these realms. In the final section, the essay offers five possible approaches to study journalistic authority in these new environments by drawing from a variety of disciplines. These disciplines include linguistics and ethnic studies, public sphere theory on counterpublics, field theory, and network analysis from sociology and science and technology studies. A table at the end diagrams potential plans of action for scholars interested in uncovering the various dynamics of authority in newly reconfigured information worlds.

Journalistic Authority: An Institutional Approach

Ten years ago, a journalist responded with exasperation to a query about the ability of citizens to report the news: "Someone's gotta be in control here," he said as he extolled the importance of a healthy fourth estate (Robinson 2007). In six words, he encapsulated the core tension that many traditional journalists at the time were feeling as digital interactivity tampered with their authority to be the world's official storytellers. At the time, I focused on this tension from the perspective of journalists and in consideration of the press as an institution. Historically, there have been several assumptions about a collective commitment to the institutional authority of the press. The first is the uniform professionalism of the industry, which has abided by a set of norms and practices, such as objectivity, that help maintain journalism's exclusivity. The second is a general consensus among common sources—typically those in power—about their symbiotic relationships with journalists: both sides strive to keep themselves "in power," though in different ways. The third assumption is that audiences themselves respect the authority of journalists. Indeed, literature about the press defines authority as "the power possessed by journalists and journalistic organizations that allows them to present their interpretations of reality as accurate, truthful and of political importance" (Anderson 2008, 250). Cook (2005) and others have noted how this power has endured over time to make the press an institution in terms of not only helping people to know what to think about and guide them in their behavior, but also in terms of reifying the status quo and entrenching a dominant power structure in society.

Thinking in these institutionalist terms, one can perceive the notion of power in this construct as largely ancillary and bound up in the concept of authority itself. The opportunity for outsiders to produce journalistic content opens up the potential for authority burglaries. Power is equated to *capability*, i.e., a successful wielding of power results in authority. For the institutionalists who have engaged with the concept of power and authority at length, authority is "legitimatized power" enacted via roles and backed by institutions. The process of gaining legitimation typically involves the evoking of "moral symbols, sacred emblems, legal formulae" (Mills 1959) that help formulate the power–dependency arrangement. Because certain norms and practices are being adhered to or referenced, the action can be legitimate and authority is achieved (Emerson 1962). Emerson describes authority as something attached more to a role rather than something any particular individual embodies. According to this scholarship about the press, authority depends on some kind of dominance over others, usually over time, as long as that person acts in commitment to an agreed-upon set of rules or standards of some collective.

The Internet appeared to offer the opportunity to challenge authoritative systems and, conceivably, to allow more people to gain authority that was perhaps even external to any collective. With its interactive capabilities and the ability of everyone "formerly of the audience" to become "mass self-communicators," digital interactivity could give voice to the voiceless (Castells 2012; Gillmor 2006; Rosen 2006). Many scholars trumpeted the opportunities for marginalized peoples to amplify their voices in mainstream information streams using blogs or Twitter. Digital technology obliterates the need for a centralized informational core. Any hierarchal flows would give way to a digitally enhanced, global, networked infrastructure (Benkler 2006; Castells 2009; Pfister 2014). In these networks, no control center exists among constellations of interconnected points (called nodes) and actors, who have varying amounts of influence. Castells has been bullish on the ability of digital networks to welcome new, successful nodes and delete old, outmoded, and inefficient ones. For Castells, power in the networked society could be reconfigured among actors. In an increasingly networked society, people can form effective collectives that bypass traditional power structures and attain new levels of authority. Mass interactivity might diffuse how power flows and, in the process, deconstruct authoritative realities.

Researchers talked to bloggers and analyzed content that was not produced by journalists. They hypothesized that the "control" to which journalists were clinging seemed tenuous. Some of these hypotheses proved true. Successful blogging enhances feelings of empowerment and communal belonging (Robinson and DeShano 2011a, 2011b). Blogging saves the public sphere by "flooding the zone with public discourse" (Pfister 2014, 60). Bloggers, external as they are to the "pack," utilize digital functionality, such as Google search, linking, immediate posting, and commenting to move the public discourse away from vertical power flows and create new, more meaningful kinds of argument (Pfister 2014, 66). More engagement on social media platforms, such as Twitter, and especially in relation to following and engaging with journalists, correlates to higher involvement civically offline. The interactivity accesses power for some of these people, according to some studies (Gil de Zúñiga et al. 2012; Xu and Feng 2014). "Regular" citizens become sources in the news more often because of their participation online (Van Leuven et al. 2014). Citizens help change the tone of information flow, making it more emotive and affective (Papacharissi 2014), and also providing a ubiquitous, "ambient" journalism (Hermida 2010). The Internet encourages multiway reciprocity of news engagement (Lewis et al. 2014) that opens the general culture of the press and creates a greater sense of immersion, participation, and connectivity for audiences (Robinson 2012). C. W. Anderson (2013), in his account of the Philadelphia news ecosystems, and others (Boyles 2014; Etling et al. 2014; Graeff et al. 2014; Kelly 2010; Robinson and Schwartz 2014) began mapping the evolving media ecologies as new players changed information flows. These studies, especially when considered in aggregate, imply a slow disintegration of the reign of journalists over information authority (although many of these studies do not address the question of authority explicitly).

Offline Hierarchies and Online Networks: Toward a Critical Framework

In 2011, I embarked on a massive, multiyear, multimethod project to better understand the composition of these ecologies and who and what comprised the information flow in local communities. After questioning journalists, active bloggers, and "regular" people about their participation in online arenas, it was clear that I needed to discuss the power to produce authority. Authority may have been undermined, but power had not been. In my datasets, not everyone was participating, even though they had the technological capability to do so. Others failed to gain any visibility or achieve any influence as content producers. Those who were attaining status because of their blogging or posting reflected savvy understandings of how to network, whom to befriend, and what was popular to say. Those who did not attain status often represented some marginalized demographic that was low-income, poorly educated, an ethnicity other than white, an alternative ideology, or some orientation that was not mainstream. Meanwhile, others adeptly managed information and worked these new networks, e.g., becoming a regular source for reporters. I set out to document how power dynamics work to bestow *authority* on individuals and withdraw it from them—both within institutions, such as the press, but also outside of formal organizations.

The following five questions persistently emerged as I worked through the data.

1. *How does offline privilege affect online contributions?* In interviews with content producers online, multifaceted questions about civic involvement, feelings of belonging, and standards of practice were posed to document whether their forays as public communicators meant they had become better citizens. We did not delve deeper into their capacity for production, their social positions, their connections that allowed them to achieve those feelings of belonging, and the very confidence it takes to hit that submit button. For example, DeShano and I looked at why people didn't contribute in public realms, although they were motivated to do so, and wrote an article (Robinson and DeShano 2011b) on the tensions that emerged about feeling intimidated in these realms. We did not fully reveal why those tensions existed in the first place, or even discern why our white participants went on to contribute more than our black ones did. How do these bloggers and others exercise social, political, and other kinds of capital, how do they gain and lose that capital, and how does their power wax and wane over time, and according to what factors? One's background (such as familiarity with technology), available resources, social networks, daily news and other habits, professional training, and other situations impact one's level of success.

2. *How does this privilege translate into authority that influences online activities?* The majority of participants in my study who contribute frequently online hold offline identities as activists or politicians, police officers, teachers, or union representatives. They are people who have attained authority in other realms before entering the information world. Indeed, it is these associations that have enabled their content production, provided them with the information and knowledge of topics about which to write, and the audience and commentary to bolster their authority. Pfister (2014) studied top bloggers, such as John Marshall and Glenn Reynolds, but didn't explore who Marshall and his other case bloggers were, whom they represented, or what roles they assumed. Hindman (2008) did so and found that of the top ten political bloggers, all but one was white, all but one was female, half were journalists or former journalists, eight had graduated from elite colleges, and seven had graduate degrees.[1] In fact, Hindman's major thesis argues that the notion of multiple perspectives and amplified voices online is largely a myth. The authority of journalists may be constantly challenged in these spaces, because people criticize posted news articles. However, these people tend to occupy positions of authority comfortably, whether they are white, educated, male, or otherwise in a domineering class. Those minority voices who do attempt to be heard face challenges. One biracial blogger, who wrote about racial achievement disparities, said she shut down because she didn't feel influential, had too few followers, and became frustrated with the one-sided, vitriolic, public dialogue. At the same time, nearly all of the highly connected activists in my datasets (not only white, educated people) were gaining huge followings in social media spaces and becoming much more influential[2] than the journalists covering the same topics. Authority bestowed in the offline world transfers to the online world in a way that can trump whatever ephemeral relationships might be resulting in new authorities in digital spaces.
3. *What are the dominating networks and who is isolated in mediated information flows?* Many scholars have noted that citizen contributors cite traditional media sources, and that broadcasters and legacy news organizations still set the agenda. Digital technology allows for new associations via social networks that change how authority over information emerges. Consider viral posts by fairly obscure people who nonetheless have networked connections with influencers of multiple communities. A successful white blogger began as an unknown (new to town, with few connections) but tapped authoritative information (direct

quotes from school board members at public meetings, links to government databases, etc.) to build a following. The man, a doctoral candidate in education history, parlayed this strategy into becoming a frequently cited source for journalists. Local media aggregators regularly listed his blog, and he campaigned successfully for the local school board. How did this digital networking increase the authoritative status of someone who began with no power? We can begin to understand this advancement by recognizing his privilege—white, credentialed, and savvy about how K–12 education systems are networked.

4. *Which digital platforms are being used and how do these platforms mimic or undermine offline communities and infrastructures and affect what appears in information flows?* The location of information production can affect the role of the producer in that environment. Journalists write stories adorned with their byline that appear under a branded masthead; the inability of the public to "talk back" contributes to the sense of exclusivity over the production of information. Today, commenting sections foster "talking back." Reporters on their own social media platforms work to use this interactivity to their advantage by milking networks and listening in on conversations. But reporters in my studies were challenged in these spaces. Niche-community actors can take over a reporter's Facebook post. For example, a reporter posted the speech of a school board candidate only to have a group of active citizens who opposed the candidate spend several days on the platform calling the candidate's words and the reporter's story into question. The reporter posted an email from the candidate in response to the Facebook conversation. This presents a new pattern of information flow involving the news institution but also has implications for who maintains control of that information production.

5. *What does the way in which content is produced because of these dynamics mean for the construction of a shared value system?* We know from institutionalism research that legitimation results from the support of a community with shared values, and also that authority translates from one group to another. If the content does not adhere to traditionally accepted values in those groups, it runs the risk of not being accepted. But much research has shown how the values that media disseminate privilege some and marginalize others, reifying hegemonic constructs, such as white privilege and the power structures that go along with them (Bonilla-Silva 2006; Shah and Thornton 2004, 1994; Squires and Jackson 2010). For example, Balaji shows how media coverage in the United States about the Haitian earthquake wove a narrative of "pity"

that served to demean "dysfunctional, dependent" black people as subordinate to the more resourceful white people: "I argue that the mediated discourse of pity exposes the subtle—and not so subtle—power relations existing between whites and blacks ..." (Balaji 2011, 52). Negative frames that portray black people and other minorities as dangerous, irresponsible, and failures also hurt achievement among minorities (Shah and Thornton 2004; Squires and Jackson 2010) and keep white people in power (Dixon and Azocar 2006; Reid-Brinkley 2012). For these people, journalism has not really been authoritative, given how journalists' performative discourse has tended to exclude them from the ritualized meanings that have come from news narratives. Focus groups conducted in 2015 with citizens of color demonstrated that without representation in mainstream newsrooms or in "normal" (that is, not negative) coverage, black participants said they checked out and refused to watch. Indeed, blogs and social media have allowed the formation of new, alternative spaces for content production for those who feel not heard in dominant information flows. When comparing these texts with traditional media, there are very stark differences in the values put forth by mainstream writers, such as journalists or white, hegemonic bloggers and those typically viewed as "other." How can these writers attain the authority necessary to share those values more effectively?

Journalistic Authority: A Working Through

Several dimensions of authority become challenged in this environment. First, consider that authority entails "control"—especially for traditional journalists. Journalistic "authority" had meant that journalists could set the world's agendas, provide knowledge for its citizens to build capital, and help achieve the aims of those in power. If journalism is waning in its ability to set the agenda because of a proliferation of content or because journalists are losing influence and access as sources bypass the press, does authority disperse, or does the concept become even more powerful? First, a world of digital networks that redistributes control and power across linked nodes might diffuse "authority" as well. Second, authority demands territorial boundaries—a "presiding over a societal sector" (Cook 2005, 81) to the exclusion of others. This dimension assumes that the press occupies a hierarchal status above the masses. The presence of new, nonprofessional actors operating with different norms and standards means a sharing of authoritative space in media ecologies. In a horizontally aligned network that creates an abundance of information from a multitude of producers, is authority over information even possible? Third, we must consider how scholarly ideas of authority have historically argued for exercising that power over time (Cook 2005; Sparrow 1999; Weber 1978). The longer journalists spent on their beat, the more access they attained and the more expert and authoritative they grew. Today, reporters are not as present, but others have appeared as new kinds of "experts" with different ways of proving their credibility. Activists who are vastly networked are influential in a way they never were before. Noting the vast amount of scholarship that shows how ephemeral and fickle yet powerful and efficacious relationships online can be, one wonders if "longevity" remains a key dimension of authority. People's expectations of "presence" and endurance have shifted as URLs vanish, new relationships manifest in the briefest of postings, and blogs thrive or idle. Given these shifting parameters, the very notion of journalistic authority itself needs to be flexible.

At the same time, we must also recognize that authority is still evident in traditional relationships and is not present in some online interactions. One's social, economic, and professional privilege dictates the position of the content producer and the resulting direction of flows within these networks in a way that introduces and perpetuates dominant value systems. The capability of power became moot when the effective connections could not be achieved. Even when a prolific activist in my data appeared highly influential in his network, the community was sometimes an isolate; without a bridge to more mainstream audiences, any authority this activist had attained was limited to a silo of discourse. Thus, another complication is added to the idea of the authority of content producers in online worlds: authority can ebb and flow, collecting in some online spaces while failing to materialize in others, depending on circumstances surrounding the content (as detailed in the above questions).

Matt Carlson's forthcoming *Journalistic Authority* offers an updated model of authority as something constantly reconstructed, undermined, and strategized, and also something relational. Journalists cannot be authoritative in a vacuum, especially given digital capabilities, he contends. He argues that blogs, sites such as the SCOTUS blog, and other citizen-enabled platforms have challenged the dominant position of the press in public information realms. Yet his model also considers how these production transformations remain mired in institutional and organizational dynamics. Carlson writes that understanding journalistic authority today means appreciating the *interactions* around journalistic text. Furthermore, he makes it clear that explorations of authority at any given time must attend to the role of nonjournalists in the creation and alteration of journalistic authority because of its performative, discursive nature.

To advance Carlson's important ideas, scholars need to consider not only how content producers are interacting with the content and audiences, but also how they are interacting with other power players, and how, where, and when other power players are interacting with that copy (or not). Appreciating a producer's social status as well as documenting his or her networks and relationship linkages around power sources—both online and offline—would make for a highly meaningful conceptualization of authority. In addition, the recognition of even the most tenuous of links to brokers and other influencers in digitally enhanced social networks can prompt a keener understanding of how someone might hold the capacity to improve his or her bid for information authority—or might not. This would mean adopting an intentional gaze toward offline and online power relationships in journalistic production work.

Journalistic Authority: Research Frameworks

But how do online researchers accomplish this? The next paragraphs propose analytical approaches that help expose how power dynamics are working to bolster or subvert information authority. Critical discourse analysis (CDA) from sociolinguistics, counterpublics from rhetoric and public sphere areas, field theory and network analysis from sociology, and science and technology studies all offer interdisciplinary techniques to uncover these opportunities and constraints. Table 11.1, Research Frameworks, details the approaches at the end of this section.

Critical Discourse Analysis

Media scholars studying race have long turned to CDA to appreciate how journalism implements societal power dynamics via semantics and language (Fairclough 2010; Fairclough et al. 2011; Wodak and Meyer 2009). CDA introduces a critical component to the analysis, encouraging the researcher to situate the findings according to societal circumstances of domination and inequality:

> Why does this topic or this information get so much (or so little) attention? Does this topic or this word challenge or maintain stereotypes or prejudices about minorities? Who are speaking and who are (or are not) allowed to give their opinion? Whose interests are defended? From whose perspective is this report written? Is discrimination or racism denied, mitigated, or trivialized? (van Dijk 1991, xii)

Those studying online production could use CDA to appreciate how the text represents a site of struggle and even boundary work over journalistic authority. For studies of online content producers, interview transcripts represent the text that the researcher interrogates for perspective, word choice, metaphors, and other language that represent power relationships at work. Such analysis uncovers ties of mutual dependence (Emerson 1962), for example, that would help situate what is happening within a power structure so we can understand whether, how, and why the content they produce attains authoritative status within an information community.

Counterpublics

Some might consider those seeking to challenge journalistic authority online an attempt to establish or support a "counterpublic." Scholars have found that marginalized groups, such as right-to-die advocates (McDorman 2001), Egyptian literary fanatics (Elsadda 2010), and antiglobalization activists (Downey and Fenton 2003), have discovered the Internet as a place to bond, organize (Palczewski 2001), and otherwise seek authority either within microcommunities or in mainstream ones. Eckert and Chadha found that their German Muslim bloggers turned to blogs as a way to establish "subaltern counterpublics" (Fraser 1990) that allow for "parallel discursive arenas" in which to develop culture and collective identities, be subversive, and rail against mainstream power. Focused as it is on counterpublics, the research did not delve into issues of authority, but the findings support the idea that these content producers sought to establish an authority to represent the Muslim perspective. The problem is that some of this work glosses over differences in online productions. Are Muslim bloggers, Egyptian literary fans, and right-to-die advocates homogenous, or are some of them more effective than others? In a 2002 essay, public sphere theorist Catherine Squires wrote, "Instead of 'the' black public sphere or counterpublic, one should speak of multiple black public spheres constituted by groups that share a common racial makeup but perhaps do not share the same class, gender, ethnic, or ideological standpoints" (Squires 2002, 452). In each of these online communities, a series of power relationships form that are internal to the group and external to the mainstream sphere. By examining the roles of the power agents within these counterpublics, the stratification of niche blogging communities themselves is revealed. We can appreciate how power-dependence relations structure in online content production environments and add to the evolving definition of "expert." Finally, the counterpublic space must be scrutinized: the platform matters in terms of the extent of the authority trying to be achieved.[3]

Field Theory

We call people of color "marginalized" because they symbolically exist on the fringes of the "mainstream" community, of the public sphere, and of information (and political, etc.) fields. Very few in "marginalized" communities occupy a social position from which they can symbolically dominate. One reason for this can be found in field theory. In sociology, field theory overlays any emergent phenomenon—such as morphing journalistic authority—with a macro understanding of evolving economic and other kinds of structures. According to French social theorist Pierre Bourdieu (1995, 40–41),

> A field is a structured social space, a field of forces, a force field. It contains people who dominate and others who are dominated. Constant permanent relationships of inequality operate inside this space, which at the same time becomes a space in which the various actors struggle for the transformation or preservation of the field. All the individuals in this universe bring to the competition all the (relative) power at their disposal. It is this power that defines their position in the field and, as a result, their strategies.

In *On Television,* Bourdieu detailed how journalism as a field included specific actors and *excluded* many. Television pundits, for example, were selected not only because of their "credentials" but also because they played by the rules of the field. They tended to be older, highly educated, and well paid, could speak on anything, react to all kinds of queries or accusations, and were always available for reportorial deadlines. This means that others—perhaps more qualified, with more diverse perspective, younger, of color, female—were unknown to the field and, thus, excluded (Bourdieu 1999). Sources called upon by journalists are not "expert" as individuals per se, but rather actors within a larger field system who can symbolically dominate only because of their position (Bourdieu 1995, 31).

This is important to think about when we think about who is not being quoted by journalists, who is not getting journalism jobs, and who is not gaining authority in information worlds, especially online ones. When journalists talk about issues involving race, for example, and they put forth unintentionally racist stereotypes, they offer their "othering" as part of an established "doxa" that helps categorize the world in constant bids for what Bourdieu called "symbolic royalty," during which periodic acts of "symbolic violence"—such as racist or superficial characterizations in text—occurs (1995, 38–39). In a 2006 essay, Rodney Benson called on researchers looking at media audience to nuance their understanding of economics and class by investigating how cultural capital can mitigate any utopic implications of digital technologies (Benson 2006). Field theory forces one to consider not merely the role of "blogger" per se, but also the competitive relationships, bids for dominance, relative isolation within the field, and other factors of the challenger in the online production environment.[4]

Network Analysis

Online journalistic platforms represent new networking opportunities. Mapping the activity on these platforms using different network analyses from sociology can be a useful starting point to trace new citizen actors in the media ecology and understand their available paths for authority in their particular information realms. Network analysis calculations identify who is most central to a digital social network or who is most connected to reveal the major influencers—that is, those who have authority. Already scholars are distinguishing levels of influence online among online contributors—such as broadcasters versus print journalists (Graeff et al. 2014; Xu and Feng 2014) or citizen bloggers versus traditional journalists (Kelly 2010). For example, Hindman analyzed nearly three million Web pages to show that the idealized notion of new voices—especially from marginalized citizens—never materialized online and that the majority of contributors have been political elites, mostly familiar activists, pundits, and politicians (Hindman 2008). Network analysis maps the relational ties between content producers and/or their sources and can assess a blogger or citizen contributor's effectiveness as an authoritative producer in an information network. Then, considering that network in an overall media ecology could grant context to the study in a way similar to what C. W. Anderson documented in Philadelphia using a more qualitative-based "network ethnography" that offered a snapshot of information evolutions flowing through a new media ecology. How do the linking habits of online producers affect authoritative status in an ecology? How do connections with traditional media outlets impede or enhance a new blogger's bid for legitimation? Can niche bloggers ever attain authority in more mainstream information worlds? A variety of quantitative and qualitative network analyses might provide some insight to such questions.

Science and Technology Studies (STS)

Some media researchers draw from science and technology studies (STS) as a way to articulate what's happening in online media production. Fortunati and Sarrica (2010) espouse a "sociotechnical system" paradigm for media research, arguing that "a technical system and a social system are always present and simultaneously operating: The global result of the system activity depends upon their interrelation" (248). Media studies

Table 11.1
Research Frameworks

Some Recommended Approaches	Parent Discipline	Considerations for Production Authority	Action Plan for Studies of Authority
Critical Discourse Analysis (CDA)	Sociolinguistics	Historical inequalities, audience relations, power influences, absences, reporters' attitudes	• Collect demographics on producer (including education, family background, socioeconomic status, race, political leanings) (Who is speaking?) • Learn social background (professional, social, civic networks)
Counterpublics	Public Sphere Studies, Rhetoric, Communication	Places of production, audience dynamics, evolving standards/practices (compared to journalists')	• Find out audience metrics (Who is listening?) • Discover social media, news, and other habits • Document routines of production (Do they exclude some groups of people?)
Field Theory	Sociology	Producer's routines, socioeconomic status, ideology, network community bridges/ connections, influences	• Uncover motivations, strategies for information production • Appreciate historical context (including past discretions, working relationships, influential decisions) • Research (social, economic, political, network) obstacles to advancement in influence • Question actions/attitudes re. race/class/privilege
Network Analysis	Sociology, Information Sciences	Social/political/economic connections; sharing/ linkage/citation patterns, brokerage, other influences; audience metrics	• Ask questions of ideology or ascertain from content analyses • Conduct network analysis of links, sharing patterns, citations, post frequencies, etc. • Consider name generator survey of most trusted information sources for producers and/or those sources in the story
SocioTechnical Theories, Actor–Network Theory, Medium Theory	Science and Technology Studies	Actant/technology considerations, platforms as actors, boundaries of technology/fields/ workspace, etc.	• Follow links and paths of content in ecology; track the information flow • Query philosophy of interactivity, journalism as a profession, role of citizens in democracy • Identify major influencers for the producer and the copy through interview, survey, network techniques

scholars have adopted STS elements to highlight the role that technologies themselves can cause systems to hiccup, hierarchies to falter, and journalists to reconceptualize their jobs. See for example, Matthew Powers's piece (2012) on "technologically specific forms of work" (25) that indicate "exemplars of continuity," "threats to be subordinated," and "possibilities for journalistic reinvention" (24). Seth Lewis and Oscar Westlund (2014) argue for a "sociotechnical emphasis in journalism studies" (3) that highlights "human actors (e.g., journalists, technology specialists, and businesspeople); technological actants (e.g., algorithms, networks, and content management systems); and audiences (e.g., assemblages of audiences distinct to certain platforms, devices, or applications)—all potentially intertwined in the activities that constitute cross-media news work" (2). Finally, Matt Carlson (2015) offers STS's emphasis on boundary work as a framework for looking at journalism's construction of authority in particular. As digital networks proliferate, boundaries are constructed or demolished as actors create alliances and challenge authorities. Places of journalistic production—professional and amateur—host this boundary struggle for control over information. Studying the content producers with an STS approach highlights the roles all the production actors and actants—from journalist to viral YouTube post—play in relation to one another. STS provides a framework for investigating how the technical aspects of online work may be modifying the values and knowledge of digital authors (Berger and Luckmann 1966), affecting the conditions of legitimacy.

Conclusions

If we accept that journalistic authority depends upon relationships for sustenance, then any study of authority in digital settings would benefit from a broad appreciation of all the relationships that may be affecting the effectiveness of that power. Offline associations of producers should be attended to as part of a critical approach. Is a blogger still challenging reportorial authority if no one reads his posts? If a frequent Facebook poster has thousands of "friends" but few of them are in positions of power, can that person's authority bloom? How can journalists with no offline networks of color use social media to increase their authority over racial issues? Even in a networked culture, society remains stratified. The linkages between offline structures and online opportunities must be problematized when we research information producers in reconfigured networks.

Key to this paradigmatic approach is a focus on power relations. Beckett and Mansell (2008, 101) recommend scholars critically approach sites of inquiry and to be particularly concerned with "power, its redistribution, and its consequences for those engaged in the production and consumption of news." If the press has traditionally operated as a cog serving the needs of those in power, then what are the roles of these contributors in reinforcing the status quo or in challenging it? How are their offline power statuses influencing copy? According to Pfister (2014, 47): "Networked sensibilities hybridize, rather than supplant, modern sensibilities." The structures and their constraints bleed into virtual production and consumption spaces. Indeed, some scholars studying online spaces have shown that not much changes in virtual environments in terms of authority and power dynamics (Hindman 2008).

Obviously this holistic undertaking is not always the goal of a research project or within the resources of researchers in any given study. But if we are to take up authority and understand its dimensions, the concept must come tied to power relationships. What role(s) is the content producer playing in the mediated ecosystem? How are the actors who create that authorial relationship situated with each other? Is one occupying a more privileged space? Is the audience granting that authority able to do so freely without institutional constraints and past expectations of hegemonic norms? Are the practices and strategies by which that authority is achieved—evidence sharing, quotations, links—inclusive or exclusive to marginalized populations? Do these authoritative strategies reflect an assumed cultural, power, economic and social status quo without acknowledging the existence of alternative realities and counter publics? The exploration of public content producers—journalists, bloggers, Facebook posters, tweeters, commenters—warrants an understanding of the relational aspects at work in bids for information authority. As we pay attention to how production norms develop among new actors, values ("objectivity" but others as well) emerge that are reflections of larger, historical power structures. Hierarchal tendencies via networked relationships uphold journalistic authority and can be constraining for nonprofessionals in some spaces but not others. These networks can also provide opportunities for those activists, politicians, and others who already hold information authority in certain offline realms.

Online journalism research has spent nearly 20 years building a formidable foundation of scholarship that draws from traditional thinking about old media and adapts for digital technologies. The field has always been interdisciplinary when thinking about media phenomena. This essay encourages scholars to build on that tradition by adopting frameworks that offer "big-picture" analysis for the information-authority transitions we are seeing. These can uncover hierarchies and the privilege they contain, unveil problematic values and the power structures they were born from, and work toward developing emergent conceptualizations of authority.

Notes

1. See also work by Jen Schradie, particularly Schradie 2012, 2011.

2. We did name-generator surveys and network analyses on one whole information exchange network in one community that quantitatively demonstrated level of influence.

3. Squires (2002) created a very interesting typology about the different counterpublic spaces operating according to different motivations and goals.

4. Rodney Benson's (2013) book comparing US immigration news to French coverage demonstrates how field theory can be used to distinguish power relations in media sociological study. He points out that field theory "is crucially concerned with how media often serve to reinforce dominant systems of power. Yet compared to hegemony, the field framework offers the advantage of paying closer attention to distinctions in forms of power, how these may vary both within a society and cross-nationally, and how they might be mobilized for democratic purposes" (195).

12 A History of Innovation and Entrepreneurialism in Journalism

Mirjam Prenger and Mark Deuze

Given the profoundly precarious condition of the news industry and the corresponding casualization of the journalism labor market, it should come as no surprise that a significant focus in the field of journalism studies is directed toward innovation and entrepreneurialism. In research as well as teaching, the "newness of the new" gets understandably overemphasized in an attempt to prepare students for precarity while supplying the industry with some much-needed perspective.

There are two issues with this approach. The first is that this focus runs the risk of ignoring the past, as innovation has been key to structural developments in journalism. Additionally, journalism has, in its innovative uses of technologies, pushed groundbreaking developments in other fields, such as telecommunications. Entrepreneurship is at the heart of breakthroughs in journalism, most notably when it comes to the introduction of new genres and news formats, investigative styles and techniques, and the development of an occupational ideology that can be both a flag behind which to rally in defense of tradition and routine, as well as providing fuel to release forces of change.

A second issue with the contemporary spotlight on entrepreneurialism and innovation is that it comes with a barely contained normative agenda, in that innovation and professionals becoming entrepreneurial tends to be seen as a good thing—marking a benevolent force. To this, one has to add some conceptual confusion: when exactly is something considered to be entrepreneurial or innovative, how does one "do" entrepreneurship, at what level of analysis does innovation lie (individual, organizational, product, or process)? Epistemological challenges further amplify these wide-ranging questions, as innovation is invariably a moving object, raising the issue of how to adequately study something so dynamic.

A general solution tends to be to treat entrepreneurialism and innovation in strictly managerial, economic, and business terms, as these fields have arguably developed the most sophisticated discourse and conceptual toolkit around such issues. At the same time, the history of research on innovation in journalism studies tends to follow a neatly boundaried institutional agenda, focusing on legacy news organizations and the content they produce. Although the worldwide media landscape is changing and working relationships are in a state of flux with the dominance of atypical media work, researchers still predominantly chart the professional cultures of news reporters working in an institutional editorial setting.

With these approaches, a whole dimension of research gets lost that is central to the object of journalism studies: professionalization, the development of a professional identity, of a news culture (particular to a country, a news organization, or division), and of an occupational ideology that works in different ways for a wide variety of practitioners professionally involved with gathering, selecting, editing, publishing, and publicizing news. Beyond the business and culture of legacy news organizations there is a wealth of questions waiting to be answered: what do ideological concepts (such as objectivity, autonomy, and being ethical) mean for particular journalists in specific circumstances in the context of entrepreneurial and transformative conditions; what do objectivity and other ideological values mean to those who either suppress or inspire innovation, creativity, and entrepreneurship in their work; and what are the implications of entrepreneurialism and innovation for the way journalists both inside and outside of professional news organizations see themselves and their role in society?

Our chapter intends to map these questions using both a historical and a contemporary setting for the investigation of entrepreneurship and innovation in journalism. The emergence of a new journalistic genre on television in the 1950s and 1960s is compared with the emergence of the current startup culture in journalism. This comparison is used to highlight particular challenges and opportunities for doing journalism studies in a dynamic field.

A Historical Dimension: The Emergence of Current Affairs Television

One could consider the 1950s and 1960s of the twentieth century a significant period of transformative innovation and change in professional journalism, specifically as it grew and matured on the television screen (Conway 2009). In this period, journalism in various (yet similar) countries made significant strides, developing a new form, voice, and approach while maintaining and enhancing core professional ideals, making this period an excellent benchmark for comparison with the turmoil of today's media landscape. A comparative analysis of the manner in which current affairs programs on television in the United States, Great Britain, and the Netherlands came of age and innovated in the 1950s and 1960s, shows that there are some very interesting parallels (Prenger 2014). These parallels help illuminate the factors which played a role in the innovation of journalism in the past, begging the question to what degree these factors still play a role today.

If we were to sketch the early history and subsequent coming-of-age of current affairs programming on television, we could draw a rather linear picture—something that has often been done by media historians (Bliss 1991; Smith 1998; Hilmes 2003). The temptation to do so is understandable: at a glance, the history of current affairs programming worldwide looks very straightforward.

The beginning can be pinpointed in the United States, with the start of the CBS program *See It Now* in November 1951. The program was the product of two ambitious men: the renowned radio journalist Edward R. Murrow and the creative and entrepreneurial television producer Fred Friendly. Inspired by the popular magazine *Life* and the radio (and movie theater) news series *The March of Time*, it was Friendly's ambition to create a news magazine on television that combined the possibilities of serious journalism, radio reporting, documentary film, and live television. The result was a weekly television program presented by Murrow that focused on current affairs, alternated with more lighthearted topics.

Acclaim for *See It Now* was immediate. Critics applauded the program because it demonstrated what television could achieve when imagination and journalistic ambition were combined. "Murrow and Friendly exploited to the full the drama and excitement inherent in the news," *Variety* wrote (Persico 1988, 305). The program focused on the gritty reality of life, and reporters and film crews were encouraged to follow news stories as they unrolled, instead of presenting them as prescripted reports. Among other innovations, Murrow and Friendly introduced the "cross-cut interview." By experimenting with the possibilities that television offered, the pair stumbled across the technique of counterpointing extracts from contrasting statements of people with different views, thereby achieving the essence of genuine controversy.

The real breakthrough came with the famous broadcast in May 1954 in which *See It Now* exposed the demagogic Senator Joseph McCarthy. Critics wrote, "no greater feat of journalistic enterprise has occurred in modern times"(Leab 1981, 20). Since that broadcast, a critical approach, a focus on serious topics, and a point of view became distinctive features of the program, and thereby of current affairs programming.

The approach proved to be inspirational. In Great Britain, the BBC had been experimenting with different ways to present current affairs on television. In 1953 the broadcaster launched *Panorama*, introduced as a biweekly "magazine of informed comment." Lacking a clear identity and sense of purpose, the program floundered. This changed when the spirited television producer Grace Wyndham Goldie took over the helm in 1955. She appointed a young 25-year-old editor-in-chief, hired a slew of experienced and outspoken reporters with a background in politics, and asked a popular radio journalist to become the presenter. Both Wyndham Goldie and her editor-in-chief set *See It Now* as an example, and experimented with ways to make the filmed reports more dynamic and the live studio interviews more confrontational.

Although the format of *Panorama* was slightly different from *See It Now*, the approach was similar. The British news magazine presented itself as a weekly "window on the world," critically examining current affairs inside and outside of Great Britain. The relaunched *Panorama* was an immediate success and the program became the flagship of the BBC in the second half of the 1950s, a position it would hold for a long time. The reporters were presented as personalities, and encouraged to voice their judgments on the topics they were investigating. Current affairs thus had become a television genre explicitly licensed to deal in values and interpretation.

Panorama became the main source of inspiration for current affairs programs on television in the rest of Europe. In the Netherlands, two news magazines were initiated simultaneously in 1960 by competing broadcasters: *Achter het nieuws* (*Behind the News*) by the Socialist broadcasting organization VARA, and *Brandpunt* (*Focus*) by the Catholic broadcasting organization KRO. Initially the focus was on lighthearted, nonconfrontational topics. But this changed in 1962 when both programs appointed new editors-in-chief who set *Panorama* as an example. They changed the tone and content of the Dutch current affairs magazines, adopting a much more hard-hitting approach, confronting politicians and other authorities, covering topics which had previously been considered taboo, and investigating misdoings. The critics and the public applauded the new direction of the programs. Television journalism had come of age.

When presented in such a chronological fashion, the history of television journalism seems simple: an innovation which started in one part of the world was copied and rearticulated, slowly making its way to other parts of the world. But this linear presentation glosses over interesting similarities across the national contexts.

For instance, when looking at the parallels between these different histories, it is noticeable that in each country the current affairs programs were introduced ten years after the start of television in that country. And it is striking that each of the programs—be it *See It Now, Panorama, Behind the News,* or *Focus*—needed about two years to find its feet and become critical and confrontational, even though inspirational examples were at hand. So it seems that factors other than just copycat behavior played a significant role in the coming of age of television journalism. Three factors stand out: public expectations, the competition between broadcasters, and the profile of the producers and editors-in-chief who helped bring about change.

Looking at public expectations, it is clear that they were not being met sufficiently at the time each current affairs program first hit the television screen. In the United States in the early 1950s there was much criticism concerning the quality of television programs in general. Television shows were deemed too commercial, violent, or immoral, and there was too much content that did not live up to public service standards. Television news programs were criticized for being superficial, concentrating on images instead of journalistic relevance. In reaction, most television broadcasters willingly adopted the Television Code, a set of ethical standards regulating the content of their programs, at the end of 1951. At the same time, *See It Now* was introduced. The enthusiastic reception of the program illustrated how much the public and the critics had been waiting for programs of substance that exploited the possibilities of the new medium. The confrontational *See It Now* broadcast concerning Joseph McCarthy in 1954 met even more critical acclaim. Television critics were full of praise, and 75,000 viewers wrote or called in, most of them voicing their approval of *See It Now* and its critical stance.

The same kind of reaction was noticeable in Great Britain. When *Panorama* started out in 1953, its lack of focus and relevance irritated many viewers. "*Panorama* is a perfect illustration of what is wrong with television," a television critic wrote (Wyndham Goldie 1977, 190). The new medium that had promised so much seemed to be stifled by old-fashioned BBC rules and routines. The *BBC News*, for example, consisted of a news bulletin read by an anonymous presenter and an image of the BBC logo on the television screen. When *Panorama* changed its format in 1955, the reactions were very positive. "If it keeps up last night's form it may become the most important live news magazine of the week," a critic wrote (Wyndham

Goldie 1977, 191). Millions of British viewers switched on their television sets each week to watch the new *Panorama*.

Impatience with the slow speed with which television journalism was evolving was tangible in the Netherlands as well. In 1961, television critics loudly complained that the Dutch current affairs programs were too immature, too dull, and not audacious enough. Where were the bold, critical, and entrepreneurial journalists who were willing to act as public watchdogs and prepared to make the programs much more dynamic and confrontational? Critics and public alike met the change in tone and approach of *Behind the News* and *Focus* in 1962 with open arms. *Focus* soon thereafter won the national prize for the best and most innovative television program.

Competition between broadcasters also played a major role in the coming of age and innovation of television journalism. As stated, all of the current affairs programs mentioned began at a similar moment: about ten years after the start of television in each country. This was also the moment that television began to break through nationally and surpassed radio as the most important medium.

In the United States, for instance, the sale of television sets exploded in 1951 and the major networks competed fiercely for a position in this new market. News and current affairs programs were considered to be strategically interesting genres with which the networks could attract new, higher-educated viewers. Hence the launch of *See It Now* by CBS in November 1951.

In Great Britain, 1953 was the breakthrough year, when the coronation of Queen Elizabeth attracted 20 million viewers, even though there were only 2 million television sets available. It made television an instant success, paving the way for the arrival of commercial television, ITV, in 1955. This caused great upheaval within the BBC, where it was feared that the corporation would lose out in the competition with ITV. Action was called for. It is significant that the newly restyled *Panorama* under guidance of Wyndham Goldie was relaunched just days before ITV began broadcasting.

The same pattern was visible in the Netherlands. Dutch television started growing in popularity in the early 1960s. The breakthrough moment came in the fall of 1962 with a marathon benefit show on television, watched by the majority of the Dutch public who donated money by the millions. It was a wake-up call for the public broadcasting organizations, making them realize how influential television had become. And it gave the politicians and other interest groups who were lobbying for the introduction of commercial television new energy. Their lobby did not succeed, as the Netherlands retained a public broadcasting model. But the threat of commercial television clearly influenced the public broadcasting organizations, which started focusing more on serious television genres with which they could accentuate their public service mission.

These broadcasting organizations, representing different ideological movements, also competed among each other for public approval. In this light, it is significant that when one of the major broadcasters, VARA, started with its current affairs program *Behind the News* in 1960; the other major player, KRO, soon thereafter launched *Focus*. And when *Focus* changed its format and became much more hard-hitting in 1962, *Behind the News* quickly followed suit.

What is clear is that the broadcasting organizations and networks in each country strategically programmed the current affairs programs, using them as a means to an end in order to heighten the prestige of the broadcasting organizations. The aim was to gain public and even political approval, and to compete with other broadcasters. The organizations and networks appointed new producers and editors-in-chief and gave them a blank check to innovate, as long as it resulted in publicly acclaimed programs with a high profile.

It is interesting to note that the producers and editors-in-chief who set about changing and innovating television journalism all seemed to fit the same mold. They generally did not have a background in newspaper journalism (although there were exceptions) and they tended to treat the new medium on its own terms, rather than imposing those brought over from another medium. They also held outspoken views on the journalistic mission of their programs. And they all had rather imposing personalities, which they used to gain authority within their network or broadcasting organization, and among their editorial staff. "When he came in, it was like the Red Sea parting," a *See It Now* team member noted about Murrow (Persico 1988, 419). Similarly, the Dutch editor-in-chief of *Behind the News* was renowned for his curt and sometimes authoritarian style of communication. And working together with Wyndham Goldie at the BBC was both a nightmare and a pleasure, a member of *Panorama*'s staff declared afterwards. She demanded total commitment. At the same time, Wyndham Goldie protected and defended her team. The British producer understood quite well how television worked, her assistant remembered: "that once you had the staff, the money, and the studio, you could tell other people to bugger off" (Lindley 2002, 37). This was important, since innovation and experiments with new ways of practicing journalism provoked backlashes and criticism, however great the general appreciation of the results.

What is perhaps most striking is that all the producers and editors-in-chief concerned were not afraid to editorialize. They believed that journalism had a duty to investigate current issues and wrong doings, and presented the results with a sharp and critical point of view. This was a deviation from the dominant paradigm of neutral and nonconfrontational reporting. The producers and editors-in-chief also had a competitive mindset. Their ambition to produce the best current affairs program in the nation strongly motivated the rest of their editorial staff. The resulting editorial culture was one that fostered the pushing back of boundaries and the exploration of new territories.

In summary, it is clear that within different historical media contexts a similar combination of internal and external forces helped bring about change and resulted in the innovation of television journalism. The external forces, in the form of public dissatisfaction and threats to the position of the broadcasters, paved the way for ideologically driven producers and editors-in-chief who were given room to experiment and explore new paths. There was a profound emotional commitment to improve television journalism, freeing it from the formats which were derived from print and radio, exploiting the possibilities of the new medium, and making use of (and sometimes stretching) the technological means that were available. In that sense, innovation went hand in hand with a passionate view on what journalism could be if it lived up to its promises.

A Contemporary Dimension: Emergence of a Global Start-up Culture

The contemporary discussion on innovation and transformation in the news industry tends to be dominated by the role of technology, specifically the Internet. This technological determinism belies global trends in the profession, showing a continuous growth of independent businesses and freelance entrepreneurship despite (or inspired by) the ongoing economic crisis. In this crisis, news organizations have seen major budget cuts, redundancies, reorganizations, and considerable downsizing. Responding to technological disruptions and changing audience practices, production practices are undergoing rapid change. The emergence of the enterprising professional in journalism is a relatively recent phenomenon, starting with the rapid growth of freelancing since the early 1990s, culminating in today's celebration of "entrepreneurial" journalism.

Although professional innovation seems to be particularly challenging for those trying to make a living as journalists working without the benefits of steady employment, nothing is further from the truth. Faced with difficult and disruptive challenges on many fronts, the news business increasingly demands its workers to shoulder the responsibility of the company. Managers and employers increasingly stress the importance of "enterprise" as an individual rather than organizational or firm-based attribute (du Gay 1996; Witschge 2012). Trends such as the integration of the business and editorial sides of the news organization, the ongoing convergence of print, broadcasting, and online news divisions in to digital journalism enterprises, and the introduction of projectized work styles show that such hybridized working practices are not particular to freelance journalists (Deuze 2007).

Shifting the notion of enterprise—with its connotations of efficiency, productivity, empowerment, and autonomy—from the level of the company to the individual, it becomes part of the professional identity of each and every worker, contingently employed or not. This shift reconstitutes "workers as more adaptable, flexible, and willing to move between activities and assignments and to take responsibility for their own actions and their successes and failures" (Storey, Salaman, and Platman 2005, 1036). At the same time, studies among professional journalists report increasingly stressful workplaces, rates of burnout rising, and people (especially younger journalists) leaving legacy news rooms (Deuze 2014). In this enterprising economy, entrepreneurial journalists increasingly start their own companies—somewhat similar to their colleagues elsewhere in the creative sector starting boutique advertising agencies or independent record labels—forming editorial or reportorial collectives as well as business start-ups. The emergence of a start-up culture is global: since the early years of the twenty-first century, new independent (generally small-scale and online-only) journalism companies have formed around the world.

As with the long history of innovation in journalism, the emerging start-up culture can be plotted along a relatively straightforward timeline: starting with freelancing as a career choice for senior reporters and a common practice for experts in broadcasting and correspondents in magazine publishing, moving to freelancing as a mainstreamed practice, and subsequently leading to reporters increasingly setting up shop with colleagues in editorial collectives and news startups. Start-ups are financed in different ways, generally influenced by factors particular to the national context—for example the dominant presence of private funding agencies and venture capitalists in Asia and the US, or state subsidies for innovation in journalism in Europe. Or their start is attributable to distinct individuals—as many start-ups initially got off the ground with large financial injections of personal funds by their founders. More recently, crowdfunding and media-savvy marketing campaigns contributed to the rise of start-ups generally.

The literature generally celebrates this kind of entrepreneurialism, or treats it as a business case study (Naldi and Picard 2012; Bruno and Kleis Nielsen 2013). But beyond this deceptively straightforward timeline are interesting similarities across the various comparable national contexts—variables that stand out in explaining the choice for entrepreneurship, the role of the professionals involved, and the potential for success down the line. Comparing new online start-ups with the emergence of public affairs television as a distinct journalistic form in the mid-twentieth century, we find three key issues are at work: frustrated expectations of professional journalism among audiences and journalists alike, heightened competition between existing news industries, and a particular personality profile of the reporters and editors involved in new journalistic enterprises determining their visibility and (early) survival.

Much has been written about the waning of public trust in institutions generally, and in the press specifically—which can be seen as "suffering from a loss in public trust and confidence" (Witschge and Nygren 2009, 41). Global PR firm Edelman conducts annual surveys on trust and credibility among college-educated, middle-class, and media-savvy adults (the primary audience for professional journalism) in 18 countries. What the firm found over time is a gradual erosion of trust in governments, traditional institutions, and the media, in favor of nongovernmental organizations and peer people. Put in the context of what Ulrich Beck (2000, 150) has considered an increasingly antihierarchical age, in which traditional institutions (including the state, the Church, and the press) are "zombie containers" without meaning, people turn to each other rather than to established experts—parents, priests, professors, or presidents—for guidance. At the same time, the media landscape is fragmenting as people snack for news and entertainment from a digital smorgasbord rather than patiently consuming whatever a handful of mass media choose to dish up. In this context there is room for innovation for niche media, specialized and personalized media, and media that provide particular services to specific people. The fact that so many (online) start-ups got their start in such a short time in so many different markets worldwide is testament to this development.

It is not just public expectations that spur innovation across the industry—a shift toward independent work is also driven by frustration about the news industry among many journalists themselves. In the often overzealous conceptualization of journalism as a more or less consistent field (with a relatively homogeneous professional population), it becomes all too easy to forget that journalists, like their colleagues in other working environments in the creative industries, experience conflict and rivalry as intrinsic if not essential ingredients in the way they do their work. There is competition and conflict between employees and freelancers, among independent journalists, between reporters and editors, between television and print (online) divisions of the industry, and so on. Indeed, if anything, the conflicts within, between, and across media organizations can be better explained (and found) by looking at the contested relationships between creativity and creative control, rather than between creativity and the market (Lampel, Lant, and Shamsie 2000).

Competition at times fuels innovation, particularly in the context of a long-term process of disruptive and discontinuous change. Beyond this, it is noteworthy that the founding histories of many start-ups include narratives of frustrated newspaper editors striking out on their own, freelancers setting up their own shop with a critical eye toward the industry that used to employ them, and newcomers seeing more (creative and market) value in starting their own media platforms. Competition between legacy news organizations all experiencing aging and declining audiences in turn inspires innovation, either through the acquisition of external businesses or new managerial initiatives intended to create and shape a start-up culture in the newsroom. This can be seen as an extension of the need for all businesses to embrace a dual management process: to protect and enhance the existing way of doing things, as well as to experiment and explore new business models, new creative cycles or productivity routines, and so on. Under conditions of increasing competition both at home and abroad, online as well as offline, and the introduction of many new players in the media field (including journalist-hiring companies such as Google, Facebook, Apple, LinkedIn, Snapchat, and Twitter), entrepreneurial journalism gets expression both in new independent businesses as well as start-up units within existing industries.

The third variable of profound influence in the emergence and initial success of entrepreneurial ventures in journalism is that of the particular kind of people involved. The key professionals leading the movement, and giving entrepreneurship a voice, tend to be those enjoying a strong reputation in the field, having certain stand-out personality traits such as extraversion, imagination, and an openness to experience, often coupled with a charismatic authority. Another noticeable element that is generally found among successful start-up founders is their deference to the basic values of the occupational ideology of journalism when proposing and defending their initiatives: truth and objectivity (or the antithesis: providing a subjective voice), ethics, public service, breaking news (or its opposite: producing slow news), and autonomy.

A strong peer reputation tends to be derived from a previous career as an editor (such as Rob Wijnberg at the Dutch *Correspondent*, Edwy Plenel at the French *Mediapart*, Andrew Jaspan at the Australian *Conversation*), or senior investigative reporter (including Guia Baggi at Italian *IRPI*, John F. Harris and Jim VandeHei at American *Politico*, Juanita León at Colombian *La Silla Vacía*). These journalists earned respect and admiration (and also resentment and criticism) from their competitor-colleagues even before they ventured into the world of news start-ups. Their respective clout is derived from having proven themselves in professional terms as mutually recognized by their peers.

Personality traits are a significant part in the makeup of those who start or choose to work at a news start-up. Founders often tend to be quite outspoken professionals, passionately voicing their enthusiasm for the new business and, correspondingly, an often-scathing critique of the existing news industry. It is striking to see that such critiques are generally grounded in the most traditional, old-school values of the profession. Legacy news operations are attacked for not doing any "real" journalism anymore—as they have to consider the market and advertisers, are limiting the ability of their reporters to do their work autonomously, are overcommitted to breaking (and short-form) news while curtailing efforts toward investigative reporting, or are too close to their political sources. Listening to start-up founders, one is struck by a fascinating paradox: they proclaim to embrace and produce a "new" kind of journalism while referencing "old" values as the source of their insights and practices.

What stands out in the "small" history of entrepreneurship and innovation at the start of the twenty-first century must be the recognition of the emotional relationship people have with news, and, more importantly, the relationship that journalists have with (doing) journalism. As we noted when looking at the emergence of current affairs television, transformative practices in journalism go hand in hand with a deeply affective interpretation of what journalism is (or should be). In this context it is important to note the fact that journalism as a distinct form of affective labor has received scant attention in the literature, even though scholars of media work signal the pitfalls of emotional labor in other disciplines, such as the production of digital games and advertising. The benchmark study here is Andrew Ross's (2003) ethnography of the New York–based new media company Razorfish. The fact that the working environment looks and feels nothing like a typical office job contributes to people working incredibly long hours, invading and disrupting their nonwork lives. The strength of this account is that it foregrounds the participants' complex negotiations of how the meanings, values, and experiences of work and labor are changed and unsettled. Contemporary ethnographic work in newsrooms is rising, yet still pays little attention to the emotional and affective dimensions of newswork, and work on independent newsworkers, editorial collectives, and news start-ups remains scarce.

Discussion and Conclusion

The literature on innovation and transformation in journalism generally focuses on the disruptive role of technology, specifically the Internet, highlights the culture of legacy news organization responding to (and resisting) change (exemplary cases include Ryfe 2012; Usher 2014), or discusses new forms of journalism largely in terms of business models and opportunities. Based on our historical and contemporary fieldwork, an additional model for theorizing and studying journalism innovation should include the factors of public and journalistic dissatisfaction and unrest, competition fueled by a sense of urgency, and personality traits and the affective dimension of newswork. Underlying all these factors is the mobilizing power of the values of the profession.

When focusing on the role of technology, one is at risk of exaggerating the influence technology has on (journalistic) innovation and disruption, and missing out on the impact that emerging journalistic practices have on the development of innovative technologies. Technology plays a role in facilitating change, but on the whole we do not find convincing evidence to conclude it induces change. The Internet and digital technologies, for instance, have been around for quite some time, but the blossoming of the start-up culture and the various innovations—and sometimes radical transformations—visible among legacy media is of a much more recent date. On a similar note, the innovation of television journalism in the past was not caused by technological innovation; rather technological innovation was accelerated because there was a need to transform television journalism.

True, technology can have a disruptive effect, but there is a tendency to exaggerate its influence, including the tendency of technomyopia (i.e., as people tend to overestimate the short-term impact of technology but underestimate the long-term impact) in both business and academia. As a disruptive force, the Internet changed the manner in which the public consume media, forcing media to adapt. However, the main disruption of the current media landscape has been caused by the steep decline of newspaper advertising revenue, combined with dwindling readership—both trends that started well before the World Wide Web was introduced.

Equally important is the realization that innovation and disruption are not new. This may have been stated many times before, but it is still worth underlining. There is nothing necessarily new about innovation and disruption in journalism. The transformation of journalism through succeeding media—print, radio, television, Internet, mobile—is an ongoing story. And what are now deemed to be "old" legacy media were once "new" media, disrupting the media landscape at some point in time. Therefore, it is a fallacy to make a distinction between old and new media, and to focus on their differences. It is much more fruitful to look at patterns that run across different media and across different phases in the development of journalism, and at the factors which play a role in inducing change at distinct moments or in particular settings.

As we have suggested, similar issues seem to be at work when considering innovation and entrepreneurialism as structural conditions of newswork over time. The frustrated expectations of professional journalism among audiences and journalists alike should be taken into consideration. To what degree does public dissatisfaction with the content, style, tone, and approach of (mainstream) media play a role in the need and urge to innovate? How is that dissatisfaction noticed and noticeable? And which kind of initiative receives public and critical acclaim, thereby highlighting what public and critics alike find lacking in the content that is generally on offer? What, in turn, produces what Pablo Boczkowski and Eugenia Mitchelstein (2013) call the news gap between the media and the public across seven countries? These are questions that are well worth taking into account, since the frustrated expectations act as a push factor, driving the audience away from the media that do not meet their (changing) demands, while at the same time pulling them toward media and start-ups that do provide the required journalistic content. For the media themselves, these changing demands clearly provide an impetus to innovate and change.

Dissatisfaction and unrest among journalists are equally worth investigating. Change comes about when professionals think they can improve the way things are done, when they feel they can make a difference and have a stake in the process of transformation in professional journalism. In that sense, dissatisfaction with the current situation is a key prerequisite for innovation. Within an editorial setting this can be a productive energy, if there is room and a (strategic) need for change. But if there is insufficient autonomy to induce creative change within a medium, the drive to search for other, more independent ways to innovate journalism will increase. Therefore, dissatisfaction can also be seen as one of the motors for an entrepreneurial culture in which journalists strike out on their own.

Competition between existing news industries clearly is of importance when studying the innovation and transformation of journalism. Competition has always played a role in journalism, even in less competitive media landscapes with strong public broadcasting systems. In times of heightened competition, caused by economic factors or sometimes—in the case of public broadcasting organizations—by political change, there is an increased need for a strong journalistic profile in order to beat the competition and retain agency. Strategic choices tend to underlie the willingness to change and innovate. In that sense, innovation can be seen as a reaction to and the result of power struggles. It is crucial to take this broader media and power context into account when researching the transformation of journalism. Transformation within news industries does not happen by itself, but is always provoked by pressures both inside and outside of organizations. These factors influence the timing as well as the form of journalism innovation.

For innovation that takes place outside of the news industries, for instance within start-ups, competition is also a driving force. It is usually not competition on economic terms, since most startups have little economic clout. But there is a clear aim to compete in terms of being better at producing "real" journalism, and being more innovative and in tune with public needs than the major news industries and legacy media. In that sense, the competition is value based and affect driven.

The particular personality profile of the reporters and editors involved in innovating journalism, as well as starting new journalistic enterprises, is a third factor that is worth researching. Not only do they generally hold strong views on what "real" journalism should look like, referring to the basic values of the occupational ideology of journalism, they also tend to possess character traits which enable them to motivate and inspire the people around them. Their backgrounds can vary, but generally they have a strong peer reputation, which also helps inspire confidence among their staff, as well as among (certain segments of) the public, that the innovations they strive for strengthen and enhance the journalistic values.

What is striking, is that for a profession that has often been criticized for not being a true profession (Anderson 2014), journalism has a surprisingly strong occupational value system, based on the cornerstones of truth, objectivity, ethics, public service, and autonomy. This occupational ideology seems to be a crucial driving force behind all factors concerned with innovation and entrepreneurialism. It is used as a critical benchmark (also by the public), as a justification, as a protection, and as a way to create cohesion within a group—whether that group is the staff of a large national newspaper or a small-scale editorial collective of collaborating freelancers. One could argue that this occupational value system allows journalists to function in environments and workplaces that can be far from ideal. The consensual occupational self-image can also lead to a reluctance of journalists to change and adapt to new work realities, when they feel these undermine their professional values. In all instances, the occupational ideology of journalism shows itself as a important and influential force of both resistance and transformation. Any research into innovation, transformation, and entrepreneurialism should take the profound role of the occupational ideology into account.

At the same time, one has to be wary of taking the values at face value or assuming that they are set in stone. The interpretation of the occupational values changes over time, and is influenced by specific national (as well as organization-specific) journalistic cultures. The challenge is to investigate what journalists and the public really mean with such catch-all terms as objectivity, autonomy, fourth estate, and public service. What is their function, how are they interpreted, and how does that change over time, as well as within specific situations? Getting a grip on how the occupational ideology works should remain one of the major demands of journalism studies. In this sense, just as journalism faces issues today that are anything but new, the field of journalism studies would be wise to recognize its own legacy, thus keeping alive its own grand narratives—while opening these approaches up to the disruptive, transformative, and precarious nature of the profession as it operates today.

13 A Manifesto of Failure for Digital Journalism

Karin Wahl-Jorgensen

This chapter argues for the need to pay attention to failure in the study of digital journalism. The field of journalism studies has frequently focused on new technology over old; on success to the detriment of failure; on innovation over resistance to change; and on the cutting-edge over the conservative. Yet such an emphasis may not be consistent with understanding the plethora of actual practices, and may therefore constitute an epistemological blind spot. This is not to suggest that failure has been entirely ignored in journalism studies and related areas. Here, the chapter suggests, we can learn from the growing body of work on journalism practice in today's complex media ecology, which has traced difficulties in adapting to new realities. Further, we can enrich our methods for studying failure by drawing on the insights of other social science fields which have long taken it seriously, as an integral part of social life in general and organizational change in particular, including complexity theory and the sociology of scientific knowledge. On the basis of insights from work done within and outside the discipline, this chapter seeks to set out a research agenda based on taking failure seriously. Such a research agenda, the chapter argues, needs to pay attention to power relations within and between news organizations, and the ways in which particular—and often less privileged—forms of news practice might be more likely to fail. Inattention to marginal and unfashionable practices is part of a broader problem of neglecting failure, but one which it most urgent to address.

"Studying Up" and the Excitement of the New

Our understanding of practices of digital journalism should be viewed within the context of scholarly work in journalism studies, and the forms of knowledge it has produced. Journalism studies has tended to privilege "studying up" or engaging in "elite research" (Conti and O'Neil 2007), by paying a disproportionate amount of attention to elite and successful individuals, news organizations, and journalistic practices within them (see Wahl-Jorgensen 2009). This is, for example, in contrast to the discipline of anthropology, where the tradition of "studying down" (Nader 1969) and directing the scholarly gaze at "the most other of others" (Hannerz 1986) has received much critical attention. This critical engagement has occasioned a turn to a broader range of subjects and power relations among scholars using ethnographic methods, particularly influential in the context of the emergence of science and technology studies (see Stryker and Gon 2014).

The practice of studying up in journalism, while not uniform or uncontested, has profoundly shaped what types of professional practice are best documented and which are neglected—and therefore, on the knowledge produced within the discipline. Studies have tended to focus on work in large, elite, and often national television and newspaper newsrooms. The classic UK and US studies of journalism practices (e.g., Gans 1980; Tuchman 1978; Schlesinger 1978) took place at national or metropolitan broadcasters and newspapers. The emphasis on the routines, cultures, practices, and processes of elite, well-resourced national newsrooms might serve to ignore those of less glamorous, successful, and innovative journalistic workplaces which are nevertheless dominant in terms of both the number of newsworkers employed by such organizations, the quantity of content output, and the audiences for their output. The scholarly neglect of a majority of the occupation it proclaims to study is particularly problematic because the practices and resources of journalists, and their ability to adapt to changing circumstances, vary hugely depending on economic, political, technological, and social conditions. It means that we have gathered an impressive body of evidence about particular privileged forms of practice, while neglecting others (see also Wahl-Jorgensen 2009).

Excitement about New Technologies

The trend of studying up has taken a particular twist in the digital era: research on digital journalism often appears to be pervaded by an infatuation with the possibilities of technological change brought about by the convergent media ecology. In his foundational exploration of processes of online news innovation, Boczkowski (2004), writing more than a decade ago, pointed to "(1) the predominance of accounts that concentrate on the effects of technological change and pay much less attention to the processes generating them and (2) the pervasiveness of analyses that underscore the revolutionary characters of online technologies and the web and overlook the more evolutionary ways in which people often incorporate new artifacts into their lives" (Boczkowski 2004, 2). This long-standing set of preoccupations has translated, on the one hand, into research which focuses on the successful adaptation to change which often (but not always) takes place at resourceful and already-privileged news organizations. On the other hand, scholars have also turned their attention to the perceived empowerment of the audience in the face of a profound renegotiation of their relationship to news producers.

What unites this work, despite the diversity in the objects of study, theoretical frameworks, and methodological approaches, is an interest in the cutting-edge, successful, and ground-breaking. Implicit in this writing is a normative optimism regarding the future of journalism as an institution, as well as the forms of citizenship as facilitated in and through engagement with new technologies. It should be stressed that this orientation is not unique to the study of journalism, but has rather been a long-standing feature of writing about media. Writers have heralded the emergence of new technologies from the telegraph to Twitter, which have frequently been seen to offer solutions to social, economic, and political problems (Carey and Quirk 1970; Standage 1998). It is characteristic of a broader technological utopianism which has also characterized social scientific engagement with technological change in general (Kasson 1991), in areas ranging from education (Surrey and Farquhar 1997) to food production (Stock and Carolan 2013) and home improvement (Sivek 2011). Similarly, the emphasis on the success of relatively privileged and successful groups, organizations, and practices is reproduced across a variety of disciplines. The field of media studies has tended to focus on successful or "cult" media texts to the detriment of unsuccessful texts (Wahl-Jorgensen 2014). Work on social movements has tended to examine large and highly visible movements (such as the Occupy movement and the large-scale uprisings of the Arab Spring), with less attention given to local/small-scale/unsuccessful activisms (personal conversation, Lina Dencik, February 2014).

The work published in key journals devoted to work in digital journalism demonstrates these trends: most of this work is focused on technological innovations and how they are being adopted by innovative news organizations, journalists, and audiences. For example, Twitter—as a tool for both professional and amateur journalists—dominates contemporary scholarship about online journalism, with the most widely cited contributions examining journalists' use of the social media platform as a reporting tool (Vis 2013), the normalization of Twitter among journalists (Lasorsa et al. 2012), and its role as a form of ambient journalism (Hermida 2010).

Other emerging technologies that have been widely charted include live-blogging, mobile news, and videojournalism, as well as the place of a broader range of social media in publishing and sharing information for both citizens and journalists—including YouTube, Facebook, and Instagram. Such work views technological change as providing both a challenge and a set of resources for journalists and audiences. As such, it is more likely to offer an analysis of how to use them to *enhance* professional practice than thinking about the ways in which particular corners of the profession may be structurally inhibited from taking advantage of new opportunities, and how they may exacerbate already existing inequalities, to mention just a few research questions which have rarely been pursued.

Central to this body of work, and its preoccupation with new technologies, is an interest in the ability of these technologies to bring about a better future both for journalism and for its audience. For example, in the most-cited article published in *Journalism Practice*, Alfred Hermida (2010), while distancing himself from "hyperbole" surrounding Twitter, takes an optimistic view of the ways in which journalists and citizens can benefit from the new technology:

> I suggest that microblogging systems that enable millions of people to communicate instantly, share, and discuss events are an expression of collective intelligence. This paper examines microblogging as a new media technology that enables citizens to "obtain immediate access to information held by all or at least most, and in which each person can instantly add to that knowledge" (Sunstein 2006, 219). It argues that new parajournalism forms such as microblogging are "awareness systems," providing journalists with more complex ways of understanding and reporting on the subtleties of public communication. (Hermida 2010, 298)

Hermida's argument is emblematic of an engagement with emerging technologies which reads these technologies in the context of the new opportunities they genuinely provide. This is not problematic in and of itself. However, the sheer volume and dominance of work which takes an optimistic view of the potential of technological change may limit our ability to see underlying problems in several ways. First of all, such work frequently sidesteps the well-documented and profound crisis in the business model and long-term survival prospects for the profession. Second, it embodies the view that the history of technological change invariably involves a progression toward a brighter future for journalism. This suggests an adherence to a version of what James Carey (1974) has referred to as a Whig interpretation of journalism history (Butterfield 1965; Carey 1974). For Butterfield (1965), a Whig interpretation of history implied the production of knowledge about particular aspects of the past to emphasize a positive view or ratification of the present based on stories of success, and on that basis to project optimism about the future. Along those lines, Butterfield (1965, 1) pointed to the tendency "in many historians to write on the side of Protestants and Whigs, to praise revolutions provided they have been successful, to emphasize certain principles of progress in the past, and to produce a story which is the ratification if not the glorification of the present." What we might call the Whig interpretation of the history of digital journalism examines particular aspects of technological change to propose an exciting present, projecting it into a bright future, even if this vision often flies in the face of more troubling evidence. It does so through the same means as the Whig interpretation of history that Butterfield critiqued: the selective emphasis on studying particular objects and practices (successful ones) and the neglect of failure.

New Technologies and Empowerment? From "The People Formerly Known as the Audience" to Citizen Journalism

The Whig interpretation of digital journalism also underpins a key strain of work on the changing relationship between news organizations and their audiences. One striking feature of this body of work is that it has moved the lens of scholarly scrutiny away from the practices of elite news organizations. Instead, it has turned its attention toward the blurring boundaries between the production and consumption of journalism and its consequences for what is understood by journalism and its role in a democratic society. The turn to giving attention to the audience is in itself an important development. The audience has tended to be neglected in the conventional "newsroom-centric" research tradition (Madianou 2009; Wahl-Jorgensen 2009), focused on the practices of professional journalists. Perhaps the best-known articulation of this shift came from Jay Rosen (2006), the media critic and academic who coined the phrase, "The people formerly known as the audience." They are:

> … those who *were* on the receiving end of a media system that ran one way, in a broadcasting pattern, with high entry fees and a few firms competing to speak very loudly while the rest of the population listened in isolation from one another—and who *today* are not in a situation like that *at all*." (Rosen 2006)

Rosen goes on to describe the liberating possibilities afforded by the digital era, including the emergence of the blog, podcasting, video production, and audience agenda-setting. He concludes that a "highly centralized media system had connected people 'up' to big social agencies and centers of power but not 'across' to each other. Now the horizontal flow, citizen-to-citizen, is as real and consequential as the vertical one" (Rosen 2006). For Rosen and other observers excited about the transformative implications of new technologies, these changes did not merely challenge the dominance of mainstream media organizations but also activated and harnessed the creativity of what Jenkins and Deuze (2008) referred to as "grassroots media production" (see also Shirky 2008; 2010):

> The "People Formerly Known as the Audience," to borrow Jay Rosen's evocative phrase, are not remaining hidden in the traditional backwaters of grassroots media production (pirate radio, community television, newsletters); rather, their work becomes increasingly central to the contemporary mediascape with YouTube videos themselves developing cult followings, getting referenced through mainstream media, and provoking their own parodies and appropriations. Grassroots intermediaries are creating new value (and perhaps damaging older meanings) around branded content which they "spread"—legally and illegally—across the mediascape. (Jenkins and Deuze 2008, 9)

For Deuze (2008, 848), an emerging form of "liquid journalism truly works in the service of the network society, deeply respects the rights and privileges of each and every consumer-citizen to be a maker and user of his own news, and enthusiastically embraces its role as, to paraphrase James Carey, an amplifier of the conversation society has with itself." Deuze's position not only gives agency to contemporary "liquid journalism" but also ascribes it a positive normative role in enhancing democracy.

The celebration of the liberated "consumer-citizen" or "citizen-journalist" can be seen as part of a broader strain within media studies scholarship which has embraced (often rather more uncritically than journalism studies colleagues) the liberatory and creative potential of new technologies. This is exemplified by the work of scholar-advocates like Clay Shirky (e.g., 2008, 2010) and David Gauntlett (e.g., 2011, 2015), for whom the birth of social media and Web 2.0 heralds an unprecedented age of creativity. Shirky (2010) argued that the networked age enables twenty-first-century individuals to spend their leisure time more productively, creatively, and altruistically:

> This linking together in turn lets us tap our cognitive surplus, the trillion hours a year of free time the educated population of the planet has to spend doing things they care about. In the twentieth century, the bulk of that time was spent watching television, but our cognitive surplus is so enormous that diverting even a tiny fraction of time from consumption to participation can create enormous positive effects.

Shirky demonstrates the assumption that technological change has taken us from a dark age of passive consumption of television into a future of creative participation. As critics have pointed out, the empirical evidence of the transformative nature of this relationship has been somewhat more mixed, demonstrating ingrained resistance among journalists to the incursions of the audience, in the context of a sweeping range of changes and threats to their professional identities (e.g., Anderson 2011b, 532–533).

More recent work in journalism studies has focused on emerging amateur/citizen practices (e.g., Allan and Thorsen 2009; Allan 2013; Andén-Papadopoulos 2011). Such work recognizes the ways in which emerging technologies, in concert with the challenging business climate for legacy journalism, might undermine professional practice as the resources and the authority of legacy news organizations are dwindling. Yet it is also discerns emerging approaches to journalistic storytelling which might open up new forms of identification and solidarity (e.g., Wahl-Jorgensen forthcoming; Chouliaraki and Blaagaard 2013).

This work does not sit outside or neglect the current challenges to the profession. Rather, much of this research simply—and, many would argue, rightly—assumes that new technologies, like Twitter, are interesting in and of themselves. As such, this work cannot be seen as guilty of technological determinism—and indeed often demonstrates an understanding that the adoption, appropriation, and use of particular technologies are contingent on, and interact with, a broader array of political, economic, and social circumstances. It reminds us that it is far more useful to see technologies as possessing particular affordances—forms of action it makes possible—and to understand how these affordances might shape their use in interaction with particular sociocultural contexts (Papacharissi 2014). As Mitchelstein and Boczkowski (2009, 566) summarized the consensus, it has been to "to reject deterministic explanations and instead propose that technological innovations are mediated and shaped by initial conditions and contextual characteristics." Nonetheless, scholars representing this more measured approach have tended to strike a cautiously optimistic tone in analyzing how innovation enables new practices and forms of participation. Papacharissi exemplifies this tone in her compelling work, *Affective Publics* (2014). The book focuses on the role of social media—particularly Twitter—in shaping the affective news streams of social movements that both mobilize and organize the activities and discourses of these movements. Papacharissi (2014) concludes her book as follows:

> Structures of feeling open up and sustain discursive spaces where stories can be told. There are particular storytelling practices that become prevalent in the discursive spaces presented by convergent and spreadable media, and these practices invite certain varieties of engagement. Networked framing and networked gatekeeping

explain how interconnected people collaboratively curate and co-create narratives. (Papacharissi 2004, kindle location 2669)

The book breaks important ground in analyzing the place of affect in public discourse, demonstrating how new forms of storytelling emerge on the basis of the affordances of social media. Yet it is also symptomatic of the ways in which the focus on technological innovation through amateur practices shifts the attention away from a crisis in news organizations. The preponderance of these more optimistic readings—while both necessary and worthwhile—can make it difficult to see the ways in which journalism is *not* progressing and is, in fact, in a state of ever-deepening crisis amid signs of systemic failure.

The Whig interpretation of the history of digital journalism has tangible consequences for our knowledge of the profession. First of all, the celebration of audience empowerment and engagement neglects the continued importance of legacy media in terms of both the demographics and self-understanding of the profession, and the consumption behavior and preferences of audience members (Blumler and Cushion 2013).

The emphasis on new technologies and their successful adoption within resource-rich, elite, and empowered organizations, as well as in the newly empowered audience, makes sense given the political economy of and power relations within the academy: researchers may be more likely to gain cultural capital in the form of institutional approval and prestige, grant money, high-profile publications, and promotions from a study of the successful use of new technologies than from examining more marginalized and failing media practices which might also be more complicated to access and analyze. Yet such complication might be exactly what is necessary given, as Simon Cottle wrote back in 2000, that we cannot "presume a generalized view of 'journalism' as an undifferentiated culture or shared professional canon" (Cottle 2000, 24).

This is all the more important because, as a profession and a failure-prone system deeply embedded in social structures, journalism is perhaps more vulnerable and prone to failure than ever before; certainly no less so than other key institutions that have failed in high-profile and devastating ways in recent years, ranging from financial institutions to nuclear power plants. It is not only marginal and poorly resourced organizations and practices which fail—historical events from the sinking of the *Titanic* to the collapse of Lehman Brothers amply demonstrate this. However, the systematic neglect of such poorly resourced organizations and the ways they fail has led to an important gap in our knowledge of journalism. Our lack of attention to concrete manifestations of the diversity of practices constitutes an epistemological blind spot or a way in which the dominant paradigm of the discipline—one underpinned by an understandable excitement with the potential of new technologies—might have steered us down a blind alley in the landscape of knowledge production. This, in turn, hints at a broader need for a reflexive turn across humanities and social sciences to consider how our dominant paradigms, produced by the power relations of the institutions and societies in which we are so firmly embedded, might be rendering invisible the problems that should, in fact, be at the top of our agendas. In aid of such a reflexive turn, I would here make the plea for the importance of bringing the study of the varied facets of journalism's failure to the forefront. Here, we can learn from the growing body of work in the field which *has* taken seriously failure in journalism.

Taking Failure Seriously: Journalism Studies and Failed Practices and Institutions

Research on failure in journalism has usually looked at one of three things: (1) the quality of journalistic texts, (2) the business model, and finally, and relatively rarely, (3) failure understood through the lived experience of journalists. First of all, work examining failure with respect to the quality of journalistic texts has tended to be informed by concerns over the hegemonic nature of news production processes. It has understood failure in terms of journalists' shortcomings in terms of providing detailed, impartial, and accurate coverage of particular events or story types, such as the events unfolding around the 9/11 terror attacks, (Carey 2002), crisis and environmental journalism (Chase 1973), the justifications for the 2003 Iraq War (Kristensen and Ørsten 2007; Bennett et al. 2007; Bruns 2008), and the reporting of Hurricane Katrina (Bennett et al. 2007), to mention just a few examples.

Second, the failure of the business model—particularly in the context of the rise of digital journalism—has been studied in detail from an *institutional* and/or *economic* perspective. Research has demonstrated that "Survival Is Success" (e.g., Bruno and Kleis Nielsen 2012; see also Lee-Wright, Philips, and Witschge 2012), and showed that it is exceedingly difficult for online news operations to turn a profit in a challenging landscape of declining circulations, audience figures, and advertising revenue (see Mitchelstein and Boczkowski 2009, for an overview). Indeed, Bruns (2008) has explicitly ascribed the rise of citizen journalism to the failure of legacy media to adapt to the digital age: "[T]his shift is one of journalism's own making, as the industry's failure to update its products for a new,

Internet- and convergence-driven environment has alienated younger audiences" (Bruns 2008, 173).

The emphasis on "updating" is one of many metaphors for a central concern with what Anderson (2013) has referred to as the project of "rebuilding the news"—of adaptation to a changing and challenging news ecology (see also Anderson et al. 2012). Chadwick and Collister (2014) have suggested that "the most convincing stories about contemporary journalism are about institutional adaptation, even if they are the least told." Work on adaption is closely tied to an interest in innovation (e.g., Aitamurto and Lewis 2013; Grueskin et al. 2013; Steensen 2009).

These metaphors point to the fact that this area of research is frequently—whether explicitly or not—shaped by Schumpeter's "creative destruction" paradigm (Schlesinger and Doyle 2015; see also Bruno and Kleis Nielsen 2012), which provides a useful language for understanding how the institutional failures of journalism may give rise to rebirth or, in Schlesinger and Doyle's (2015) language, "recomposition." Schumpeter, in articulating the creative destruction paradigm, suggested that processes of renewal operate through destruction of "old" technologies and modes of production:

> [T]he same process of industrial mutation … that incessantly revolutionizes the economic structure from within, incessantly destroying the old one, incessantly creating a new one. This process of Creative Destruction is the essential fact about capitalism. It is what capitalism consists in and what every capitalist concern has got to live in. (1950, 83)

It is clear why the language of creative destruction is particularly helpful for making sense of the challenges facing journalism in the digital era, and it is indeed central to the discourse of many writers engaging with the crisis in the media industry (e.g., Bruno and Kleis Nielsen 2012). At the same time, because accounts of the crisis are often accompanied by the idea that it will result in productive renewal, such language may underwrite a continued Whig interpretation of journalism history which assumes that the future will always be brighter and better.

A third line of research moves closer to investigating the questions around the relationship between failure and the marginal which are at the heart of this chapter. This research focuses on the lived experience of journalists who may be struggling to adapt, who work for old-fashioned and, indeed, unfashionable, non-elite news organizations. This body of research remains indebted to a creative destruction paradigm, taking an interest in "the struggle of the media industry to reinvent itself in order to adapt to structural social changes and overcome the slow demise of its traditional business model" (Domingo 2011; see also Domingo and Paterson 2011; Paterson and Domingo 2008; Usher 2010). But it moves beyond "the tyranny of technology" to suggest that adaptation to change—while urgently necessary—can and will be painful, difficult, and uneven in its success, and that failure of various sorts is not just an option, but a likely outcome. This body of research often draws on ethnographic or interview methods to gain insights into the concrete manifestations and complexities in the experiences of change, adaptation, and failure (e.g., Anderson 2011a, 2013). Such work acknowledges the fact that in their everyday work, journalists "live out a tension between tradition and change" (Mitchelstein and Boczkowski 2009, 562). For example, Boczkowski's (2004) early work, *Digitizing the News*, examined how a variety of US news organizations adapted to processes of convergence. One of his case studies focused on HoustonChronicle.com's "Virtual Voyager" project, which sought to foster vicarious experiences among its users (Boczkowski 2004, 105). Boczkowki concluded that the "unfolding of Virtual Voyager exhibited a seemingly contradictory trajectory: successful with users and industry colleagues, it nonetheless resulted in a commercial failure. These were not contradictory outcomes, but rather the two sides of the same innovation coin" (2004, 137). Through careful ethnographic work and examinations of the information architecture of the site, Boczkowski foregrounded the ways in which this and other experiments with innovation were shaped by existing practices, but also had to negotiate complex terrains of audience expectations, needs and literacies, and commercial pressures.

Such work has highlighted the fact that adaptation to technological change does not occur in a vacuum but is rather strongly shaped by material circumstances, including the crisis in the business model of journalism. For example, Ekdale et al. (2014) carried out in-depth interviews with journalists at an independently owned media company in a mid-sized US city that had recently gone through a round of lay-offs. They found that the growing "culture of job insecurity" in the newsroom hindered adaptation to change "as those who fear their jobs are in danger are unlikely to risk altering well-understood practices, while many others who perceive job security would rather accommodate than initiate change." Similarly, David Ryfe's book, *Can Journalism Survive?* (2012), based on long-term ethnographic work at US regional newsrooms, presents a devastating picture of a profession in terminal decline, in part due to its inability to respond to the challenges of the Internet:

What did I find? The short answer is that journalists have not adapted very well. For the most part, they continue to gather the same sorts of information, from the same sorts of people, and package it in the same forms they have used for decades. Newspapers have the same look and feel they have had since the 1930s, and newspaper websites still look uncomfortably like newspapers. When journalists have tried to break from tradition, their efforts largely have come to naught. I know of no recent innovations in news that were invented in a metro daily newsroom, and no newsroom, to my knowledge, has adopted the new innovations in a comprehensive way. (2012, 3)

Ryfe's (2012) pessimistic assessment makes for oddly refreshing reading, not just because he draws on detailed ethnographic work, but also because he examines the nitty-gritty detail of the daily lives of, and the newsroom cultures inhabited by, journalists working for metropolitan dailies—newspapers which have long been such a central part of our news media landscape, but which are now dying a rapid death, the experience of which remains poorly understood.

At the same time, it is also important to pay attention to the failure of the analytical category of the newsroom in and of itself. Whereas traditional ethnographies in journalism have frequently been characterized by "newsroom-centricity" (Wahl-Jorgensen 2009), it is important to understand that news organizations can no longer be seen as isolated and singular institutions, but rather as part of a larger journalistic "ecosystem" (Anderson 2013). Further, news production can no longer be understood as concentrated in the material space of the newsroom and carried out by specialized professionals, but should rather be seen as a radically diverse, networked, and dispersed set of activities carried out by a broad range of organizations, groups, and individuals in many different places and on varied platforms. Usher (2015) studied the decline of the newsroom itself—as a space for newswork, but also as a material symbol of the centrality of journalism—as newspapers are increasingly moving from large, landmark, city-center buildings to smaller, anonymous, and often suburban spaces. She focused on how journalists have experienced the loss of the material space of the newspaper building:

These moves matter because they are testaments to the changing fortunes of the news industry. For journalists, what it means to leave behind tradition, move from downtown, and rethink space can have implications on what news they think they can cover. To a larger public, these moves might well be the first time that there has been a noticeable, public symbol that traditional journalism is in decline. (Usher 2015, 2)

Usher (2015) reminds us that the ways in which the profession responds to change must be mapped with an acute awareness of how the daily work of journalists, their routines, self-understandings, resources, and constraints vary, and shape both their failures and their successes.

Approaches to Understanding Failure

Here, we can draw on the work of sociologists, political scientists, and organizational theorists who have studied failure in complex systems, ranging from nation-states to nuclear power stations and banks (e.g., Malpas and Wickham 1995; Perrow 2011). Such research assumes that complex systems and organizations can and will fail on a regular basis. Clarke and Perrow's (1996) widely cited piece on "Prosaic Organizational Failure" opened as follows: "Organizations fail often and they fail in important ways. Police departments become corrupt. Banks invest unwisely. Schools do not educate. Investment houses make bad bets." They explain the importance of this agenda in suggesting that complex, "highly interactive systems increasingly insinuate themselves into society. The justifications that attend those systems often mask the failures we need to see more clearly" (Clarke and Perrow 1996, 1040–1055). Scholars who have studied failure—including in areas of market failure and political economy—have demonstrated that rather than assuming it is "an aberration; a temporary breakdown of the social system" (Malpas and Wickham 1995, 38), we need to take seriously and understand it as "an ubiquitous and central feature of social life" (Malpas and Wickham 1995, 37). Failure affects institutions of all types and sizes, and needs to be carefully studied through a variety of theoretical and methodological means developed for that purpose. Along those lines, the observations made by Little (2012) about political theorists' understanding of political failure could be similarly applied to work about journalism:

> [I]f forms of failure are likely outcomes from political practice in complex societies, how is this propensity for failure reflected in political theory itself? How do political theorists talk about failure? The short answer is that they rarely do and, when they do, it often involves rather simplistic, pejorative understandings of failure. (Little 2012, 6)

As this chapter has discussed, forms of failure in contemporary news organizations are complex and frequently overlapping, arising out of the nexus of challenges to conventional models of journalism in the context of networked society (Benkler 2006). They include the failure to thrive financially, attract audiences, implement technological innovation, provide "quality" journalism, and follow ethical standards. We need new tools and approaches to understand these complex forms of failure, but the development of these tools also requires a more fundamental perceptual shift: it requires us to "see" failure in the first place, and to view it as a viable and important object of study. As such, we can meaningfully analyze the woes of journalism through the lens of complexity theory—a framework often drawn on by social scientists seeking to understand failure. This framework recognizes that "dynamic, nonlinear systems, change, instability, and disequilibrium are the norm, not the stability and equilibrium assumed in traditional mechanistic models" (Sanderson 2009, 705; Urry 2005). Proponents of complexity theory suggest instead that "failures are seen as an emergent property of complexity" (Dekker et al. 2011, 939). As Dekker et al. (2011) conclude, to understand failure requires us to "gather multiple narratives from different perspectives inside of the complex system, which give partially overlapping and partially contradictory accounts of how emergent outcomes come about. The complexity perspective dispenses with the notion that there are easy answers to a complex systems event—supposedly within reach of the one with the best method or most objective investigative viewpoint. It allows us to invite more voices into the conversation, and to celebrate their diversity and contributions" (Dekker et al. 2011, 944).

This perspective provides helpful methodological guidance for journalism scholars in reminding us, first of all, that failure is inevitable and does not arise out of simple mistakes caused by the actions of individuals, but rather are the outcome of complex processes. What the chapter has highlighted is that we can only understand the complexities of both adaptation to technological change *and* failure if we systematically shift our scholarly gaze away from the elite news organizations that are the powerhouses of successful innovation, and toward the marginal and unfashionable, unfashionable and poorly resourced players in the media ecosystem which are nonetheless just as urgently seeking to respond to the profound changes impacting the industry. In his response to my chapter, Schudson suggested that so-called "ice cream gazettes"—poorly resourced local media outfits which may have limited and parochial news agendas—are not worthy of study because they do not provide the high-quality journalism that citizens need in a democratic society. However, I would argue that ice cream gazettes are, in fact, *more* representative of the material circumstances of the profession today, and the kinds of journalism that are possible as a result. There are many more ice cream gazettes than elite newspapers, they employ a far larger number of journalists, and are read by more people than the *New York Times*. They also represent the very real and embodied struggle for survival that is at the heart of the everyday experience of journalists, and without understanding their occasional successes and regular failures, we know little about what is going on within the profession we proclaim to study.

Conclusion: Failure is Inevitable

What, then, does the attention to failure and marginalized practices mean for our research? It includes taking a deliberate interest in failures, strugglers, and those who resist innovation and change in organizations, and understand *why* it is so difficult to change and what that has to do with the power relations that often remain invisible in scholarship. We need to recognize that failure is inevitable, and more likely than success. This is true for any complex social systems, but more so in the case of journalism. As we have seen, the emphasis on creative destruction shows that failure itself can be productive of success, but that it is frequently a prerequisite for renewal.

How, then, does such a preoccupation shape the kinds of questions we should ask and what kinds of methodologies we should deploy? First of all, it means that we should identify the news organizations that are struggling, and view them as worthy of study. Second, it means understanding the place of these news organizations within local and geographically bounded ecologies which are, at the same time, embedded in larger national and global contexts. This might entail the use of methods including those of political economy, network analysis, and historiography—methods providing us with a bigger picture that enables us to situate the practices of particular institutions within a larger system. Third, drawing on the insights of complexity theory, it means using ethnographic methods of mapping the lived experience of journalists and other newsworkers at all levels of the organization, paying attention to the concrete textures of individual practices and accounts, but also understanding how these are produced by power relations and institutional cultures.

Finally, attention to failure also means an increased reflexivity around power relations and political economy in shaping the object of our study as well as scholarly work itself. It entails a renewed commitment to the margins of media practice, and to seeing and understanding failure as well as success.

Commentary: The Journalism Studies Tree

Michael Schudson

Each of the papers I have been asked to discuss places on the agenda of journalism studies a different issue. All three issues bear witness to some common difficulties in thinking through the digital transformation in journalism. I will focus on the larger questions that these papers, each in a different way, urge us to confront.

Each paper is a plea for more attention in journalism studies to a particular problem. Karin Wahl-Jorgensen notes that research in this field is skewed strongly toward a small set of elite news organizations and that thousands of others have no resident PhD students knocking on their doors or gaining access for a newsroom study. Why should we neglect those that do not stand out? Why should we look at successes or apparent successes and ignore the fact that in this domain of commercial enterprise as in so many others, failure is much more the common experience than success—what sort of a field is this that ignores the common experience?

Sue Robinson sees a different gaping hole in the corpus of journalism studies: it omits or underplays the enduring, fundamental social inequalities that news production and news content do little more than reproduce. Her concern is that the connections of broad and deep social inequalities of race, class, and gender to news production get little attention from the studies that attract the greatest notice in our field. Wealthy white males are here, there, and everywhere—but why do we so often take them for granted or take their dominance as background, not foreground, to our studies? How can we—yet again—marginalize the experience of women, people of color, and people of modest means?

Mark Deuze and Mirjam Prenger worry that we pay too much attention to innovation in journalism today, and, worse, that we tie it too closely to technology as the chief driver of change. They urge a shift of attention and offer a nice example of nontechnological innovation—the creation in the

1950s and 1960s of magazine-style, hard-hitting "current affairs" television programming in the United States, Britain, and the Netherlands.

All three papers reflect on the shortcomings of the field of journalism studies. What is striking to me about this feature of the papers, and of the volume as a whole, is that they all take for granted that there *is* a field of journalism studies! Indeed there is, but to me, this is an entirely unexpected and still surprising development. It is as if some large, flowering tree arose in my front yard while I was taking a nap. Some of my former PhD students are leaders in constructing this field, maintaining an academic enterprise that I myself had not conceived or imagined. What is it doing in the middle of my front lawn?

Well, this is how intellectual life in universities grows and thrives. It is how critical conversations develop. It is also how discourses subdivide and then develop amnesia about adjacent bodies of work. And it calls for exactly the sort of critical second thoughts these papers provide.

The journalism studies tree in my yard is surprising but not unwelcome. Yes, in the multiplication of subfields, whether in communication or sociology or other disciplines, there is a threat to general intellectual cohesion. The danger of overspecialization is obvious, but so are the advantages of creating a critical and supportive intellectual subcommunity.

With that in mind, I would just note about all three papers that more connections to other relevant literature in other domains of communication studies or sociology or political theory or elsewhere might have been helpful. There are, after all, other studies of forms of failure—Diane Vaughan's study of organizational failure in *The Challenger Disaster*, or Charles Bosk's *Forgive and Remember: Managing Medical Failure*, two exemplary works of sociology. Vaughan examines how a kind of normalization of shortcomings becomes part of organizational routine with, in this case, disastrous consequences. Bosk shows how the everyday failures on a surgical ward are incorporated into medical routines. All organizations fail in various ways. Those that are spectacularly prominent may have the most spectacular failures—the *New York Times* failed to detect reporter Jayson Blair's deceptions in its own pages; the *Washington Post* and the *New York Times* both failed to blow the whistle on the dubious Bush administration claims that Iraq held weapons of mass destruction in the run-up to the Iraq War. These were gigantic failures because these news organizations have also produced huge achievements, have the highest aspirations for their work, and embody remarkable traditions of professionalism. How do we think about their sorts of failures and the failures of lesser organizations? Fourteen hundred other American newspapers also missed the story of the r by Donald Matheson, documenting the emergence in British n between 1880 and 1930 of a distinctive journalistic "voice"—as o the miscellany of voices that had dominated and for which the were not much more than scribes.[4]

ard in 2016 not to be overwhelmed by the technological advances elated "apps" and extensions that seem to pour out weekly in and ournalism, but these are by no means the only sites of innovation. t as the new hardware of the digital age is for transforming the nalists, professional and amateur alike, communicate, the new that makes use of the hardware and changes our language and our for gathering knowledge relevant to us is of mammoth impor- itself. The software is technology-related and technology-enabled but at heart it is literary or artistic or presentational innovation.

shake my head in wonder at this journalism studies tree in the s a fine growth, on balance, and offers shade for our reflections on life that has grown more complex, more disturbing, and more fas- than ever.

Hall, "Encoding/Decoding," in Stuart Hall, Dorothy Hobson, Andrew Love, Willis, eds., *Culture, Media, Language* (London: Hutchinson, 1980): 128–138.

ey Benson, "Institutional Forms of Media Ownership and Their Modes of in Martin Eide, Leif Ove Larsen, Helle Sjovaag, eds., *The Journalistic Institu- amined* (Intellect, forthcoming).

ael Schudson, "Question Authority: A History of the News Interview" in Schudson, *The Power of News* (Cambridge, MA: Harvard University Press, 2–93. (Originally published in *Media, Culture & Society* 16 [October, 1994]).

ld Matheson, "The Birth of News Discourse: Changes in News Language in Newspapers, 1880–1930," *Media, Culture & Society* 22 (2000): 557–573.

nonexistent weapons of mass destruction but no one ever expected them to get it.

On broad social inequalities and the role of the media in perpetuating them, the literature is almost endless, including all the many studies of racial and gender stereotyping in media industries and media products, from journalism to television entertainment, to video games, to film, and much more. There was no need for Sue Robinson to systematically review that literature. Still, I was surprised that she did not discuss at all one of the most productive areas of research over several decades, the work on audiences that grew out of British cultural studies. That body of work insisted that the communicative process in mass media includes a phase of "encoding" by producers and also a phase of "decoding" by audiences, as Stuart Hall insisted.[1] Audiences sometimes accept the "preferred reading" that the producers intended but quite often arrive at a "negotiated meaning" and sometimes "oppositional" readings as well. To neglect this is to arrive at an old and familiar trouble—if you assume that wealthy white males control society's show of meanings because they largely bankroll the show, you miss the plurality of meanings available in a liberal society and the plurality of interpretations that audiences engender and that human agency enables, and you have no way to understand how opinion ever changes. Without some attention to the multiplicity of meanings even a single text affords, we would have no explanation of the upheavals in eastern Europe after 1989 or the US civil rights movement or the women's movement and the extraordinary differences they have made in so many lives. Without recognition of multiple and conflicting views even among powerful white males, there would be little hope of understanding how cigarette smoking, once normative, came to be despised among educated people in industrial democracies around the world, or how homosexuality, once widely understood as immoral or unnatural, came to be accepted as one of the ordinary expressions of human sexuality. How do these things happen if the entrenched powers are so fully supported by the chief institutions of opinion formation?

Dominant meanings are in some measure derivative of who has the biggest Swiss bank account, who owns the most vacation homes, and who are the CEOs of the biggest corporations, but it is not a closed system. And some of the intervening institutions—like journalism—have in any case never fully compelled belief. The public never granted sovereignty to reporters and editors as democratic peoples do a government official or a public school teacher. Journalism cannot execute us or imprison us or even give us a failing grade. What we grant to the media is a provisional and partial control over defining everyday reality so long as they do so in collaboration with legitimate authorities, who have grown in the digital world where millions of Americans read non-US news sources and where many millions more take their cues on what to read not from one of three networks nor from the dominant daily newspaper in the metropolitan area where they live but from their friends on social media. The news media are well aware in their daily routines that they are as dependent on the opinions their audiences already hold as those audiences are dependent on the information and ideas the media purvey. What Rodney Benson calls "audience adjustment"—the phenomenon that news content adapts to consumers' established preferences—can be more important in explaining news content than the political views and predilections of media owners.[2]

This is not to discount Robinson's primary concern—that power matters and that a study of journalism in the digital age cannot leave the study of power to the side as if Facebook's easy accessibility erased all inequality. But just how do the power relations in and around journalism change with the digital transformation, if they do at all? The more precise question may be this: how does the collapse of the advertising-based business model, precipitated by the flow of advertisements and a variety of other personal and commercial economic transactions to online sales, alter the political and economic power of and within news organizations?

In Wahl-Jorgensen's focus on "success" and "failure," I wonder if these terms don't require more subtle definition in relation to the digital transformation of journalism. How does the meaning of "success" and "failure" change—if at all—in a digital era? What is failure in a small start-up that no one really expected to last compared to retrenchment in a major daily newspaper that no one ever expected to fail? How, if at all, does the digital transformation change the meaning of success and failure?

Wahl-Jorgenson's paper refers to my remarks at the Northwestern conference about the *Ice Cream Gazette*. Let me explain: in 2009 I drove across the United States from San Diego to New York. In every city or small town where I stopped along the way, I picked up the local newspaper. Someone of a different frame of mind might have found this exhilarating—after all, the first discovery here is that most every town had a local newspaper to sell to me. (On another car trip in 2014, I kept buying, but I found it a very dispiriting experience—it made me think that a dying newspaper business might not produce much of a gap in the educational and political experience of Americans.)

I remember particularly one Ohio news
July, it was hot, and this paper had a big fe
about why they liked ice cream, complete w
It turns out that the town's residents br
a refreshing treat on a hot day. That con
newsprint.

Okay, so it was a slow news day. Still, I
failure, a sad use of dead trees. I don't know
thought; maybe they judged the ice cream
tured about a dozen individuals who other
names in print. Community inclusion is one
sometimes seek to achieve and sometimes
do not think this brand of community bu
does much to ennoble journalism. And I ra
deserves from scholars the same kind of atte
organizations.

Failure, Wahl-Jorgensen tells us, is "ine
—"more likely than success." This may be tru
to justify studying "the failure, the underdogs,
online journalism." The *Ice Cream Gazettes* of
are typically small businesses dedicated to pr
small profit, and maybe contributing to the
Little League team. They will never unseat a lo
They are unlikely to innovate. They will rarely
or anything except maybe the local high schoo
work habits in an internship or after-school job

But are they failures? Perhaps not in their ov
own terms? Winning a prize? Growing circ
Producing investigative work that has a real p
Producing emotionally touching work that pr
refreshment from the ordinary course of the day

Some of the most important changes in jou
been closely tied to technological change. Deuze
television's magazine-style "current affairs" prog
petition, rivalry, and leadership. What it was, ess
of a new genre or style of presenting news that m
available technology. Another important exampl
of interviewing in the late nineteenth century as
tice, first in the United States and within a few de
parts of Europe, too.[3] Yet another development is

fine pap
journalis
opposed
reporters

It is h
or tech-
around
Importa
way jou
software
resource
tance in
change,

I stil
yard. It
a public
cinating

Notes

1. Stuar
and Pau

2. Rodn
Power,"
tion Ree

3. Mich
Michael
1995):

4. Don
British

Postscript: The Who, What, When, Where, Why, and How of Journalism and Journalism Studies

Michael X. Delli Carpini

I keep six honest-serving men
(They taught me all I knew)
Their names are What and Why and When
And How and Where and Who
—Rudyard Kipling (1902, 63)

That "journalism is in crisis" has become something of a mantra for many scholars, critics, and practitioners of the profession. Certainly a case could be made for this view, at least in the United States as it regards the "legacy" media of print and broadcast news. In the last three decades over 300 daily newspapers have closed, newspaper circulation has dropped by 35 percent, the number of professional journalists has declined by over 40 percent, and revenues are at the same level as 1950, when the population was half what it is today and the economy was one seventh its current size (Newspaper Association of America; American Society of News Editors). Over this period the trends for local and national television news are equally grim, with nightly viewership dropping by over 50 percent and the average age of viewers rising to over 60 (Potter 2008). And trust in the news media in general is at an all-time low (Gallup 2015).

At the same time, the very cultural, political, economic, and technological changes often blamed for journalism's impending demise have arguably led to a plethora of new sources of public information and analyses, beginning with cable news and talk and extending to online news, blogs and microblogs, citizen journalists (and random acts of citizen journalism), crowdsourcing, online access to international media, and popular culture genres such as satirical news. In addition, trends in legacy news found in the United States are countered by those in many other parts of the world, where news production and consumption is on the rise. In this light the crisis of journalism is perhaps better thought of as an historical moment in which journalism worldwide is in a state of major transition, the implications of which remain very much in doubt (Williams and Delli Carpini 2011; Chadwick 2013).

Understanding these changes and their implications is the motivating force behind this collection of essays and the conference from which they emerged. In this postscript I give my thoughts on what we learn from them. To do so I organize my comments around the classic "five Ws and the H" that were taught in many journalism courses and textbooks beginning in the early twentieth century, and that have come to represent what we mean by professional news reporting. In doing so I use this trope to explicate what these essays tell us about both the changing nature of our object of study (i.e., journalism), and the changes in journalism studies that are (or should be) occurring as a result.[1]

Who?

The "who" of journalism can refer to several different categories of people. For journalists themselves it refers mainly to the subjects and sources of their reporting, but as scholars of journalism we would add the roles of news producers and consumers. Who filled each of these roles was deceptively simple for most of the twentieth century. US news was produced by professional journalists working within established, easily identifiable, and almost exclusively privately and increasingly corporately-owned news organizations. As "a matter of journalistic convention" (Sigal 1986, 12), both the subjects and the sources of news consisted largely of "Knowns" (government officials and other elites, experts, and celebrities). "Unknowns," or ordinary people, were the subject of only one in five news stories, and rarer still the source of news (Gans 1980), relegated instead to the role of relatively passive news consumers. This media system or "regime" emerged from the political, economic, cultural, and technological environments of the time, and was maintained and naturalized by the norms, routines, and institutions of twentieth-century professional journalism (Williams and Delli Carpini 2011, chapters 2 and 3).

The essays in this volume point to a gradual unraveling of this system, and in doing so complicate both the practice and the study of these always oversimplified and socially constructed roles. Most obvious in this unraveling is the blurring of boundaries between news producers, consumers, sources, and subjects. As Boczkowski and Mitchelstein, quoting Picard (2014) note, "the traditional functions of bearing witness, holding to account opinion leadership, and shaming are no longer provided solely by the news media." They go on to identify a number of still sporadic innovations in newsroom processes and practices that expand the definition of sources to include social media, user-generated content, and the recycling of topics and stories culled from other online sources. Similarly, Ananny's examples of how contemporary journalists struggle with the pros and cons of filling the "absences" created by elite silences (in one case, by enlisting "followers" to become "a verification network"), with profound implications for whose voices are "heard, made actionable, and [gain] power." And as Wahl-Jorgensen observes (quoting from Jenkins and Deuze),

> The "People Formerly Known as the Audience," to borrow Jay Rosen's evocative phrase, are not remaining hidden in the traditional backwaters of grassroots media production (pirate radio, community television, newsletters); rather, their work becomes increasingly central to the contemporary mediascape with YouTube videos themselves developing cult followings, getting referenced through mainstream media, and provoking their own parodies and appropriations. Grassroots intermediaries are creating new value (and perhaps damaging older meanings) around branded content which they "spread"—legally and illegally—across the mediascape.

This relatively optimistic view is tempered, however, by Robinson's thoughtful call for scholars to carefully and systematically consider how these changes may be dependent upon, and thus reproduce, inequitable power relationships found in the offline world of both journalism and society writ large.

While the blurring of boundaries among news subjects, sources, producers, and consumers is arguably the most dramatic result of the changing character of the information environment, it has implications for the "who" of journalism even when these boundaries have remained largely intact. This is clear for professional journalists, as several of the essays in this volume attest. For example, Hindman argues that "new forms of data about the audience" require that journalists pay "strenuous attention to data on audience behavior." Both Singer, and Prenger and Deuze explore how notions of "entrepreneurship" are changing the ways in which journalists do their jobs. And Boczkowski and Mitchelstein note how "online news has increased the pressure on reporters to perform multiple tasks and combine several media formats."

To this I would add that the role of news consumer has changed dramatically as well. The explosion in the number and types of information choices makes the "job" of being an informed, engaged citizen more complex and time consuming, the results of which remain unclear. Will it lead to widening information gaps (Prior 2007) or to greater opportunities for less politically-motivated citizens to encounter useful information (Baum

2003)? Will it lead to the fragmentation of audiences into ideologically consonant "echo chambers" (Sunstein 2007; Jamieson and Cappella 2010) or to opportunities for creating new forms of community and "connected action" (Benkler 2006; Shirky 2008; Bennett and Segerberg 2012)? Will it lead to the loss of any meaningful distinction between fact and rumor (Thorson 2016) or to new possibilities for achieving collective wisdom (Noveck 2015)? These are unanswered questions with likely context dependent answers that we as scholars have a responsibility to answer.

Finally, the still evolving changes occurring in the contemporary media environment also suggest the need to expand the "who" of journalism studies. I mean this in several ways. Most obviously (and clear from the essays in this volume and the work they cite) is the need to include the many nonprofessional, semiprofessional, and independent journalists who now populate the newsmaking environment. But beyond this we also need to consider the many other actors that are increasingly involved in news and information production, from computer programmers to popular entertainers. As both Schudson and Dutton suggest in their respective commentaries, we also need to expand the "who" we draw on—both across disciplines and within the subdisciplines of communication and media studies—as we develop our theories, methods, and interpretations. And, as Stroud and Benson point out, we need to consider what our roles as scholars are, not simply as students of journalism, but as potential partners in the creation of useful, useable, and accessible public information.

What?

The "what" of journalism traditionally refers to "what happened"—the recounting of events that make up a news story. Despite journalists' commitments to notions of "objectivity," the socially constructed nature of this process has long been understood. As Carlin Romano wrote three decades ago:

> What does the press cover? The off-the-cuff answers come quickly. What it feels like covering. What sells papers. What the competition is covering. What it can get into the paper by 10 p.m. What it has always covered. A few more dignified answers come to mind. The news. Facts. What's important. (1986, 38)

The changes illuminated in this volume have direct implications for the "what" of traditional journalism, though whether the resulting information environment is an improvement or degradation remains uncertain. The changes to the practices of journalism, challenges to journalists' professional dynamics, and the reconfiguration of journalists' authority noted in different ways by Boczkowski and Mitchelstein, Annany, and Robinson clearly make Romano's quip that the press covers "what they can get into the paper by 10 p.m." obsolete, and more provocatively, perhaps his more serious claim that they cover "the facts" as well. The greater access and attention to, and use of, audience data and social media content noted by Hindman, Boczkowski and Mitchelstein, and Annany, coupled with the economic pressures highlighted by Singer, Prenger and Deuze, and Pickard also suggest the possibility of a shift in emphasis in what constitutes news. In Romano's language, this shift might signal an even greater move away from covering "what journalists feel like covering" (and perhaps "what is important") to "what sells." Less cynically, it could also signal a shift to "what people are interested in," and perhaps even to "what works" through "adherence to evidence-based practices likely to advance civic goals" (Hindman) and partnerships between journalists and academics (Stroud).

If we expand the "who" of journalism studies beyond the usual suspects (i.e., professional journalists) as I suggested earlier, the issue of the "what" of journalism becomes even more complicated, requiring a rethinking of what we even mean by "news." The essays by Singer, and by Prenger and Deuze suggest a possible move away from "what the competition is covering" and "what it always covered" to greater efforts to distinguish the form and substance of one "news product" from another through various forms of innovation. Here Nielson's essay is particularly instructive, suggesting that technological advances have the potential to change the status of news (provided by both traditional and new journalism entities) beyond Park's (1940) classic view that it "does not so much inform as orient the public, giving each and all notice as to what is going on." Drawing on the work of William James, Nielson argues that in the current information environment, we can no longer think of news as providing only one form of information, but three: "news-as-impressions," characterized by "decontextualized snippets of information presented via headline services, news alerts, live tickers, and a variety of new digital intermediaries including search engines, social media, and messaging apps"; "news-as-items," represented by the more traditional "self-contained, discrete articles and news stories bundled together in a newspaper, a broadcast stream, on a website, or in an app"; and "news-about-relations," constituting deeper knowledge emerging from the combination of "elements of long-form 'contextual' or 'explanatory' forms of journalism well known from some twentieth-century newspapers, magazines, and current affairs programs ... with new forms of data journalism, visualization, and interactivity enabled and empowered by digital technologies."

Lewis's and Zamith's use of Becker's (2008) concept of "art worlds" takes us yet a step further in this conceptual unpacking of the "what" of a more expansive, networked, fluid, and multiple notion of journalism(s). The distinct journalistic products produced by these overlapping "journalism worlds" (among them the hypothesized worlds of "ambient journalism," "data journalism," and "algorithmic journalism"),

> are the result of the combined labor of a large set of social actors and technological actants that is more heterogeneous than typically is acknowledged in the literature. ... [are] enabled by conventions, which both facilitate and constrain the creation of particular works ... [and] are constantly in flux, with the valuation of particular actors, works, and forms of labor differing between worlds, even as they contribute to the general understanding of what we call journalism.

While Lewis's and Zamith's use the notion of journalism worlds to expand and differentiate the distinct ways journalism and technology can intersect and coexist, their logic can be, I think, easily extended to additional worlds such as "citizen journalism," "satirical journalism," "alternative journalism," "local journalism," and "global journalism," each with their own overlapping and changing actors, conventions, valuations, and products.

Two other uses of the "what" of journalism seem appropriate as we consider the changes wrought by the contemporary information environment. First, and as noted to varying degrees in the essays by Boczkowski and Mitchelstein, Benson, Lewis and Zamith, Prenger and Deuze, and Wahl-Jorgensen, the convergence of both offline and online, and of text, images, and sound, create a different type of boundary blurring and as a result a different type of news product that have implications for how news is constructed and received (as one extreme example, consider the experimentation by the *New York Times* with the use of virtual reality).

Second, it is important that as scholars we keep in mind the "what" of our research includes not only an expanded notion of what constitutes news and those who produce and consume it, but the larger (and also changing and diverse) social, political, and institutional/organizational contexts in which news production and consumption occurs, a point foregrounded in the essays by Pickard, Robinson, Boczkowski and Mitchelstein, Anderson, Benson, Singer, Prenger and Deuze, and Wahl-Jorgensen, as well as in the commentaries by Schudson, Papacharissi, Dutton, and Neuman.

When?

As Michael Schudson (1986) has noted, the "when" of late-twentieth-century journalistic practice had several meanings, including what has happened in the past 24 hours, when is a journalist's deadline for submitting a story, are we ahead of our competition, and what of note happened on this date in the past. The first three of these four meanings have shifted in recent years (as they have in the past), the result of the various technological, economic, political, and cultural developments documented in this volume. The twice-a-day news cycle and the related notion of regular, rhythmic deadlines have given way to near instantaneous releasing of "breaking news" and constant updates. A "scoop" is measured in seconds, and often limited to the more mundane aspects of an evolving story. At the same time, control over the life cycle of specific stories has been wrested away from journalists. In short, "time" as it relates to journalism has both collapsed and expanded in ways less clearly under the control of journalists and the organizations for which they work.

The implications of this time shift are significant for both the "who" and the "what" of journalism discussed earlier. As Boczkowski and Mitchelstein note, information-gathering processes have changed, leading to less original reporting, "a culture of immediacy and speed," and an endangering of "fact checking and in-depth reporting." At the same time the simultaneous collapsing and expanding of time allows for an expansion of journalistic content that on the one hand merely provides an "acquaintance with" the topic at hand, while on the other can provide deeper "knowledge about" it than ever before (Nielson). It also expands the range of sources and subjects as journalists struggle to get information as quickly as possible and also fill the time and space available in moments when more obviously relevant or trustworthy information is absent (Annany). The greater ease with which information from the past can be accessed today also creates more opportunities for looking backward, sometimes for no other reason than to point out inconsistencies, but also to put current happenings in more nuanced, historical contexts.

For journalism scholars, however, "when" takes on a somewhat different meaning, as a number of the essays in this volume make clear. In the search for what is new about the current information environment, it can be easy to miss what is, in fact, the same. It is also easy to miss when what is different represents an evolution of, rather than a radical departure from, the past. Boczkowski and Mitchelstein draw on 25 years of scholarship on online journalism to put the current state of play in context, quoting

Cawley's (2012) argument that "the application of a strong historical perspective to scholarship on online news is necessary to gauge the depth of any changes from and the strength of continuity with print and broadcast news." Anderson states that "we need to see the newsroom as a space and an organizational assemblage that travels through time. This sense of the historical context is particularly important in the digital age, an age in which both scholars and everyday media users seem to be living within a culture of technology without a history." Benson observes that both actor–network theory and network society theory provide useful descriptions (though importantly, not generalizable explanations) of contemporary journalism in part by historicizing the present. Pickard looks to past efforts at subsidizing the media's public interest obligations as a way to reimagine what may be possible today. Stroud draws on John Dewey's 1892 experiment with *Thought News* to make the case for contemporary collaborations between academics and journalists. Hindman puts his call for the use of new and more detailed audience data to create a more citizen-centered notion of journalism ethics in historical context by noting that "print journalists have for centuries been essentially ignorant of the audience their specific stories receive." Singer sees the emergence of entrepreneurial journalists as being connected to issues of "autonomy or professional independence" that stretch back over 100 years. Nielsen discusses the potential of the digital media environment to expand the kind of knowledge produced by journalism by drawing on the work of the nineteenth-century philosopher and psychologist William James. Annany draws on examples as distant as the Sumerians to place the notion of "whitespaces" in context. Prenger and Deuze place current discussions of journalistic innovation in the context of the emergence of television in the 1950s and 1960s. Wahl-Jorgenson turns to past examples of failure and the lessons learned from them to bolster her call for more attention to failure in the digital age. And Dutton reminds us that "when" also requires looking into the future, but doing so in a way that avoids what he calls "innovation amnesia."

Where?

In traditional news stories the dateline serves the important function of informing the reader that the journalist reporting is located in the place from which the story emanated (or some other relevant site), thus adding an authoritativeness to what follows; a similar function is often served by the background shots in television news. "Where" also tells us what parts of the nation and the globe (according to journalists, news organizations, and elites at least) we should attend to.

As Dan Hallin has argued, however, "after the dateline, many stories drop the question 'where' altogether, or give it the most perfunctory answer" (1986a, p. 109). Unfortunately, while the contemporary information environment has a number of implications for these and other issues related to "where," a similar critique could be made of journalism scholars. While a handful of the essays in this volume make brief mention of journalism outside the US, mention US journalism's coverage of other parts of the world (most notably, Benson), cite non-US research, or make explicit appeals for more comparative research, by and large the discussions in these essays are geographically "decontextualized," at least as far as issues of international, comparative, and/or global differences are concerned.

This is unfortunate in that changes of the sort documented in these essays have important consequences for the "where" of US journalism, and the "where" of journalism can provide useful contexts for teasing out what we mean by capital "J" Journalism in the twentieth century. Most obviously, the declining resources and personnel of legacy news organizations in the United States has meant the closing of most foreign bureaus, in turn affecting how much, in what ways, and from where news from outside our borders is covered. This includes an acceleration of the long-standing limitations that come with "fly-in" news coverage (Hallin 1986a; 1986b), such as greater dependence on "official" sources, less understanding of subtle context, ignoring large portions of the world except for when natural disasters, wars, or revolutions occur, etc. But they also include an increasing dependence (on the part of journalists *and* news consumers) on nontraditional sources such as Twitter, Facebook, YouTube, and other social media reports provided by "ordinary people." Also important is the growing role of non-traditional news organizations, governments, private foundations, NGOs, and advocacy groups in filling the gap in foreign and global news coverage, either through direct dissemination via digital technologies, or by providing stories and/or funding to traditional news organizations.

Close analyses of the economics, institutional structures, journalistic norms and practices, news content, and consumers of non-US journalism can also be highly instructive regarding how different mixes of economics, politics, culture, and technology can produce different types of journalism and different types of larger information environments. In addition, globalization, coupled with new technologies, radically alters where news is produced and who consumes it, blurring national and regional borders of ownership and audiences in ways that demand theorizing and empirical research on this emerging "global journalism."

Just as there is much that we can learn from national, international, and global research and comparisons, so too can we learn from greater attention to the local, even the hyperlocal. This argument is implicit in several of the essays in this volume, particularly those using or calling for a greater use of comparative case studies and ethnographic methods in journalism research (e.g., Anderson, Benson, Dutton, Robinson, Singer, Prenger and Deuze, Wahl-Jorgensen). Local comparisons (both inside and outside the US) can provide some of the same research-design advantages as national ones, but also "control for" or hold constant some important macro contextual factors, allowing one to tease out the influences of other, more nuanced sources of variation (e.g., ownership, news room structures, use of technology, entrepreneurship, etc.).

Beyond these global, national, and local meanings, "where?" takes on additional importance for both journalists and journalism scholars in the contemporary information environment. Where are journalists working: On location or online? Within a traditional news organization, some other organization, or independently? (Boczkowski and Mitchelstein, Prenger and Deuze, Singer). Where do corporate owners (from traditional media organizations to Internet providers such as Google) and/or their targeted audiences reside, and how does it influence what they cover and how they cover it? Where do the subjects and sources of news come from, geographically but also in terms of race, ethnicity, gender, class, religion, and other groupings integrally related to issues of authority, power, and privilege (Robinson). Even, where do journalism scholars reside and, perhaps more importantly, conduct their research?

How and Why?

James Carey (1986) entitled his essay on the "Why and How" of news reporting "The Dark Continent of American Journalism." This has been less true, as reading his essay makes clear, for the "how" than the "why," but only if the former is defined narrowly and descriptively. Anything more in-depth than this, and "how merges into why: a description becomes an explanation" (149). Carey further notes that the thin version of "how" and the often missing "why" is the result of the usual limitations of time and space faced by reporters in their daily jobs. Ultimately he argues that "how and why" should be thought of as "properties of the whole, not the part"

(150), meaning of the entirety of coverage of an issue or event rather than a single story or report:

> If a story can be kept alive in the news long enough, it can be fleshed out and rounded off. Journalists devote much of their energy to precisely that: keeping significant events afloat long enough so that interpretation, explanation, and thick description can be added as part of the ongoing development. Alas management and the marketing department devote much of their energy to precisely the opposite—making each front page look like a new chapter in human history. (150–151)

Not surprisingly, there is much we can still learn from Carey's assessment, even though the contours of journalism have changed significantly over the past three decades. First and most consonant with his use of "how and why" is considering the changes that have occurred and how they might affect journalists' ability to answer these questions. As with all the issues addressed in this postscript (and the essays in this volume), speculative answers push in two directions. On the one hand, the increased demand on professional journalists' time created by larger news holes, increased responsibilities, and fewer resources and staff (e.g., Boczkowski and Mitchelstein) suggests that the "how and why" of a story are less likely than ever to be answered. On the other hand, online news essentially has no time or space limits, and new online data sources (and new tools for analyzing them) provide opportunities for journalists to explore the "how and why" either through original analyses or links to existing information. Add to this the explosion in alternative sources of news (entrepreneurial journalists, citizen journalists, random acts of journalism, direct accounts from participants, and so forth), the networked, interactive, and iterative nature of contemporary news construction, dissemination, and consumption, and the growth in often ideologically driven online and cable talk and commentary, and the opportunities for discussing and debating (if not determining) the "how and why" of events increases significantly. This expansion of "journalists" and "journalism" beyond the usual suspects also takes control of when a story begins and ends out of the hands of professionals and the organizations they work for, thus often extending the life cycle of stories and the time this allows for in-depth, often disparate, explanations. These competing pressures help explain Prenger's and Deuze's view that the new information environment creates new opportunities for journalists to produce information that provides both simple "acquaintance with' and deeper "knowledge about." Understanding this dynamic can also benefit from Lewis's and Zamith's conceptualization of news as emerging from different, networked, and fluid "journalism worlds."

Asking "how and why" is also relevant to the larger issues addressed in this volume. "How have economic, political, cultural, and technological changes affected the practice, products, status, influence, even definition of journalism?" "Why did these changes occur and why did journalism (and other actors) react as it did?" Here Benson's critique of the "'new descriptivist' media research" and call for more "systematic critique, patterned observation, and generalizable explanation" seems to parallel Carey's own regarding journalism, though we should keep Dutton's warnings about the risks of over- or premature generalizations foregrounded. In this regard we can learn from his observation that explanation (i.e., the "how and why") is a quality of the collective enterprise rather than an individual story; that it is a collaborative effort that occurs over time and through the sharing of information. The same could be said about journalism scholarship.

"How and why" are also questions we should ask ourselves as scholars, not only as it relates to our object of study, but regarding the work we do. How should we study contemporary journalism? The essays in this volume suggest a range of interdisciplinary theoretical and empirical approaches, including but not limited to critical discourse analysis, the study of counterpublics, field theory, network analysis, and actor–network theory (e.g., Robinson); ethnography (e.g., Anderson; Singer; Wahl-Jorgensen); historiography (e.g., Anderson; Pickard; Prenger and Deuze); political economy (e.g., Pickard; Singer; Wahl-Jorgensen); field experiments (e.g., Stroud); audience analysis (e.g., Hindman) and comparative case studies (Benson), including examples of failure (e.g., Wahl-Jorgensen). More generally, Boczkowski and Mitchelstein make a strong case (reinforced in Dutton's commentary) for "two-way" exchanges of theories and findings with other subdisciplines in our field, and other disciplines doing related work, some of which are "the very same disciplines from which [journalism studies] once emerged."

In the multimethod and multidisciplinary spirit of these recommendations, I would also suggest more openness to data and research that draws on both traditional (e.g., experiments, surveys, human-coded content analyses) and newer (e.g., computational science approaches to networks, "big data" analyses, computer-assisted content analysis) quantitative methods; methods that are downplayed in this volume. And of course while this ecumenical approach to theory and methods is undoubtedly the right way to go, it is incumbent on us to build bridges rather than walls across these various epistemologies and ontologies if we are to move, as Benson urges us, from description to explanation.

Finally, we should be asking ourselves "why" we study journalism, a question I turn to in my concluding thoughts.

Concluding Thoughts: What If?

There are many personal and professional answers to the question "'why' do we study journalism," but my answer emerges by putting the conclusions drawn by Boczkowski and Mitchelstein and Pickard in conversation with each other. As noted earlier, the former (along with the commentary by Schudson) makes a strong case that journalism scholarship would benefit from, and provide benefits through, conversations with other subdisciplines and disciplines. In his essay, Pickard notes critically that while "journalism scholarship often contains tacit normative implications … they are not a central concern."

Both these arguments (echoed in various ways by other authors in this volume) have great merit. Certainly there are aspects of journalism, especially in its expanded contemporary meaning, that have similarities to the objects of study in other areas of research: information science, organizational theory, public opinion research, network analysis, political economy, policy studies, etc. And certainly there are things to be learned and taught through interdisciplinary sharing and collaboration. But Pickard (and Dutton) remind us that we study journalism because we believe that it *matters*—to the lives of individuals in their private, professional, and civic lives, to the local, national, and global communities to which we belong, and ultimately to the functioning of democracy—in ways that are unique.

The assumption that journalism matters animates (implicitly and often explicitly) the questions addressed by the authors in this volume. Who should count as "a journalist" in today's digital, networked, and fluid information environment (Lewis and Zamith)? Whose voices and concerns were absent in the journalism of the twentieth century, and do the new contours that are emerging mitigate, exacerbate, or replicate these inequities (Robinson; Annany)? What journalistic values (e.g., "citizen participation, interorganizational collaboration, standards of digital evidence, ideas about the audience, and journalists understanding of 'the public'") are most appropriate for the digital age (Anderson) and which values from earlier eras need to be preserved? What conditions and contexts might allow for effective journalistic innovation and entrepreneurialism while avoiding the more exploitative and problematic aspects of this growing trend (Boczkowski and Mitchelstein; Prenger and Deuze; Singer)? When is it appropriate for the state to intervene on behalf of the public interest in order to address the market failures associated with corporate journalism (Pickard)? When does a focus on variation across different journalistic contexts and locales provide useful and usable information, and when does it inhibit our ability to make meaningful generalizations (Benson)? Where should we focus our attention so as to increase the chance of improving journalism in ways that benefit those on the margins (Robinson)? Where are innovations in the funding, practice, dissemination, or reception of journalism occurring, and what lessons can we learn from them (Singer; Prenger and Deuze; Wahl-Jorgensen)? How might academics partner with journalists and news organizations to identify ways to identify informative and "artful" ways of presenting public affairs content (Stroud), and assess the democratic effectiveness of what is produced, disseminated, received, and used by citizens (Benson)? How can we encourage a journalism capable of providing citizens with an appropriate mix of familiarity, facts, and deeper knowledge about important issues of the day (Nielsen)? Why is there a gap between the news preferences of consumers and those of editors and journalists and how can we address this in ways that are democratically useful (Hindman)? Why do journalism scholars attend to elite media at the expense of both journalism and citizens who are "at the margins" (Wahl-Jorgensen)?

These and the many other questions asked or implied by the authors in this volume require theory and empirical evidence to answer. As such, answering them would benefit from more dialogue and sharing across researchers, subdisciplines, and disciplines. But it is their normative importance that should be the foundation of the collective enterprise of *both* journalism and journalism studies. Our metric of success should include more than thick description, important as this is. It should include more than generalizable explanation, though Benson correctly notes the need for it. Ultimately it should include impact; evidence that what we learn can be used to improve the information environment of individuals and communities in democratically useful ways. This, I think, will require collaboration and dialogue with not only a wider and more diverse circle of scholars, but with journalists (professional or not), the wide array of other information providers that have emerged in recent decades, and the people for whom this information is intended to serve. To do this effectively, we need to add one more question to the classic who, what, when, where, why and how. Not a question to be used in the construction of individual news stories, but rather to serve as a reminder to journalists and journalism scholars about the reason for our work, and the opportunities presented in the current moment. That question is “what if”?

Note

1. My use of the “five Ws and the H” is also a modest homage to Karl Manoff and Michael Schudson’s influential 1986 collection of essays entitled *Reading the News*, which took a similar approach.

Contributors

Mike Ananny is Assistant Professor at Annenberg School for Communication and Journalism, University of Southern California.

Chris Anderson is Associate Professor of Media Culture at City University of New York.

Rodney Benson is Professor of Media, Culture and Communication at New York University.

Pablo J. Boczkowski is Professor of Communication Studies at Northwestern University.

Michael X. Delli Carpini is Professor and Walter H. Annenberg Dean at Annenberg School for Communication, University of Pennsylvania.

Mark Deuze is Professor of Media Studies at University of Amsterdam.

William H. Dutton is Quello Professor of Media and Information Policy, College of Communication Arts and Sciences at Michigan State University.

Matthew Hindman is Associate Professor in the School of Media and Public Affairs at George Washington University.

Seth C. Lewis is Shirley Papé Chair in Electronic Media in the School of Journalism and Communication at University of Oregon.

Eugenia Mitchelstein is Assistant Professor of Social Sciences at Universidad de San Andrés.

W. Russell Neuman is Professor of Media Technology at New York University.

Rasmus Kleis Nielsen is Director of Research at the Reuters Institute for the Study of Journalism, University of Oxford.

Zizi Papacharissi is Professor and Head of Communication at University of Illinois at Chicago.

Victor Pickard is Associate Professor at Annenberg School for Communication, University of Pennsylvania.

Mirjam Prenger is Assistant Professor of Media and Culture at University of Amsterdam.

Sue Robinson is Associate Professor of Journalism and Mass Communication at University of Wisconsin-Madison.

Michael Schudson is Professor of Journalism at Columbia University.

Jane B. Singer is departmental Director of Research and Professor of Journalism Innovation at City, University of London.

Natalie Jomini Stroud is Associate Professor of Communication Studies at University of Texas at Austin.

Karin Wahl-Jorgensen is Professor of Journalism, Media and Cultural Studies at Cardiff University.

Rodrigo Zamith is Assistant Professor of Journalism at University of Massachusetts at Amherst.

References

Aalberg, T., and James Curran, eds. 2011. *How Media Inform Democracy: A Comparative Approach*. London: Routledge.

Abbate, J. 1999. *Inventing the Internet*. Cambridge, MA: MIT Press.

Ahlers, D. 2006. "News Consumption and the New Electronic Media." *Harvard International Journal of Press/Politics* 11 (1): 29–52.

Alexander, J. C. 2011. *Performance and Power*. Cambridge: Polity.

Alexander, J. C. 2015. "The Crisis of Journalism Reconsidered: Cultural Power." *Fudan Journal of Humanities and Social Sciences* 8 (1): 9–31.

Ali, C. 2013. "Where Is Here? An Analysis of Localism in Media Policy in Three Western Democracies." Unpublished doctoral dissertation, University of Pennsylvania, Philadelphia, PA.

Allan, S. 2006. *Online News*. Maidenhead, UK: Open University Press.

Almiron, N. 2010. *Journalism in Crisis: Corporate Media and Financialization*. New York: Hampton Press.

Ananny, M. 2013a. "Breaking News Pragmatically: Some Reflections on Silence and Timing in Networked Journalism." Nieman Journalism Lab. http://www.niemanlab.org/2013/04/breaking-news-pragmatically-some-reflections-on-silence-and-timing-in-networked-journalism.

Ananny, M. 2013b. "Press-Public Collaboration as Infrastructure: Tracing News Organizations and Programming Publics in Application Programming Interfaces." *American Behavioral Scientist* 57 (5): 623–642.

Ananny, M. 2014. "Networked Press Freedom and Social Media: Tracing Historical and Contemporary Forces in Press-Public Relations." *Journal of Computer-Mediated Communication* 19 (4): 938–956.

Ananny, M., and L. Bighash. 2016. "Why Drop a Paywall? Mapping Industry Accounts of Online News Decommodification." *International Journal of Communication* 10: 3359–3380.

Ananny, M., and Kate Crawford. 2015. "A Liminal Press: Situating News App Designers within a Field of Networked News Production." *Digital Journalism* 3 (2): 192–208.

Anderson, A., Dominique Brossard, Dietram A. Scheufele, Michael A. Xenos, and Peter Ladwig. 2014. "The 'Nasty Effect': Online Incivility and Risk Perceptions of Emerging Technologies." *Journal of Computer-Mediated Communication* 19 (3): 373–387.

Anderson, C. W. 2008. "Journalism: Expertise, Authority, and Power in Democratic Life." In *The Media and Social Theory*, ed. Jason Toynbee and David Hesmondhalgh. New York: Routledge, 248–264.

Anderson, C. W. 2011. "Between Creative and Quantified Audiences: Web Metrics and Changing Patterns of Newswork in Local US Newsrooms." *Journalism* 12 (5): 550–566.

Anderson, C. W. 2011b. "Deliberative, Agonistic, and Algorithmic Audiences: Journalism's Vision of Its Public in an Age of Audience Transparency." *International Journal of Communication* 5:19.

Anderson, C. W. 2013. "Towards a Sociology of Computational and Algorithmic Journalism." *New Media & Society* 15 (7): 1005–1021.

Anderson, C. W. 2013. *Rebuilding the News: Metropolitan Journalism in the Digital Age*. Philadelphia: Temple University Press.

Anderson, C. W. 2014. "The Sociology of the Professions and the Problem of Journalism Education." *Radical Teacher* 99: 62–68.

Anderson, C. W. 2015. "Media Scholarship and Digital Time." Culture Digitally. Online at http://culturedigitally.org/2015/06/media-scholarship-and-digital-time-2/. Accessed July 15, 2015.

Anderson, C. W., and Daniel Kreiss. 2013. "Black Boxes as Capacities for and Constraints on Action." *Qualitative Sociology* 36:365–382.

Anderson, C. W., and Juliette De Maeyer. 2015. "Objects of Journalism and the News." *Journalism* 16 (1): 3–9.

Anderson, C. W., Emily Bell, and Clay Shirky. 2012. "Post-Industrial Journalism: Adapting to the Present." Report for the Tow Center for Digital Journalism. http://towcenter.org/research/post-industrial-journalism-adapting-to-the-present-2.

Anderson, Chris. 2008. "The End of Theory: The Data Deluge Makes the Scientific Method Obsolete." *Wired Magazine*, June 23. http://archive.wired.com/science/discoveries/magazine/16-07/pb_theory.

Anderson, D. A. 1975. "Libel and Press Self-Censorship." *Texas Law Review* 53:422–481.

Anderson, M., and Andrea Caumont. 2014. "How Social Media Is Reshaping News." Pew Research Center, September 24. http://www.pewresearch.org/fact-tank/2014/09/24/how-social-media-is-reshaping-news/.

Anthonissen, C. 2003. "The Sounds of Silence in the Media: Censorship and Self-Censorship." In *Handbook of Communication in the Public Sphere*, ed. R. Wodak and V. Koller. Paris: Mouton de Gruyter.

Armstrong, J. 2015. "Media Blackout: Would I be Happier If I Didn't Read the News?" *The Guardian*, March 14. http://www.theguardian.com/media/2015/mar/14/would-i-be-happier-if-i-didnt-read-the-news.

Ashuri, T. 2014. "When Online News Was New: Online Technology Use and Constitution of Structures in Journalism." *Journalism Studies* 17 (3): 301–318.

Asur, S., B. A. Huberman, G. Szabo, and C. Wang. 2011. *Trends in Social Media: Persistence and Decay*. Paper presented at the AAAI Conference on Weblogs and Social Media.

Aumente, J. 1987. *New Electronic Pathways: Videotex, Teletext, and Online Databases*. London: Sage.

Babineaux, Ryan, and John Krumboltz. 2013. *Fail Fast, Fail Often: How Losing Can Help You Win*. New York: Penguin.

Bagdikian, Ben H. 2000. *The Media Monopoly*. 6th ed. Boston: Beacon Press.

Baines, David, and Clara Kennedy. 2010. "An Education for Independence? Should Entrepreneurial Skills be an Essential Part of the Journalist's Toolbox?" *Journalism Practice* 4 (1): 97–113.

Balaji, M. 2011. "Racializing Pity: The Haiti Earthquake and the Plight of 'Others.'" *Critical Studies in Media Communication* 28: 50–67.

Ball, J. 2014. "EU's Right to Be Forgotten: Guardian articles Have Been Hidden by Google." *The Guardian*, July 2. http://www.theguardian.com/commentisfree/2014/jul/02/eu-right-to-be-forgotten-guardian-google.

Baker, C. E. 1989. *Human Liberty and Freedom of Speech*. Oxford: Oxford University Press.

Baker, C. Edwin. 2002. *Media, Markets, and Democracy*. New York: Cambridge University Press.

Bakker, P. 2014. "Mr. Gates Returns: Curation, Community Management and Other New Roles for Journalists." *Journalism Studies* 15 (5): 596–606.

Baruah, B., and Anthony Ward. 2015. "Metamorphosis of Intrapreneurship as an Effective Organizational Strategy." *International Entrepreneurship and Management Journal* 11 (4): 811–822.

Bardoel, J., and Mark Deuze. 2001. "Network Journalism: Converging Competences of Old and New Media Professionals." *Australian Journalism Review* 23 (2): 91–103.

Barnhurst, K. G. 2010. "The Form of Reports on US Newspaper Internet Sites, an Update." *Journalism Studies* 11 (4): 555–566.

Barnhurst, K. G. 2012. "The Content of Online News in the Mainstream US Press, 2001–2010." In *Communication @ the Center*, ed. Steve Jones, 231–253. New York: Hampton Press.

Barnhurst, K.G., and Diana Mutz. 1997. "American Journalism and the Decline in Event-Centered Reporting." *Journal of Communication* 47 (4): 27–5.

Barnouw, E. 1968. *The Golden Web: A History of Broadcasting in the United States: Vol. 2, 1933 to 1953*. New York: Oxford University Press.

Baum, M. 2003. *Soft News Goes to War: Public Opinion and American Foreign Policy in the New Media Age*. Princeton, NJ: Princeton University Press.

Baumer, E. P. S., P. Adams, V. D. Khovanskaya, T. C. Liao, M. E. Smith, V. S. Sosik, and K. Williams. 2013. *Limiting, Leaving, and (Re)lapsing: An Exploration of Facebook Non-use Practices and Experiences*. Paper presented at the SIGCHI Conference on Human Factors in Computing Systems.

Bayard, S. 2009. "News Websites in Texas and Kentucky Invoke Shield Laws for Online Commenters." Digital Media Law Project. http://www.dmlp.org/blog/2009/news-websites-texas-and-kentucky-invoke-shield-laws-online-commenters.

Beaujon, A., and J. Moos. 2013. "*New York Times, Wall Street Journal* Drop Paywalls for Storm Coverage." Poynter. http://www.poynter.org/news/mediawire/203601/new-york-times-to-drop-paywall-for-storm-coverage.

Beam, R. 2006."Organizational Goals and Priorities and the Job Satisfaction of U.S. Journalists." *Journalism & Mass Communication Quarterly* 83 (1): 169–185.

Beam, R., B. J. Brownlee, D. H. Weaver, and D. T. Di Cicco. 2009. "Journalism and Public Service in Troubled Times." *Journalism Studies* 10 (6): 734–753.

Beck, U. 2000. *The Brave New World of Work*. Cambridge: Polity.

Becker, H. 2008. *Art Worlds*. Berkeley: University of California Press. (1982).

Becker, H. 2008. "Epilogue to the 25th Anniversary Edition: A Dialogue on the Ideas of 'World' and 'Field,' with Alain Pessin." In *Art Worlds*, 372–386. Berkeley, CA: University of California Press.

Beckett, C. 2015. "How Journalism Is Turning Emotional and What That Means for the Future of News." On Medium. https://medium.com/@CharlieBeckett/this-blog-is-based-on-a-talk-charlie-beckett-gave-at-the-2015-british-science-festival-in-bradford-7236691301ab.

Beckett, C., and R. Mansell. 2008. "Crossing Boundaries: New Media and Networked Journalism." *Communication, Culture & Critique* 1:92–104.

Belair-Gagnon, Valerie. 2015. *Social Media at BBC News: The Re-Making of Crisis Reporting*. London: Routledge.

Bell, Emily. 2014. "Journalism Startups Aren't a Revolution If They're Filled with All These White Men." *The Guardian*, March 12. http://www.theguardian.com/commentisfree/2014/mar/12/journalism-startups-diversity-ezra-klein-nate-silver.

Bell, Emily. 2014. "Silicon Valley and Journalism: Make up or Break Up?" Presented at the Reuters Memorial Lecture 2015, Reuters Institute for the Study of Journalism. http://reutersinstitute.politics.ox.ac.uk/events/silicon-valley-and-journalism-make-or-break.

Benkler, Y. 2006. *The Wealth of Networks: How Social Production Transforms Markets and Freedom*. New Haven: Yale University Press.

Bennett, S. 2015. 2015. "What Are the Best Times to Post on #Facebook, #Twitter and #Instagram?" *Adweek*, January 6. http://www.adweek.com/socialtimes/best-time-to-post-social-media/504222.

Bennett, W. L. 2005. "Social Movements beyond Borders: Understanding Two Eras of Transnational Activism." In *Transnational Protest and Global Activism*, ed. D. Della Porta and S. Tarrow, 203–226. Lanham, MD: Rowman & Littlefield Publishers, Inc.

Bennett, W. L. 2011. *News: The Politics of Illusion*. 9th ed. Chicago: University of Chicago Press.

Bennett, W. Lance, and Shanto Iyengar. 2008. "A New Era of Minimal Effects? The Changing Foundations of Political Communication." *Journal of Communication* 58 (4): 707–731.

Bennett, W. Lance, and Alexandra Segerberg. 2012. "The Logic of Connective Action: Digital Media and the Personalization of Contentious Politics." *Information, Communication & Society* 15 (5).

Benson, R. 2004. "Bringing the Sociology of Media Back In." *Political Communication* 21 (3): 275–292.

Benson, R. 2010. "Futures of the News: International Considerations and Other Reflections." In *New Media, Old News*, ed. N. Fenton, 187–200. London: Sage.

Benson, R. 2013a. *Shaping Immigration News: A French-American Comparison*. Cambridge: Cambridge University Press.

Benson, R. 2013b. "On the Explanatory and Political Uses of Journalism History." *American Journalism* 30 (1): 4–14.

Benson, R. 2014. "Strategy Follows Structure: A Media Sociology Manifesto." In *Media Sociology: A Reappraisal*, ed. S. Waisbord, 25–45. Cambridge: Polity.

Benson, R. 2014. "Challenging the 'New Descriptivism': Restoring Explanation, Evaluation, and Theoretical Dialogue to Communication Research." Presentation at Qualitative Political Communication Pre-Conference, International Communication Association, Seattle, May 22.

Benson, R., and Erik Neveu, eds. 2005. *Bourdieu and the Journalistic Field*. Malden, MA: Polity.

Benson, R., and Matthew Powers. 2011. *Public Media and Political Independence*. Washington, DC: Free Press.

Bercovici, J. 2014. "The *Playboy* Interview: A Candid Conversation with Gawker's Nick Denton." *Playboy*, February 2. http://playboysfw.kinja.com/the-playboy-interview-a-candid-conversation-with-gawke-1527302145.

Berger, P., and Thomas Luckmann. 1966. *The Social Construction of Reality: A Treatise in the Sociology of Knowledge*. New York: Vintage Books.

Berry, D. 2012. *Journalism, Ethics and Society*. Farnham, UK: Ashgate Publishing, Ltd.

Bertrand, C. J. 2000. *Media Ethics & Accountability Systems*. Piscataway, NJ: Transaction Publishers.

Best, Joel, ed. 1989. *Images of Issue: Typifying Contemporary Social Problems*. New York: Aldine de Gruyter.

Bhattacharya, D., and S. Ram. 2012. *Sharing News Articles Using 140 Characters: A Diffusion Analysis on Twitter*. Paper presented at the IEEE Conference on Applications of Social Network Analysis.

Bickford, S. 1996. *The Dissonance of Democracy: Listening, Conflict, and Citizenship*. Ithaca, NY: Cornell University Press.

Bijsterveld, K. 2008. *Mechanical Sound: Technology, Culture, and Public Problems of Noise in the Twentieth Century*. Cambridge, MA: MIT Press.

Black, J., B. Steele, and R. D. Barney. 1999. *Doing Ethics in Journalism*. Upper Saddle River, NJ: Prentice Hall.

Bliss Jr., Edward. 1991. *Now the News: The Story of Broadcast Journalism*. New York: Columbia University Press.

Block, Jörn H., and Andreas Landgraf. 2016. "Transition from Part-time Entrepreneurship to Full-time Entrepreneurship: The Role of Financial and Non-financial Motives." *International Entrepreneurship and Management Journal* 12 (1): 259–282.

Blom, Robin, and Lucinda D. Davenport. 2012. "Searching for the Core of Journalism Education: Program Directors Disagree on Curriculum Priorities." *Journalism & Mass Communication Educator* 67 (1): 70–86.

Blumer, Herbert. 1986. *Symbolic Interactionism: Perspective and Method*. Berkeley: University of California Press.

Blumler, Jay G. 1980. "Mass Communication Research in Europe: Some Origins and Prospects." *Media Culture & Society* 2:367–376.

Blumler, Jay G., and Stephen Cushion. 2014. "Normative Perspectives on Journalism Studies: Stock-taking and Future Directions." *Journalism* 15 (3): 259–272.

Boczkowski, P. 2004. *Digitizing the News: Innovation in Online Newspapers*. Cambridge, MA: MIT Press.

Boczkowski, P. 2010. *News at Work: Imitation in an Age of Information Abundance*. Chicago, IL: University of Chicago Press.

Boczkowski, P. 2014. "The Material Turn in the Study of Journalism: Some Hopeful and Cautionary Remarks from an Early Explorer." *Journalism* 16 (1): 65–68.

Boczkowski, P. J., and M. de Santos. 2007. "When More Media Equals Less News: Patterns of Content Homogenization in Argentina's Leading Print and Online Newspapers." *Political Communication* 24 (2): 167–180.

Boczkowski, P. J., and J. A. Ferris. 2005. "Multiple Media, Convergent Processes, and Divergent Products: Organizational Innovation in Digital Media Production at a European Firm." *Annals of the American Academy of Political and Social Science* 597 (1): 32–47.

Boczkowski, Pablo J., and Eugenia Mitchelstein. 2010. "Is There a Gap between the News Choices of Journalists and Consumers? A Relational and Dynamic Approach." *International Journal of Press/Politics* 15 (4): 420–440.

Boczkowski, P. J., and E. Mitchelstein. 2013. *The News Gap: When the Information Preferences of the Media and the Public Diverge*. Cambridge, MA: MIT Press.

Bonilla-Silva, E. 2006. *Racism without Racists: Color-Blind Racism and the Persistence of Racial Inequality in America*. Lanham, MD: Rowman & Littlefield.

Born, G. 2005. *Uncertain Vision: Birt, Dyke and the Reinvention of the BBC*. London: Random House UK.

Bourdieu, Pierre. 1993. *The Field of Cultural Production*. New York: Columbia University Press.

Bourdieu, P. 1999. *On Television*. New York: New Press.

Bourdieu, Pierre. 2005. "The Political Field, the Social Science Field, and the Journalistic Field." In *Bourdieu and the Journalistic Field*, ed. Rodney Benson and Erik Neveu, 29–47. Malden, MA: Polity.

Bourdieu, Pierre, and Loic Wacquant. 1992. *An Invitation to Reflexive Sociology*. Chicago: University of Chicago Press.

Boyd-Barrett, O. 2004. "Judith Miller, *The New York Times*, and the Propaganda Model." *Journalism Studies* 5 (4): 435–449.

Boyer, D. 2011. "News Agency and News Mediation in the Digital Era." *Social Anthropology* 19 (1): 6–22.

Boyles, J., 2014. "When the Newsprint Fades: How the Media Ecology of New Orleans Produces News Knowledge." Dissertation, American University.

Boyles, Jan Lauren. 2016. "The Isolation of Innovation: Restructuring the Digital Newsroom through Intrapreneurship." *Digital Journalism* 4 (2): 229–246.

Brandtzaeg, P. B., M. Lüders, J. Spangenberg, L. Rath-Wiggins, and A. Følstad. 2015. "Emerging Journalistic Verification Practices Concerning Social Media." *Journalism Practice* 10 (3): 323–342.

Braun, Joshua A. 2013. "Going Over the Top: Online Television Distribution as SocioTechnical System." *Communication, Culture & Critique* 6:432–458.

Braun, Joshua A. 2015. "News Programs: Designing MSNBC.com's Online Interfaces." *Journalism* 16 (1): 27–43.

Braun, Joshua A. 2015. *This Program Is Brought to You by…: Distributing Television News Online*. New Haven, CT: Yale University Press.

Briggs, Mark. 2012. *Entrepreneurial Journalism: How To Build What's Next for News*. Washington, DC: CQ Press.

Bro, P., and F. Wallberg. 2014. "Digital Gatekeeping: News Media versus Social Media." *Digital Journalism* 2 (3): 446–454.

Brown, W. 1998. "Freedom's Silences." In *Censorship and Silencing: Practices of Cultural Regulation*, ed. R. C. Post, 313–327. Los Angeles: Getty.

Brubaker, J. R., M. Ananny, and K. Crawford. 2014. "Departing Glances: A Sociotechnical Account of 'Leaving' Grindr." *New Media & Society* 18 (3): 373–390.

Bruno, N., and R. Kleis Nielsen. 2012. *Survival Is Success: Journalistic Online Start-Ups in Western Europe*. Reuters Institute for the Study of Journalism.

Bruns, A. 2011a. "Gatekeeping, Gatewatching, Real-time Feedback: New Challenges for Journalism." *Brazilian Journalism Research Journal* 7 (2): 117–136.

Bull, M. 2012. "The Audio-visual iPod." In *The Sound Studies Reader*, ed. J. Sterne, 197–208. New York: Routledge.

Burt, R. S. 1992. *Structural Holes: The Social Structure of Competition*. Cambridge, MA: Harvard University Press.

Burt, R. 1998. "(Un)censoring in Detail: The Fetish of Censorship in the Early Modern Past and the Postmodern Present." In *Censorship and Silencing: Practices of Cultural Regulation*, ed. R. C. Post, 17–42.

Burt, R. S. 2002. "Bridge Decay." *Social Networks* 24 (4): 333–363.

Burt, R. S., M. Kilduff, and S. Tasselli. 2013. "Social Network Analysis: Foundations and Frontiers on Advantage." *Annual Review of Psychology* 64:527–547.

Bustamante, E. 2004. "Cultural Industries in the Digital Age: Some Provisional Conclusions." *Media Culture & Society* 26 (6): 803–820.

Cage, J. 1961. *Silence: Lectures and writings*. Middletown, CT: Wesleyan.

Cain, S. 2012. *Quiet: The Power of Introverts in a World That Can't Stop Talking*. New York: Random House/Broadway Paperbacks.

Callon, Michel. 1986. "Some Elements of a Sociology of Translation: Domestication of Scallops and the Fishermen of St. Brieuc Bay." In *Power, Action and Belief: A New Sociology of Knowledge?* ed. J. Law, 196–233. London: Routledge.

Cappella, Joseph N., and Kathleen Hall Jamieson. 1997. *Spiral of Cynicism: The Press and the Public Good*. New York: Oxford University Press.

Carey, James. 1986. "Why and How? The Dark Continent of American Journalism." In *Reading the News*, ed. Robert Karl Manoff and Michael Schudson, 146–196. New York: Pantheon Press.

Carey, J. W. 1989. "A Cultural Approach to Communication." In *Communication as Culture: Essays on Media and Society*, 13–36. New York: Routledge.

Carey, J. 1989. "Technology and Ideology; The Case of the Telegraph." In *Communication as Culture: Essays on Media and Society*. London: Routledge.

Carlson, M. 2007. "Blogs and Journalistic Authority." *Journalism Studies* 8 (2): 264–279.

Carlson, Matt. 2009. "Media Criticism as Competitive Discourse: Defining Reportage of the Abu Ghraib Scandal." *Journal of Communication Inquiry* 33 (3): 258–277.

Carlson, M. 2011. *On the Condition of Anonymity: Unnamed Sources and the Battle forJournalism*. Chicago: University of Illinois Press.

Carlson, Matt. 2015. "The Robotic Reporter: Automated Journalism and the Redefinition of Labor, Compositional Forms, and Journalistic Authority." *Digital Journalism* 3 (3): 416–431.

Carlson, Matt. 2015. "When News Sites Go Native: Redefining the Advertising-Editorial Divide in Response to Native Advertising." *Journalism* 16 (7): 849–865.

Carlson, M. 2015. "Introduction: The Many Boundaries of Journalism." In *Boundaries of Journalism*, ed. Matt Carlson and Seth C. Lewis. New York: Routledge.

Carlson, M. Forthcoming. *Journalistic Authority: A Relational Approach*. New York: Columbia University Press.

Carlson, Matt, and Seth C. Lewis, eds. 2015. *Boundaries of Journalism: Professionalism, Practices and Participation*. New York: Routledge.

Carter, C., and S. Allan. 2000. "'If It Bleeds, It Leads': Ethical Questions about Popular Journalism." In *Ethics and Media Culture: Practices and Representations*, ed. David Berry, 132–153. Burlington, MA: Focal Press.

Carter, R. G. S. 2006. "Of Things Said and Unsaid: Power, Archival Silences, and Power in Silence." *Archivaria* 61:215–233.

Carvajal, M., J. A. Garcia-Aviles, and J. L. Gonzalez. 2012. "Crowdfunding and Non-profit Media: The Emergence of New Models for Public Interest Journalism." *Journalism Practice* 6 (5–6): 638–647.

Carvin, A. 2013. *Distant Witness: Social Media, the Arab Spring and a Journalism Revolution*. New York: CUNY Journalism Press.

Casella, P. A. 2012. "Breaking News or Broken News? Reporters and News Directors Clash on 'Black Hole' Live Shots." *Journalism Practice* 7 (3): 362–376.

Cassidy, W. P. 2005. "Variations on a Theme: The Professional Role Conceptions of Print and Online Newspaper Journalists." *Journalism & Mass Communication Quarterly* 82 (2): 264–280.

Cassidy, W. P. 2007. "Online News Credibility: An Examination of the Perceptions of Newspaper Journalists." *Journal of Computer-Mediated Communication* 12 (2): 478–498.

Castells, Manuel. 1996. *The Rise of the Network Society*. Oxford: Blackwell.

Castells, Manuel. 2009. *Communication Power*. New York: Oxford University Press.

Castells, M. 2012. *Networks of Outrage and Hope: Social Movements in the Internet Age*. 1st ed. Cambridge: Polity.

Castle, S., and D. Dalby. 2015. "Irish Media, Fearing Lawsuits, Steers Clear of a Billionaire." *New York Times*, May 29. http://www.nytimes.com/2015/05/30/world/europe/ireland-rte-denis-obrien.html.

Cawley, A. 2008. "News Production in an Irish Online Newsroom: Practice, Process and Culture." In *Making Online News: The Ethnography of New Media Production*, ed. C. A. Paterson and D. Domingo, 45–60. New York: Peter Lang.

Cawley, A. 2012. "Towards a Historical Perspective on Locating Online News in the News Ecology: The Case of Irish News Websites, 1994–2010." *Media History* 18 (2): 219–237.

Center for Public Integrity (CPI). 2014. "Center Wins First Pulitzer Prize: Stories Detailed Systematic Denial of Benefits to Black Lung Sufferers." April 14. http://www.publicintegrity.org/2014/04/14/14593/center-wins-first-pulitzer-prize

Chadwick, Andrew. 2013. *The Hybrid Media System*. New York: Oxford University Press.

Chaffee, S. H., and M. J. Metzger. 2001. "The End of Mass Communication?" *Mass Communication & Society* 4 (4): 365–379.

Chaffee, Steven H. 1991. *Explication*. Newbury Park, CA: Sage.

Chan-Olmsted, S. M., and L. S. Ha. 2003. "Internet Business Models for Broadcasters: How Television Stations Perceive and Integrate the Internet." *Journal of Broadcasting & Electronic Media* 47 (4): 597–617.

Christians, C. G., J. P. Ferré, and P. M. Fackler. 1993. *Good News: Social Ethics and the Press*. New York: Oxford University Press.

Christians, Clifford G., Theodore L. Glasser, Denis McQuail, Kaarle Nordenstreng, and Robert A. White. 2009. *Normative Theories of the Media: Journalism in Democratic Societies*. Urbana: University of Illinois Press.

Christin, Angele. 2014. *Counting Clicks: Commensuration in Online Journalism in the United States and France*. Paper presented at American Sociological Association annual conference, San Francisco, August.

Chomsky, D. 2006. "'An Interested Reader': Measuring Ownership Control at the *New York Times*." *Critical Studies in Mass Communication* 23 (1): 1–18.

Chung, D. S., and S. Nah. 2009. "The Effects of Interactive News Presentation on Perceived User Satisfaction of Online Community Newspapers." *Journal of Computer-Mediated Communication* 14 (4): 855–874.

Chyi, H. I. 2005. "Willingness to Pay for Online News: An Empirical Study on the Viability of the Subscription Model." *Journal of Media Economics* 18 (2): 131–142.

Chyi, Hsiang Iris. 2009. "Information Surplus in the Digital Age: Impact and Implications." In *Journalism and Citizenship: New Agendas in communication*, ed. Zizi Papacharissi, 91–107. New York: Routledge.

Chyi, Hsiang Iris, Seth C. Lewis, and Nan Zheng. 2012. "A Matter of Life and Death?: Examining How Newspapers Covered the Newspaper 'Crisis'." *Journalism Studies* 13 (3): 305–324.

Chyi, H. I., and M. J. Yang. 2009. "Is Online News an Inferior Good? Examining the Economic Nature of Online News among Users." *Journalism & Mass Communication Quarterly* 86 (3): 594–612.

Clarke, Debra. 2014. *Journalism and Political Exclusion: Social Conditions of News Production and Reception*. Montreal: McGill Queens University Press.

Coddington, Mark. 2015. "Clarifying Journalism's Quantitative Turn: A Typology for Evaluating Data Journalism, Computational Journalism, and Computer-Assisted Reporting." *Digital Journalism* 3 (3): 331–348.

Coddington, M. 2015. "The Wall Becomes a Curtain: Revisiting Journalism's News–Business Boundary." In *Boundaries of Journalism: Professionalism, Practices, and Participation*, ed. Matt Carlson and Seth C. Lewis. New York: Routledge.

Cohen, E. L. 2002. "Online Journalism as Market-driven Journalism." *Journal of Broadcasting & Electronic Media* 46 (4): 532–548.

Cohen, N., and M. Scott. 2014. "*Times* Articles Removed from Google Results in Europe." *New York Times*, October 4. http://www.nytimes.com/2014/10/04/business/media/times-articles-removed-from-google-results-in-europe.html.

Colapinto, Cinzia, and Colin Porlezza. 2012. "Innovation in Creative Industries: From the Quadruple Helix Model to the Systems Theory." *Journal of the Knowledge Economy* 3 (4): 343–353.

Communications Management Inc. 2011. "Sixty Years of Daily Newspaper Circulation Trends." http://media-cmi.com/downloads/Sixty_Years_Daily_Newspaper_Circulation_Trends_050611.pdf.

ComScore. 2015. "comScore Ranks the Top 50 U.S. Digital Media Properties for August 2014." http://www.comscore.com/Insights/Market-Rankings/comScore-Ranks-the-Top-50-US-Digital-Media-Properties-for-August-2014.

Conboy, M., and J. Steel. 2008. "The Future of Newspapers." *Journalism Studies* 9:650–661.

Compaine, Benjamin M., and Douglas Gomery. 2000. *Who Owns the Media? Competition and Concentration in the Mass Media Industry*. New York: Routledge.

Compaine, Ben, and Anne Hoag. 2012. "Factors Supporting and Hindering New Entry in Media Markets: A Study of Media Entrepreneurs." *International Journal on Media Management* 14 (1): 27–49.

Conway, Michael. 2009. *The Origins of Television News in America: The Visualizers of CBS in the 1940s*. New York: Peter Lang.

Cook, Timothy. 1998. *Governing with the News*. Chicago: University of Chicago Press.

Cooper, Mark. 2011. "The Future of Journalism: Addressing Pervasive Market Failure with Public Policy." In *Will the Last Reporter Please Turn Out the Lights*, ed. Robert McChesney and Victor Pickard, 320–339. New York: The New Press.

Codrea-Rado. Anna. 2012. "Between the Spreadsheets." *Columbia Journalism Review*, August 8. http://www.cjr.org/data_points/spreadsheets_olympics.php.

Cottle, S. 2007. "Ethnography and News Production: New(s) Developments in the Field." *Sociology Compass* 1 (1): 1–16.

Couldry, Nick. 2008. "Actor Network Theory and Media: Do They Connect and on What Terms?" In *Cultures of Connectivity*, ed. A. Hepp and R. Winter, 93–110. Creskill, NJ: The Hampton Press.

Couldry, N. 2013. *Why Media Ethics Still Matters*. Hoboken, NJ: Wiley-Blackwell.

Craig, Douglas B. 2000. *Fireside Politics: Radio and Political Culture in the United States, 1920–1940*. Baltimore: Johns Hopkins University Press.

Craig, Robert, and Karen Tracy. 1995. "Grounded Practical Theory: The Case of Intellectual Discussion." *Communication Theory* 5 (3): 248–272.

Cranberg, Gilbert, Randall Bezanson, and John Soloski. 2001. *Taking Stock: Journalism and the Publicly Traded Newspaper Company*. Ames: Iowa State University Press.

Crawford, K. 2009. "Following You: Disciplines of Listening in Social Media." *Continuum (Perth)* 23 (4): 525–535.

Crawford, Susan. 2013. *Captive Audience: The Telecom Industry and Monopoly Power in the New Gilded Age*. New Haven, CT: Yale University Press.

Creech, Brian, and Andrew L. Mendelson. 2015. "Imagining the Journalist of the Future: Technological Visions of Journalism Education and Newswork." *Communication Review* 18 (2): 142–165.

Davis, M. S. 1971. "That's Interesting! Towards a Phenomenology of Sociology and a Sociology of Phenomenology." *Philosophy of the Social Sciences* I:309–344.

Daulerio, A. J. 2012. "Gawker Will Be Conducting an Experiment, Please Enjoy Your Free Cute Cats Singing and Sideboobs." Gawker, January 23. http://gawker.com/5878065/gawker-will-be-conducting-an-experiment-please-enjoy-your-free-cute-cats-singing-and-sideboobs.

Delli Carpini, Michael X., Fay Lomax Cook, and Lawrence R. Jacobs. 2014. "Public Deliberation, Discursive Participation, and Citizen Engagement: A Review of the Empirical Literature." *Annual Review of Political Science* 7:315–344.

Dennis, E. E. (Fall 2006). "Television's Convergence Conundrum: Finding the Right Digital Strategy." *Television Quarterly* XXXVII:22–26.

Deuze, M. 2003. "The Web and Its Journalisms: Considering the Consequences of Different Types of News Media Online." *New Media & Society* 5 (2): 203–230.

Deuze, M. 2004. "What Is Multimedia Journalism?" *Journalism Studies* 5:139–152.

Deuze, M. 2005. "What Is Journalism? Professional Identity and Ideology of Journalists Reconsidered." *Journalism* 6 (4): 442–464.

Deuze, M. 2007. *Media Work*. Cambridge: Polity Press.

Deuze, Mark. 2012. *Media Life*. Cambridge: Polity Press.

Deuze, M. 2014. "Journalism, Media Life, and the Entrepreneurial Society." *Australian Journalism Review* 36 (2): 119–130.

Deuze, M., A. Bruns, and C. Neuberger. 2007. "Preparing for an Age of Participatory News." *Journalism Practice* 1 (3): 322–338.

Deuze, M., and C. Dimoudi. 2002. "Online Journalists in the Netherlands: Towards a Profile of a New Profession." *Journalism* 3 (1): 85–100.

Deuze, Mark, and Tim Marjoribanks. 2009. "Newswork." *Journalism* 10 (5): 555–561.

Dewey, John. 1983. "Public Opinion." In *The Middle Works of John Dewey*, vol. 13., ed. Jo Ann Boydston. Carbondale: Southern Illinois University Press. (1922).

Dewey, John. 1984. "The Public and Its Problems." In *The Later Works of John Dewey*, vol. 2, ed. Jo Ann Boydston. Carbondale: Southern Illinois University Press. (1927).

Diakopoulos, Nicholas. 2011. "A Functional Roadmap for Innovation in Computational Journalism," April. http://www.nickdiakopoulos.com/2011/04/22/a-functional-roadmap-for-innovation-in-computational-journalism.

Dickinson, R., J. Matthews, and K. Saltzis. 2013. "Studying Journalists in Changing Times: Understanding News Work as Socially Situated Practice." *International Communication Gazette* 75 (1): 3–18.

Dickinson, Roger. 2008. "Studying the Sociology of Journalists: The Journalistic Field and the News World." *Sociology Compass* 2 (5): 1383–1399.

Diesing, Paul. 2008. *Patterns of Discovery in the Social Sciences*. New Brunswick: Aldine Transaction. (1971)

van Dijck, José. 2013. *The Culture of Connectivity: A Critical History of Social Media*. Oxford: Oxford University Press.

DiSalvo, C. 2012. *Adversarial Design*. Cambridge, MA: MIT Press.

Dixon, T. L., and C. Azocar. 2006. "The Representation of Juvenile Offenders by Race on Los Angeles Area Television News." *Howard Journal of Communications* 77:143–161.

Dobson, A. 2012. "Listening: The New Democratic Deficit." *Political Studies* 60 (4): 843–859.

Domingo, D., P. Masip, and I. Costera Meijer. 2015. "Tracing Digital News Networks: Towards an Integrated Framework of the Dynamics of News Production, Circulation and Use." *Digital Journalism* 3 (1): 53–67.

Domingo, D., T. Quandt, A. Heinonen, S. Paulussen, J. B. Singer, and M. Vujnovic. 2008. "Participatory Journalism Practices in the Media and Beyond: An International Comparative Study of Initiatives in Online Newspapers." *Journalism Practice* 2 (3).

Dörr, Konstantin Nicholas. 2015. "Mapping the Field of Algorithmic Journalism." *Digital Journalism* 4 (6): 700–722.

Doudaki, V., and L.-P. Spyridou. 2013. "Print and Online News: Remediation Practices in Content and Form." *Journalism Studies* 14 (6): 907–925.

Downey, J., and N. Fenton. 2003. "New Media, Counter Publicity and the Public Sphere." *New Media & Society* 5:185–202.

Downie, L., and M. Schudson. 2009. "The Reconstruction of American Journalism." *Columbia Journalism Review* 19:2009.

Downie, Leonard, and Michael Schudson. 2009. "The Reconstruction of American Journalism." *Columbia Journalism Review*. http://www.cjr.org/reconstruction/the_reconstruction_of_american.php.

Downs, A. 1957. "An Economic Theory of Political Action in a Democracy." *Journal of Political Economy* 65 (2): 135–150.

Drucker, Peter. 1985. *Innovation and Entrepreneurship: Practice and Principles*. New York: Harper & Row.

Dunaway, Johanna. 2008. "Markets, Ownership, and the Quality of Campaign News Coverage." *Journal of Politics* 70 (4): 1193–1202.

Du Gay, P. 1996. *Consumption and Identity at Work*. London: Sage.

Dunbar-Hester, C. 2014. *Low Power to the People: Pirates, Protest, and Politics in FM Radio Activism*. Cambridge, MA: MIT Press.

Dunne, A., and F. Raby. 2013. *Speculative Everything: Design, Fiction, and Social Dreaming*. Cambridge, MA: MIT Press.

Dutton, W. H. 1995. "Driving into the Future of Communications? Check the Rear View Mirror." In *Information Superhighways: Multimedia Users and Futures*, ed. S. J. Emmott, 79–102. New York: Academic Press.

Dutton, W. H. 2007. "Through the Network of Networks—The Fifth Estate." https://www.oii.ox.ac.uk/archive/downloads/events/2007/20071015_WD_5thEstateLecture.pdf.

Dutton, W. H. 2009. "The Fifth Estate Emerging through the Network of Networks." *Prometheus* 27 (1): 1–15.

Dutton, W. H., ed. 2013. *The Oxford Handbook of Internet Studies*. Oxford: Oxford University Press.

Dutton, W. H., and P. Jeffreys, eds. 2010. *World Wide Research: Reshaping the Sciences and Humanities*. Cambridge, MA: MIT Press.

Dutton, W. H., Sun Huan, and Weiwei Shen. 2015, "China and the Fifth Estate: Net Delusion or Democratic Potential?" In *The Impact of Social Media in Politics and Public Administrations* (online special issue), coordinated by J. Balcells, A. Battle, and A. Padró-Solanet. *IDP. Revista de Internet, Derecho y Política* 20: 1–19. http://elcrps .uoc.edu/index.php/idp/article/view/n20-dutton-huan-shen/2797.

Dyos, H. J. 1982. "Some Historical Reflections on the Quality of Urban Life. In *Exploring the Urban Past: Essays in Urban History by H. J. Dyos*, ed. D. Cannadine and D. Reeder, 56–78. Cambridge: Cambridge University Press.

Eckman, Alyssa, and Thomas Lindlof. 2003. "Negotiating the Gray Lines: An Ethnographic Case Study of Organizational Conflict between Advertorials and News." *Journalism Studies* 4 (1): 65–77.

Edmonds, Rick. 2004. "News Staffing, News Budgets and News Capacity." *Newspaper Research Journal* 25 (1): 98–109.

Edmonds, Rick, Emily Guskin, Amy Mitchell, and Mark Jurkowitz. 2013. "Newpapers Stablilizing, but Still Threatened." http://www.stateofthemedia.org/2013/ newspapers-stabilizing-but-still-threatened.

Edmonds, Rick, Emily Guskin, Tom Rosenstiel, and Amy Mitchell. 2012. "Newspapers: By the Numbers." In The Pew Research Center, "The State of the News Media 2012." http://www.stateofthemedia.org/2012/newspapers-building-digital-revenues -proves-painfully-slow/newspapers-by-the-numbers.

Eliasoph, N. 1988. "Routines and the Making of Oppositional News." *Critical Studies in Mass Communication* 5 (4): 313–334.

Eliasoph, N. 1998. *Avoiding Politics: How Americans Produce Apathy in Everyday Life*. Cambridge: Cambridge University Press.

Elliott, D., and D. Ozar. 2010. "An Explanation and a Method for the Ethics of Journalism." In *Journalism Ethics: A Philosophical Approach*, ed. Christopher Meyers, 9–24. Oxford: Oxford University Press.

Elsadda, H. 2010. "Arab Women Bloggers: The Emergence of Literary Counterpublics." *Middle East Journal of Culture and Communication* 3:312–332.

Emerson, R. M. 1962. "Power-Dependence Relations." *American Sociological Review* 27:31–41.

Emerson, T. I. 1970. *The System of Freedom of Expression*. New York: Random House.

Enda, Jodi. "Retreating from the World." 2011. *American Journalism Review*. http://ajrarchive.org/Article.asp?id=4985.

Enda, Jodi, Katerina Eva Matsa, and Jan Lauren Boyles. 2014. "America's Shifting Statehouse Press: Can New Players Compensate for Lost Legacy Reporters?" Pew Research Center: Journalism & Media, July 10. http://www.journalism.org/2014/07/10/americas-shifting-statehouse-press.

Entman, R. M. 1989. *Democracy without Citizens*. Oxford: Oxford University Press.

Epstein, E. J. 1973. *News From Nowhere: Television and the News*. New York: Random House.

Eshbaugh-Soha, Matthew, and Jeffrey S. Peake. 2011. *Breaking through the Noise: Presidential Leadership, Public Opinion, and the News Media*. Stanford, CA: Stanford University Press.

Esser, Frank, and Thomas Hanitzsch, eds. 2012. *The Handbook of Comparative Communication Research*. London: Routledge.

Ettema, James S., and Theodore Lewis Glasser. 1998. *Custodians of Conscience : Investigative Journalism and Public Virtue*. New York: Columbia University Press.

Fairclough, N. 2010. *Critical Discourse Analysis: The Critical Study of Language*. 2nd ed. Harlow, England: Routledge.

Fairclough, N., J. Mulderrig, and R. Wodak. 2011. "Critical Discourse Analysis." In *Discourse Studies: A Multidisciplinary Introduction*, ed. T. A. van Dijk. London: Sage.

Federal Communications Commission. 2015. "2015 Broadband Progress Report." https://www.fcc.gov/reports/2015-broadband-progress-report.

Federal Trade Commission. 2010. "Potential Policy Recommendations to Support the Reinvention of Journalism." https://www.ftc.gov/sites/default/files/documents/public_events/how-will-journalism-survive-internet-age/new-staff-discussion.pdf.

Fengler, Susanne, and Stephan Russ-Mohl. 2008. "Journalists and the Information-Attention Markets: Towards an Economic Theory of Journalism." *Journalism* 9 (6): 667–690.

Fenton, Natalie, ed. 2010. *New Media, Old News: Journalism & Democracy in the Digital Age*. London: SAGE.

Ferree, Myra Marx, William Gamson, Jürgen Gerhards, and Dieter Rucht. 2002. *Shaping Abortion Discourse: Democracy and the Public Sphere in Germany and the United States*. Cambridge: Cambridge University Press.

Ferrier, Michelle Barrett. 2013. "Media Entrepreneurship: Curriculum Development and Faculty Perceptions of What Students Should Know." *Journalism & Mass Communication Educator* 68 (3): 222–241.

Fidler, Roger. 2013. "News Consumption on Mobile Devices Surpasses Desktop Computers, Newspapers." Reynolds Journalism Institute, May 1. http://www.rjionline.org/news/news-consumption-mobile-devices-surpasses-desktop-computers-newspapers.

Filloux, F. 2008. "The Economics of Moving from Print to Online: Lose One Hundred, Get Back Eight." http://www.mondaynote.com/2008/09/29/the-economics-of-moving-from-print-to-online-lose-one-hundred-get-back-eight.

Fink, Katherine, and Michael Schudson. 2014. "The Rise of Contextual Journalism, 1950s–2000s." *Journalism* 15 (1): 3–20.

Fink, Katherine, and C. W. Anderson. 2015. "Data Journalism in the United States." *Journalism Studies* 16 (4): 467–481.

Fishman, M. 1980. *Manufacturing the News*. Austin: University of Texas Press.

Fiss, O. 1996. *The Irony of Free Speech*. Cambridge, MA: Harvard University Press.

Fligstein, Neil, and Doug Adam. 2012. *A Theory of Fields*. New York: Oxford University Press.

Foreman, Gene. 2015. *The Ethical Journalist: Making Responsible Decisions in the Digital Age*. 2nd ed. Malden, MA: John Wiley & Sons.

Foucault, M. 1971. "Nietzsche, Genealogy, History." In *Language, Counter-Memory, Practice: Selected Essays and Interviews*, ed. D. F. Bouchard. Ithaca, NY: Cornell University Press.

Foucault, M. 1994. *The Order of Things: An Archaeology of the Human Sciences*. New York: Vintage. (1966).

Foucault, M. 1994. "Two Lectures." In *Culture/Power/History: A Reader in Contemporary Social Theory*, ed. N. B. Dirks, G. Eley and S. B. Ortner, 200–221. Princeton, NJ: Princeton University Press.

Foxman, Maxwell. 2015. "Play the News: Fun and Games in Digital Journalism," A Tow/Knight report. http://towcenter.org/wp-content/uploads/2015/02/PlayTheNews_Foxman_TowCenter.pdf.

Fraser, N. 1990. "Rethinking the Public Sphere: A Contribution to the Critique of Actually Existing Democracy." *Social Text* 25/26:56–80.

Freedman, Des. 2008. *The Politics of Media Policy*. Cambridge: Polity Press.

Friend, C., and J. B. Singer. 2007. *Online Journalism Ethics*. Armonk, NY: M. E. Sharpe.

Frosh, P., and A. Pinchevski. 2011. "Introduction: Why Media Qitnessing? Why Now?" In *Media Witnessing: Testimony in the Age of Mass Communication*, ed. P. Frosh and A. Pinchevski, 1–22. New York: Palgrave.

Gade, Peter J. 2004. "Newspapers and Organizational Development: Management and Journalist Perceptions of Newsroom Cultural Change." *Journalism & Communication Monographs* 6 (1): 3, 5–55.

Graeff, E., M. Stempeck, and E. Zuckerman. 2014. "The Battle for 'Trayvon Martin': Mapping a Media Controversy Online and Off-line." *First Monday* 19 (2).

Galbraith, J. K. 1978. "Writing, Typing, and Economics." *Atlantic* 241 (March): 102–105. http://www.theatlantic.com/magazine/archive/1978/03/writing-typing-and-economics/305165/.

Gallup, G. H. 1928. "An Objective Method for Determining Reader Interest in the Content of a Newspaper." Doctoral thesis, University of Iowa.

Gallup. 2015. "Americans' Trust in Media Remains at Historical Low." http://www.gallup.com/poll/185927/americans-trust-media-remains-historical-low.aspx.

Gamson, William, and Andre Modigliani. 1989. "Media Discourse and Public Opinion on Nuclear Power: A Constructionist Approach." *American Journal of Sociology* 95 (1): 1–37.

Gans, Herbert J. 1974. *Popular Culture and High Culture: An Analysis and Evaluation of Taste*. New York: Basic Books.

Gans, H. 1980. *Deciding What's News: A Study of CBS Evening News, NBC Nightly News, Newsweek, and Time*. New York: Vintage.

Gans, H. J. 2007. "Everyday News, Newsworkers, and Professional Journalism." *Political Communication* 24 (2): 161–166.

García, E. P. 2008. "Print and Online Newsrooms in Argentinean Media: Autonomy and Professional Identity." In *Making Online News: The Ethnography of New Media Production*, ed. C. A. Paterson and D. Domingo, 61–75. New York: Peter Lang.

Garrison, B. 2005. "Online Newspapers." In *Online News and the Public*, ed. M. B. Salwen, B. Garrison and P. D. Driscoll, 3–46. Mahwah, NJ: Lawrence Erlbaum Associates.

Geertz, C. 1983. *Local Knowledge: Further Essays in Interpretive Anthropology*. New York: Basic.

Geertz, Clifford. 1973. *The Interpretation of Cultures*. New York: Basic Books.

Geidner, N., and D. D'Arcy. 2013. "The Effects of Micropayments on Online News Story Selection and Engagement." *New Media & Society*, October 18.

Gentzkow, M. 2007. "Valuing New Goods in a Model with Complementarity: Online Newspapers." *American Economic Review* 97 (3): 713–744.

Gentzkow, Matthew, and Jesse M. Shapiro. 2010. "What Drives Media Slant? Evidence from U.S. Daily Newspapers." *Econometrica* 78 (1): 35–71.

Gergen, K. J. 1991. *The Saturated Self: Dilemmas of Identity in Contemporary Life*. New York: Basic Books.

Gil de Zúñiga, H., N. Jung, and S. Valenzuela. 2012. "Social Media Use for News and Individuals' Social Capital, Civic Engagement and Political Participation." *Journal of Computer-Mediated Communication* 17:319–336.

Gillespie, T. 2014. "The Relevance of Algorithms." In *Media Technologies: Essays on Communication, Materiality, and Society*, ed. T. Gillespie, P. Boczkowski and K. A. Foot, 167–194. Cambridge, MA: MIT Press.

Gillespie, Tarleton, Pablo J. Boczkowski, and Kirsten A. Foot, eds. 2014. *Media Technologies: Essays on Communication, Materiality, and Society*. Cambridge, MA: MIT Press.

Gillmor, D. 2004. *We the Media: Grassroots Journalism By the People, For the People*. Sebastopol, CA: O'Reilly Media, Inc.

Gillmor, D. 2009. "Towards a Slow News Movement." *Mediactive*. http://mediactive.com/2009/11/08/toward-a-slow-news-movement.

Gitelman, L. 2014. *Paper Knowledge: Toward a Media History of Documents*. Durham, NC: Duke University Press.

Gitlin, Todd. 2001. *Media Unlimited: How the Torrent of Images and Sounds Overwhelms Our Lives*. New York: Metropolitan Books.

Glasser, T. L., and M. Gunther. 2005. "The Legacy of Autonomy in American Journalism." In *The Institutions of American Democracy: The Press*, ed. G. Overholser and K. H. Jamieson, 384–399. Oxford: Oxford University Press.

Glenn, C. 2004. *Unspoken: A Rhetoric of Silence*. Carbondale, IL: Southern Illinois University Press.

Goode, L. 2009. "Social News, Citizen Journalism and Democracy." *New Media & Society* 11 (8): 1287–1305.

Gordon, A. D., J. M. Kittross, J. C. Merrill, W. Babcock, and M. Dorsher. 2012. *Controversies in Media Ethics*. New York: Routledge.

Goyanes, M. 2014. "An Empirical Study of Factors That Influence the Willingness to Pay for Online News." *Journalism Practice* 8 (6): 742–757.

Grafström, M., and K. Windell. 2012. "Newcomers Conserving the Old: Transformation Processes in the Field of News Journalism." *Scandinavian Journal of Management* 28 (1): 65–76.

Granovetter, M. S. 1973. "The Strength of Weak Ties." *American Journal of Sociology* 78 (6):1360–1380.

Graves, Lucas. 2016. *Deciding What's True: The Fact-Checking Movement in American Journalism*. New York: Columbia University Press.

Graves, Lucas, and Magda Konieczna. 2015. "Sharing the News: Journalistic Collaboration as Field Repair." *International Journal of Communication* 9:1966–1984.

Graves, L., J. Kelly, and M. Gluck. 2010. "*Confusion Online: Faulty Metrics and the Future of Digital Journalism*." Tow Center for Digital Journalism, Columbia University Graduate School of Journalism. http://towcenter.org/research/confusion-online-faulty-metrics-and-the-future-of-digital-journalism.

Greenberger, M. 1985. *Electronic Publishing Plus*. White Plains, NY: Knowledge Industries.

Greer, J. D. 2004. "Advertising on Traditional Media Sites: Can the Traditional Business Model Be Translated to the Web?" *Social Science Journal* 41 (1): 107–113.

Grusin, R. 2010. *Premediation: Affect and mediality after 9/11*. New York: Palgrave.

Gulyas, Agnes. 2013. "The Influence of Professional Variables on Journalists' Uses and Views of Social Media: A Comparative Study of Finland, Germany, Sweden and the United Kingdom." *Digital Journalism* 1 (2): 270–285.

Gurevitch, M., Coleman, S., and Blumler, J. 2009. "Political Communication—Old and New Media Relationships." *The ANNALS of the American Academy of Political and Social Science* 625 (1): 164.

Haas, T. 2005. "From 'Public Journalism' to the 'Public's Journalism'? Rhetoric and Reality in the Discourse on Weblogs." *Journalism Studies* 6 (3): 387–396.

Hagood, M. 2011. "Quiet Comfort: Noise, Otherness, and the Mobile Production of Personal Space." *American Quarterly* 63 (3): 573–589.

Haigh, T., A. L. Russell, and W. H. Dutton. 2015. "Histories of the Internet." *Information & Culture* 50 (2): 143–159.

Hallett, Tim. 2010. "The Myth Incarnate: Recoupling Processes, Turmoil, and Inhabited Institutions in an Urban Elementary School." *American Sociological Review* 75 (1): 52–74.

Hallin, D. C. 1986a. *The "Uncensored War": The Media and Vietnam*. New York: Oxford University Press.

Hallin, Daniel. 1986b. "Where? Cartography, Community, and the Cold War." In *Reading the News*, ed. Robert Karl Manoff and Michael Schudson, 109–145. New York: Pantheon Press.

Hallin, Daniel, and Paolo Mancini. 2004. *Comparing Media Systems: Three Models of Media and Politics*. Cambridge: Cambridge University Press.

Hallin, Daniel, and Paolo Mancini. 2011. *Comparing Media Systems Beyond the Western World*. Cambridge: Cambridge University Press.

Hamilton, James T. 2006. *All the News That's Fit to Sell: How the Market Transforms Information into News*. Princeton, NJ: Princeton University Press.

Hampton, K., L. S. Goulet, C. Marlow, and L. Rainie. 2012. "Why Most Facebook Users Get More Than They Give." Pew Research Center. http://www.pewinternet.org/2012/02/03/why-most-facebook-users-get-more-than-they-give.

Hampton, K., L. Rainie, W. Lu, M. Dwyer, S. Inyoung, and K. Purcell. 2014. "Social Media and the 'Spiral of Silence': Summary of Findings." Pew Research Internet Project. http://www.pewinternet.org/2014/08/26/social-media-and-the-spiral-of-silence.

Hanitzsch, Thomas. 2007. "Deconstructing Journalism Culture: Toward a Universal Theory." *Communication Theory* 17 (4): 367–385.

Hara, K. 2009. *White*. London: Lars Muller.

Hargittai, E. 2007. "Whose Space? Differences among Users and Non-users of Social Network Sites." *Journal of Computer-Mediated Communication* 13 (1): 276–297.

Harris, M. 2014. *The End of Absence: Reclaiming What We've Lost in a World of Constant Connection*. New York: Current.

Hartley, J. 2002. *Communication, Cultural and Media Studies*. London: Routledge.

Hartley, J., J. Burgess, and A. Bruns, eds. 2013. *A Companion to New Media Dynamics*. Chichester, UK: Blackwell Publishing Ltd.

Havelock, E. 1982. *Preface to Plato*. Cambridge, MA: Belknap Press.

Hayes, Danny, and Jennifer L. Lawless. 2015. "As Local News Goes, So Goes Citizen Engagement: Media, Knowledge, and Participation in U.S. House Elections." *Journal of Politics* 77 (2): 447–462.

Hellmueller, Lea, Tim P. Vos, and Mark A. Poepsel. 2013. "Shifting Journalistic Capital? Transparency and Objectivity in the Twenty-first Century." *Journalism Studies* 14 (3): 287–304.

Heinrich, Ansgard. 2011. "Foreign Reporting in the Sphere of Network Journalism." *Journalism Practice* 6 (5–6): 766–775.

Heinrich, Ansgard. 2013. "News Making as an Interactive Practice: Global News Exchange and Network Journalism." In *Rethinking Journalism: Trust and Participation in a Transformed News Landscape*, ed. Chris Peters and Marcel Broersma, 89–100. New York: Routledge.

Hemmingway, Emma. 2008. *Into the Newsroom: Exploring the Digital Production of Regional Television News*. London: Routledge.

Hendrickson, L. 2007. "Press Protection in the Blogosphere: Applying a Functional Definition of "Press" to News Weblogs." In *Blogging, Citizenship and the Future of the Media*, ed. M. Tremayne, 187–204. New York: Routledge.

Herbert, J., and N. Thurman. 2007. "Paid Content Strategies for News Websites." *Journalism Practice* 1 (2): 208–226.

Hermans, L., and M. Vergeer. 2009. "Internet in the Daily Life of Journalists: Explaining the use of the Internet by Work-Related Characteristics and Professional Opinions." *Journal of Computer-Mediated Communication* 15 (1): 138–157.

Hermida, Alfred. 2010. "Twittering the News." *Journalism Practice* 4 (3): 297–308.

Hermida, Alfred. 2012. "Tweets and Truth: Journalism as a Discipline of Collaborative Verification." *Journalism Practice* 6 (5–6): 659–668.

Hermida, Alfred, Seth C. Lewis, and Rodrigo Zamith. 2014. "Sourcing the Arab Spring: A Case Study of Andy Carvin's Sources on Twitter During the Tunisian and Egyptian Revolutions." *Journal of Computer-Mediated Communication* 19 (3): 479–499.

Hermida, Alfred, and Neil Thurman. 2008. "A Clash of Cultures: The Integration of User-generated Content within Professional Journalistic Frameworks at British Newspaper Websites." *Journalism Practice* 2 (3): 343–356.

Herndon, Keith L. 2012. *The Decline of the Daily Newspaper: How an American Institution Lost the Online Revolution*. New York: Peter Lang International Academic Publishers.

Hesmondhalgh, David. 2006. "Bourdieu, the Media and Cultural Production." *Media Culture & Society* 28 (2): 211–231.

Hesmondhalgh, David, and Sarah Baker. 2011. *Creative Labour: Media Work in Three Cultural Industries*. London: Routledge.

Hetland, P. 2015. "Popularizing the Internet: Traveling Companions Supporting Good News." *Nordicom Review* 36 (2): 157–171.

Hilmes, Michele, ed. 2003. *The Television History Book*. London: BFI Publishing.

Hindman, Matthew. 2008. *The Myth of Digital Democracy*. Princeton, NJ: Princeton University Press.

Hindman, M. 2014. "Stickier News." Shorenstein Center on Media, Politics and Public Policy. http://shorensteincenter.org/wp-content/uploads/2015/04/Stickier-News-Matthew-Hindman.pdf.

Hollifield, C. Ann, Jan LeBlanc Wicks, George Sylvie, and Wilson Lowrey. 2015. *Media Management: A Casebook Approach*. 5th ed. New York: Routledge.

Hong, S. 2012. "Online News on Twitter: Newspapers' Social Media Adoption and Their Online Readership." *Information Economics and Policy* 24 (1): 69–74.

Howard, Alexander Benjamin. 2014. "*The Art and Science of Data-Driven Journalism*." Tow Center for Digital Journalism. http://towcenter.org/wp-content/uploads/2014/05/Tow-Center-Data-Driven-Journalism.pdf.

Howard, Phil. 2002. "Network Ethnography and the Hypermedia Organization: New Media, New Organizations, New Methods." *New Media & Society* 4 (4): 550–574.

Hujanen, J., and S. Pietikainen. 2004. "Interactive Uses of Journalism: Crossing between Technological Potential and Young People's News-using Practices." *New Media & Society* 6 (3): 383–401.

Hunter, Andrea. 2015. "Crowdfunding Independent and Freelance Journalism: Negotiating Journalistic Norms of Autonomy and Objectivity." *New Media & Society* 17 (2): 272–288.

Hunter, Anna, and François P. Nel. 2011. "Equipping the Entrepreneurial Journalist: An Exercise in Creative Enterprise." *Journalism & Mass Communication Educator* 66 (1): 9–24.

Hutchins, B. 2007. "Public Culture, Independent Online News and the *Tasmanian Times*." *Journalism* 8 (2): 205–225.

Ihlebæk, K. A., and A. H. Krumsvik. 2014. "Editorial Power and Public Participation in Online Newspapers." *Journalism*.

Innis, H. 1950. *Empire and Communications*. Toronto: University of Toronto Press.

Iyengar, Shanto, Helmut Norpoth, and Kyu S. Hahn. 2004. "Consumer Demand for Election News: The Horserace Sells." *Journal of Politics* 66 (1): 157–175.

Jacobs, Ron, and Eleanor Townsley. 2011. *The Space of Opinion*. New York: Oxford University Press.

James, William. 1890. *The Principles of Psychology*. New York: Henry Holt.

Jamieson, Kathleen Hall, and Joseph Cappella. 2010. *Echo Chamber: Rush Limbaugh and the Conservative Media Environment*. New York: Oxford University Press.

Jamieson, P., K. H. Jamieson, and D. Romer. 2003. "The Responsible Reporting of Suicide in Print Journalism." *American Behavioral Scientist* 46 (12): 1643–1660.

Jansen, Sue Curry. 2009. "Phantom Conflict: Lippmann, Dewey, and the Fate of the Public in Modern Society." *Communication and Critical/Cultural Studies* 6 (3): 221–245.

Jarvis, J. 2007. "New Rule: Cover What You Do Best and Link to the Rest." *Buzzmachine*. http://buzzmachine.com/2007/02/22/new-rule-cover-what-you-do-best-link-to-the-rest.

Jenner, Michael, Esther Thorson, and Anna Kim. 2014. "How U.S. Daily Newspapers Decide to Design and Implement Paywalls." Paper presented at the Association for Education in Journalism and Mass Communication Annual Conference, Montreal, Canada.

Jian, L., and N. Usher. 2014. "Crowd-Funded Journalism." *Journal of Computer-Mediated Communication* 19 (2): 155–170.

John, Richard R. 1995. *Spreading the News: The American Postal System from Franklin to Morse*. Cambridge, MA: Harvard University Press.

Johnson, F. 2015. "Going It Alone." *Harper's* (April): 31–40.

Jones, S. C., and T. G. Schumacher. 1992. "Muzak: On Functional Music and Power." *Critical Studies in Mass Communication* 9 (2): 156–169.

Juniper, A. 2003. *Wabi Sabi: The Japanese Art of Impermanence*. Tokyo, Japan: Tuttle.

Jurkowitz, Mark. 2014. "The Growth in Digital Reporting: What It Means for Journalism and News Consumers." The Pew Research Center, The State of the News Media 2014. http://www.journalism.org/2014/03/26/the-growth-in-digital-reporting.

Kahn, D. 1997. "John Cage: Silence and Silencing." *Musical Quarterly* 81 (4): 556–598.

Kaplan, R. 1999. *The Nothing That Is: A Natural History of Zero*. Oxford: Oxford University Press.

Karlsson, Michael. 2011. "The Immediacy of Online News, the Visibility of Journalistic Processes and a Structuring of Journalistic Authority." *Journalism* 12 (3): 279–295.

Karlsson, M., and C. Clerwall. 2012. "Patterns and Origins in the Evolution of Multimedia on Broadsheet and Tabloid News Sites: Swedish Online News, 2005–2010." *Journalism Studies* 13 (4): 550–565.

Karlsson, M., and J. Strömbäck. 2010. "Freezing the Flow of Online News: Exploring Approaches to the Study of the Liquidity of Online News." *Journalism Studies* 11 (1): 2–19.

Karpf, D. 2012. *The MoveOn Effect: The Unexpected Transformation of American Political Advocacy*. New York: Oxford University Press.

Katz, E., and P. Lazarsfeld. 1955. *Personal Influence: The Part Played by People in the Flow of Mass Communications*. New York: Free Press.

Kawahata, M. 1987. "HI-OVIS." In *Wired Cities: Shaping the Future of Communications*, ed. W. H. Dutton, J. G. Blumler and K. L. Kraemer, 179–200. Boston: G. K. Hall.

Kaye, Jeff, and Stephen Quinn. 2010. *Funding Journalism in the Digital Age: Business Models, Strategies, Issues and Trends*. New York: Peter Lang.

Keeling, K., and J. Kun. 2011. "Introduction: Listening to American Studies." *American Quarterly* 63 (3): 445–459.

Keen, A. 2007. *The Cult of the Amateur: How Today's Internet is Killing Our Culture*. New York: Doubleday.

Kellogg, Katherine C. 2009. "Operating Room: Relational Spaces and Microinstitutional Change in Surgery." *American Journal of Sociology* 115 (1): 657–711.

Kelly, J. 2010. "Parsing the Online Ecosystem: Journalism, Media, and the Blogosphere." In *Transitioned Media, The Economics of Information, Communication and Entertainment*, ed. G. Einav, 93–108. New York: Springer.

Kennedy, J. 2005. "Right to Receive Information: The Current State of the Doctrine and the Best Application for the Future." *Seton Hall Law Review* 35 (2): 789–821.

Kenny, C. 2011. *The Power of Silence: Silent Communication in Daily Life*. London: Karnac.

Kent, S. 1991. "Partitioning Space: Cross-cultural Factors Influencing Domestic Spatial Segmentation." *Environment and Behavior* 23 (4): 438–473.

Kerbel, M. R. 2000. *If It Bleeds, It Leads: An Anatomy of Television News*. New York: Basic Books.

Kernell, S. 1997. *Going Public: New Strategies of Presidential Leadership*. Washington, DC: Congressional Quarterly.

Kerrane, Kevin. 1998. *The Art of Fact: A Historical Anthology of Literary Journalism. A Touchstone Book*. New York: Touchstone.

Kipling, R. 2015. "The Elephant's Child." In *Rudyard Kipling, Just So Stories*. London: Macmillan Children's Books. (1902).

Kirk, Andy. 2012. *Data Visualization: A Successful Design Process. Community Experience Distilled*. Birmingham, UK: Packt Publishing.

Klaidman, S. 1987. *The Virtuous Journalist*. New York: Oxford University Press.

Klein, E. 2015. "This Is My Best Advice to Young Journalists. Vox, February 9. http://www.vox.com/2015/2/9/8008365/advice-to-young-journalists.

Klinenberg, Eric. 2005. "Convergence: News Production in a Digital Age." *Annals of the American Academy of Political and Social Science* 597 (4): 48–64.

Knight Foundation. 2009. Informing Communities: Sustaining Democracy in the Digital Age. http://www.knightfoundation.org/media/uploads/publication_pdfs/Knight_Commission_Report_-_Informing_Communities.pdf.

Konieczna, Magda. 2014. "A Better News Organization: Can Nonprofits Improve on the Commercial News Organizations from Which They Arose?" Unpublished dissertation, Department of Journalism and Mass Communication, University of Wisconsin-Madison.

Konieczna, M., and S. Robinson. 2014. "Emerging News Non-profits: A Case Study for Rebuilding Community Trust?" *Journalism* 15 (8): 968–986.

Krause, Monika. 2015. "Comparative Research: Beyond Linear-Causal Explanation." In *Practicing Comparison: Logics, Relations, Collaborations*, ed. J. Deville, M. Guggenheim, and Z. Hrdličková. Manchester, UK: Mattering Press.

Kreiss, Daniel. 2013. "Temporality and Political Communication Research." Qualitative Political Communication Research blog, posted December 30. https://qualpolicomm.wordpress.com/2013/12/30/temporality-and-political-communication-research.

Kreiss, D. 2015. "The Networked Democratic Spectator." *Social Media and Society* (1). http://sms.sagepub.com/content/1/1/2056305115578876.abstract.

Kreiss, Daniel, and J. S. Brennen. 2016. "Journalism and Democracy, Normative Models." In *Sage Handbook of Digital Journalism Studies*, ed. T. Witschge, C. W. Anderson, D. Domingo, and A. Hermida. New York: Sage.

Kreiss, Daniel, Laura Meadows, and John Remensperger. 2015. "Political Performance, Boundary Spaces, and Active Spectatorship: Media Production at the 2012 Democratic National Convention." *Journalism* 16 (5): 577–595.

Küng, Lucy. 2011. "Managing Strategy and Maximizing Innovations in Media Organizations." In *Managing Media Work*, ed. Mark Deuze, 43–56. Thousand Oaks, CA: Sage.

Kuratko, Donald F. 2005. "The Emergence of Entrepreneurship Education: Development, Trends and Challenges." *Entrepreneurship Theory and Practice* 29 (5): 577–598.

Kurpius, David D., Emily T. Metzgar, and Karen M. Rowley. 2010. "Sustaining Hyperlocal Media: In Search of Funding Models." *Journalism Studies* 11 (3): 359–376.

Lacey, K. 2011. "Listening Overlooked: An Audit of Listening as a Category in the Public Sphere." *Javnost—The Public* 18 (4): 5–20.

Ladany, N., C. E. Hill, B. J. Thompson, and K. M. O'Brien. 2004. "Therapist Perspectives on Using Silence in Therapy: A Qualitative Study." *Counselling & Psychotherapy Research* 4 (1): 80–89.

Lampel, Joseph, Theresa Lant, and Jamal Shamsie. 2000. "Balancing Act: Learning from Organizing Practices in Cultural Industries." *Organization Science* 11 (3): 263–269.

Lang, K., and G. E. Lang. 1953. "The Unique Perspective of Television and Its Effect: A Pilot Study." *American Sociological Review* 18 (1): 3–12.

Larsson, A. O. 2012. "Understanding Nonuse of Interactivity in Online Newspapers: Insights from Structuration Theory." *Information Society* 28 (4): 253–263.

Lasica, J. D. 2003. "Blogs and Journalism Need Each Other." *Nieman Reports* 57 (3): 70–74.

Latour, B. 2005. *Reassembling the Social: An Introduction to Actor-Network Theory*. Oxford: Oxford University Press.

Latour, Bruno, and Steve Woolgar. 1979. *Laboratory Life: The Social Construction of Scientific Facts*. Beverly Hills, CA: Sage.

Latour, Bruno. 1991. "Technology Is Society Made Durable." In *A Sociology of Monsters: Essays on Power, Technology and Domination*, Sociological Review Monograph 38, ed. J. Law, 103–132. London: Routledge.

Law, John. 2009. "Actor-Network Theory and Material Semiotics." In *The New Blackwell Companion to Social Theory*, ed. B. S. Turner, 141–158. London: Blackwell.

Lawson-Borders, G. 2006. *Media Organizations and Convergence: Case Studies of Media Convergence Pioneers*. Mahwah, NJ: Lawrence Erlbaum Associates.

Lazarsfeld, Paul F. 1941. "Remarks on Administrative and Critical Communications Research." *Studies in Philosophy and Social Science* 9 (1): 2–16.

Leab, Dan. 1987. "*See It Now*: A Legend Reassessed." In *American History/America Television. Interpreting the Video Past*, ed. John E. O'Connor. New York: Ungar.

Le Masurier, M. 2014. "What Is Slow Journalism?" *Journalism Practice* 9 (2): 138–152.

Lee, Eun-Ju. 2012. "That's Not the Way It Is: How User-Generated Comments on the News Affect Perceived Media Bias." *Journal of Computer-Mediated Communication* 18 (1): 32–45.

Lee, S. K., and J. E. Katz. 2014. "Disconnect: A Case Study of Short-term Voluntary Mobile Phone Non-use." *First Monday* 19 (12).

Lee, Eun-Ju, and Ye Weon Kim. 2015. "Effects of Infographics on News Elaboration, Acquisition, and Evaluation: Prior Knowledge and Issue Involvement as Moderators." *New Media & Society* 18 (8): 1579–1598

Lehmann, J., C. Castillo, M. Lalmas, and E. Zuckerman. 2013. "Transient News Crowds in Social Media." Paper presented at the ICWSM 2013. http://chato.cl/papers/lehmann_castillo_lalmas_zuckerman_2013_transient_news_crowds.pdf.

Lerman, J. 2013. "Big Data and Its Exclusions." *Stanford Law Review* 66:55–63.

Levendusky, Matthew. 2013. *How Partisan Media Polarize America*. Chicago: University of Chicago Press.

Levy, M. R. 1981. "Disdaining the News." *Journal of Communication* 31 (3): 24–31.

Levy, David A. L., and Rasmus Kleis Nielsen. 2010. *The Changing Business of Journalism and Its Implications for Democracy*. Oxford: Reuters Institute for the Study of Journalism. http://reutersinstitute.politics.ox.ac.uk/publication/changing-business-journalism-and-its-implications-democracy.

Lewis, C. 2007. "The Nonprofit Road." *Columbia Journalism Review* 46 (3): 32–36.

Lewis, S. C. 2012. "The Tension between Professional Control and Open Participation: Journalism and Its Boundaries." *Information Communication and Society* 15 (6): 836–866.

Lewis, Seth C. 2015. "Journalism in an Era of Big Data: Cases, Concepts, and Critiques." *Digital Journalism* 3 (3): 321–330.

Lewis, S. C., A. E. Holton, and M. Coddington. 2014. "Reciprocal Journalism." *Journalism Practice* 8 (2): 229–241.

Lewis, S. C., and N. Usher. 2013. "Open Source and Journalism: Toward New Frameworks for Imagining News Innovation." *Media Culture & Society* 35 (5): 602–619.

Lewis, S. C., and N. Usher. 2014. "Code, Collaboration, and the Future of Journalism: A Case Study of the Hacks/Hackers Global Network." *Digital Journalism* 2 (3): 383–393.

Lewis, Seth C., and Nikki Usher. 2016. "Trading Zones, Boundary Objects, and the Pursuit of News Innovation: A Case Study of Journalists and Programmers." *Convergence: The International Journal of Research Into New Media Technologies* 22 (5): 543–560. doi:10.1177/1354856515623865.

Lewis, Seth C., and Oscar Westlund. 2015. "Actors, Actants, Audiences, and Activities in Cross-Media News Work." *Digital Journalism* 3 (1): 19–37.

Lewis, Seth C., and Oscar Westlund. 2016. "Mapping the Human–Machine Divide in Journalism." In *The SAGE Handbook of Digital Journalism*, ed. Tamara Witschge, Chris W. Anderson, David Domingo, and Alfred Hermids, 341–353. London: Sage.

Leys, Colin. 2001. *Market-driven Politics: Neoliberal Democracy and the Public Interest*. London: Verso.

Lievrouw, L. A., and S. Livingstone, eds. 2002. *Handbook of New Media*. Thousand Oaks, CA: Sage.

Lindley, R. 2002. *Panorama: Fifty Years of Pride and Paranoia* . London: Politico's.

Lipari, L. 2010. "Listening, Thinking, Being." *Communication Theory* 20:348–362.

Lippmann, W. 1922. *Public Opinion*. New York: Harcourt, Brace.

Lippmann, Walter. 2011. *The Phantom Public*. New Brunswick, NJ: Transaction Publishers. (Originally published 1927.)

Lohr, Steve. 2011. "In Case You Wondered, a Real Human Being Wrote This Column." *New York Times*, September 10. http://www.nytimes.com/2011/09/11/business/computer-generated-articles-are-gaining-traction.html?_r=0.

Loosen, Wiebke, and Jan-Hinrik Schmidt. 2012. "(Re-)discovering the Audience: The Relationship between Journalism and Audience in the Networked Digital Media." *Information Communication and Society* 15 (6): 867–887.

Lowrey, W. 2006. "Mapping the Journalism-Blogging Relationship." *Journalism* 7 (4): 477–500.

Lowrey, Wilson. 2012. "Journalism Innovation and the Ecology of News Production: Institutional Tendencies." *Journalism & Communication Monographs* 14 (4): 214–287.

Lowrey, Wilson, and Peter J. Gade. 2011. *Changing the News: The Forces Shaping Journalism in Uncertain Times*. New York: Routledge.

Luiselli, V. 2014. "Relingos: The Cartography of Empty Spaces," trans. C. MacSweeney. In *Sidewalks*, 69–78. Minneapolis, MN: Coffee House Press.

McDorman, T. 2001. "Crafting a Virtual Counterpublic: Right-to-die Advocates on the Internet." In *Counterpublics and the State*, 187–209. New York: SUNY Press.

MacGregor, P. 2007. "Tracking the Online Audience: Metric Data Start a Subtle Revolution." *Journalism Studies* 8 (2): 280–298.

Madison, J. 1822. "James Madison to W. T. Barry," Chapter 18, Document 35 in *Writings* 9: 103–9, August 4. http://press-pubs.uchicago.edu/founders/documents/v1ch18s35.html.

Madrigal, A. C. 2012. "Dark Social: We Have the Whole History of the Web Wrong." *The Atlantic*. http://www.theatlantic.com/technology/archive/2012/10/dark-social-we-have-the-whole-history-of-the-web-wrong/263523.

Madrigal, Alexis C. 2013. "2013: The Year 'The Stream' Crested." *Atlantic* 12 (December). http://www.theatlantic.com/technology/archive/2013/12/2013-the-year-the-stream-crested/282202.

Maeda, J. 2006. *The Laws of Simplicity*. Cambridge, MA: MIT Press.

Mair, J., and R. L. Keeble, eds. 2014. *Data Journalism: Mapping the Future*. Suffolk, UK: Abramis.

Madianou, M., and D. Miller. 2012. "Polymedia: Towards a New Theory of Digital Media in Interpersonal Communication." *International Journal of Cultural Studies* 16 (2): 1–19.

Maitland, S. 2014. *How to Be Alone*. London: Picador.

Mancini, P. 2013. "What Scholars Can Learn from the Crisis of Journalism." *International Journal of Communication* 7:127–136.

Manovich, Lev. 2001. *The Language of New Media*. Cambridge, MA: MIT Press.

Manjoo, F. 2013. "You Won't Finish This Article." Slate, June 6. http://www.slate.com/articles/technology/technology/2013/06/how_people_read_online_why_you_won_t_finish_this_article.html.

Martin, John Levi. 2011. *The Explanation of Social Action*. New York: Oxford University Press.

Marvin, C. 1988. *When Old Technologies Were New: Thinking About Electric Communication in the Late Nineteenth Century*. Oxford: Oxford University Press.

Marx, L. 1987. "Does Improved Technology Mean Progress?" *Technology Review* (January): 33–41.

Massing, Michael. 2015a. "Digital Journalism: How Good Is It?" *New York Review of Books* 4 (June). http://www.nybooks.com/articles/archives/2015/jun/04/digital-journalism-how-good-is-it.

Massing, Michael. 2015b. "Digital Journalism: The Next Generation." *New York Review of Books* 25 (June). http://www.nybooks.com/articles/archives/2015/jun/25/digital-journalism-next-generation.

McBride, Kelly, and Tom Rosenstiel. 2014. *The New Ethics of Journalism*. Los Angeles: Sage.

McChesney, Robert W. 2013. *Digital Disconnect: How Capitalism Is Turning the Internet against Democracy*. New York: The New Press.

McChesney, Robert W. 2015. *Rich Media, Poor Democracy: Communication Politics in Dubious Times*. New York: The New Press.

McChesney, Robert W., and John Nichols. 2010. *The Death and Life of American Journalism: The Media Revolution That Will Begin the World Again*. New York: Nation Books.

McChesney, Robert W., and Victor Pickard. 2011. *Will the Last Reporter Please Turn out the Lights: The Collapse of Journalism and What Can Be Done to Fix It*. New York: The New Press.

McChesney, Robert W., and Victor Pickard. 2014. "News Media as Political Institutions." In *The Oxford Handbook of Political Communication*, ed. Kate Kenski and Kathleen Hall Jamieson. Oxford: Oxford University Press.

McChesney, Robert. 2014. *Digital Disconnect*. New York: The New Press.

McDevitt, Michael. 2003. "In Defense of Autonomy: A Critique of the Public Journalism Critique." *Journal of Communication* 53 (1): 155–164.

McCoy, M. E. 2001. "Dark Alliance: News Repair and Institutional Authority in the Age of the Internet." *Journal of Communication* 51 (1): 164–193.

McIntosh, N. 2015. "List of BBC Web Pages Which Have Been Removed from Google's Search Results." BBC, June 25. http://www.bbc.co.uk/blogs/internet/entries/1d765aa8-600b-4f32-b110-d02fbf7fd379.

McKeown, G. 2014. *Essentialism: The Disciplined Pursuit of Less*. New York: Crown Business.

McManus, John H. 1997. "Who's Responsible for Journalism?" *Journal of Mass Media Ethics* 12 (1): 5–17.

McNair, B. 2000. "Journalism and Democracy." In *The Handbook of Journalism Studies*, ed. K. Wahl-Jorgensen and T. Hanitzch, 237–249. New York: Routledge.

McNair, Brian. 2005. "What Is Journalism?" In *Making Journalists: Diverse Models, Global Issues*, ed. Hugo de Burgh, 25–43. New York: Routledge.

Meiklejohn, Alexander. 1948. *Free Speech and Its Relation to Self-government*. New York: Harper Brothers Publishers.

Mensing, Donica. 2010. "Rethinking (again) the Future of Journalism Education." *Journalism Studies* 11 (4): 511–523.

Meraz, S. 2011. "Using Time Series Analysis to Measure Intermedia Agenda-Setting Influence in Traditional Media and Political Blog Networks." *Journalism & Mass Communication Quarterly* 88 (1): 176–194.

Merrill, John Calhoun. 1993. *The Dialectic in Journalism: Toward a Responsible Use of Press Freedom*. Baton Rouge: Louisiana State University Press.

Merton, Robert. 1949. *Social Theory and Social Structure*. New York: Free Press.

Meyers, C., W. N. Wyatt, S. L. Borden, and E. Wasserman. 2012. "Professionalism, Not Professionals." *Journal of Mass Media Ethics* 27 (3): 189–205.

Mill, J. S. 1974. *On Liberty*. London: Penguin Books. (1859).

Millen, D. R., and S. M. Dray. 2000. "Information Sharing in an Online Community of Journalists." *Aslib Proceedings* 52:166–173.

Mills, C. W. 1959/2000. *The Sociological Imagination*. Oxford: Oxford University Press.

Mindich, D. T. 2000. *Just the Facts: How Objectivity Came to Define American Journalism*. New York: NYU Press.

Mitchell, A., M. Jurkowitz, J. Holcomb, J. Enda, and M. Anderson. 2013. "*Nonprofit Journalism: A Growing but Fragile Part of the US News System*." Pew Research Center's Project for Excellence in Journalism. http://www.journalism.org/2013/06/10/nonprofit-journalism.

Mitchelstein, E., and P. J. Boczkowski. 2009. "Between Tradition and Change: A Review of Recent Research on Online News Production." *Journalism: Theory, Practice & Criticism*. 10 (5): 562–586.

Molotch, H., and M. Lester. 1974. "News as Purposive Behavior: On the Strategic Use of Routine Events, Accidents, and Scandals." *American Sociological Review* 39 (1): 101–112.

Murdock, G. 1977. *Patterns of Ownership: Questions of Control*. Milton Keynes, UK: Open University Press.

Musgrave, Richard A. 1959. *The Theory of Public Finance: A Study in Public Economy*. New York: McGraw-Hill.

Mutz, Diana C., and Byron Reeves. 2005. "The New Videomalaise: Effects of Televised Incivility on Political Trust." *American Political Science Review* 99 (1): 1–15.

Mutter, A. 2014. "The Newspaper Crisis, by the Numbers." Newsosaur, July. http://newsosaur.blogspot.com/2014/07/the-newspaper-crisis-by-numbers.html.

Naldi, Lucia, and Robert G. Picard. 2012. "'Let's Start an Online News Site': Opportunities, Resources, Strategy, and Formational Myopia in Startups." *Journal of Media Business Studies* 4:47–59.

Napoli, P. M. 2010. *Audience Evolution*. New York: Columbia University Press.

Nee, Rebecca Coates. 2013. "Creative Destruction: An Exploratory Study of How Digitally Native News Nonprofits Are Innovating Online Journalism Practices." *International Journal on Media Management* 15 (1): 3–22.

Nel, François. 2010. "Where Else Is the Money? A Study of Innovation in Online Business Models at Newspapers in Britain's 66 Cities." *Journalism Practice* 4 (3): 360–372.

Nelkin, D. 1991. "AIDS and the News Media." *Milbank Quarterly* 69 (2): 293–307.

Nerone, J., and K. G. Barnhurst. 2001. "Beyond Modernism—Digital Design, Americanization and the Future of Newspaper Form." *New Media & Society* 3 (4): 467–482.

Neuman, W. R. 1986. *The Paradox of Mass Politics: Knowledge and Opinion in the American Electorate*. Cambridge, MA: Harvard University Press.

Neuman, W. R. 1990. "The Threshold of Public Attention." *Public Opinion Quarterly* 54 (2): 159–176.

Neuman, W. R., M. Just, and A. Crigler. 1992. *Common Knowledge: News and the Construction of Political Meaning*. Chicago: University of Chicago Press.

Neuman, W. Russell, Yong Jin Park, and Elliot Panek. 2012. "Tracking the Flow of Information into the Home: An Empirical Assessment of the Digital Revolution in the U.S. from 1960–2005." *International Journal of Communication* 6:1022–1041.

New York Times. 2015. "Digital Subscriptions." www.nytimes.com/content/help/account/purchases/subscriptions-and-purchases.html

New York Times. 2015 "Where the Story Ends, Times Premier Begins." http://www.nytimes.com/subscriptions/switch/lp8UXWF.html#insider.

Newman, Nic. 2015. "Media, Journalism and Technology Predictions 2015." Reuters Institute for the Study of Journalism. https://reutersinstitute.politics.ox.ac.uk/news/media-journalism-and-technology-predictions-2015.

Newman, N., W. H. Dutton, and G. Blank. 2012. "Social Media in the Changing Ecology of News: The Fourth and Fifth Estates in Britain." *International Journal of Internet Science* 7 (1): 6–22.

Newman, Nic, and David A. L. Levy. 2014. *Reuters Institute Digital News Report 2014: Tracking the Future of News*. Oxford: Reuters Institute for the Study of Journalism.

Newman, Nic, David A. L. Levy, and Rasmus Kleis Nielsen. 2015. *Reuters Institute Digital News Report 2015*. Oxford: Reuters Institute for the Study of Journalism.

Nguyen, A. 2008. "Facing 'the Fabulous Monster': The Traditional Media's Fear-driven Innovation Culture in the Development of Online News." *Journalism Studies* 9 (1): 91–104.

Nguyen, A. 2013. "Online News Audiences." In *Journalism: New Challenges*, ed. K. Fowler-Watt and S. Allan, 146–161. Bournemouth: Centre for Journalism & Communication Research Bournemouth University.

Nielsen, C. E. 2014. "Coproduction or Cohabitation: Are Anonymous Online Comments on Newspaper Websites Shaping News Content?" *New Media & Society* 16 (3): 470–487.

Nielsen, Rasmus Kleis. 2012a. "How Newspapers Began to Blog: Recognizing the Role of Technologists in Old Media Organizations' Development of New Media Technologies." *Information Communication and Society* 15 (6): 959–978.

Nielsen, Rasmus Kleis. 2012b. *Ground Wars: Personalized Communication in Political Campaigns*. Princeton, NJ: Princeton University Press.

Nielsen, Rasmus Kleis, and Geert Linnebank. 2011. "Public Support for the Media: A Six-Country Overview of Direct and Indirect Subsidies." Reuters Institute for the Study of Journalism. https://reutersinstitute.politics.ox.ac.uk/sites/default/files/Public%20support%20for%20Media_0.pdf.

Noelle-Neumann, E. 1984. *The Spiral of Silence: Public Opinion—Our Social Skin*. Chicago: University of Chicago Press.

Nonnecke, B., and J. Preece. 2000. *Lurker Demographics: Counting the Silent*. Paper presented at the ACM Conference on Computer Human Interaction, The Hague, The Netherlands, April 1–6, 2000.

Noveck, Beth. 2015. *Smart Citizens, Smarter State: The Technologies of Expertise and the Future of Governing*. Cambridge, MA: Harvard University Press.

O'Sullivan, J., and A. Heinonen. 2008. "Old Values, New Media: Journalism Role Perceptions in a Changing World." *Journalism Practice* 2 (3): 357–371.

Obstfeld, D. 2005. "Social Networks, the Tertius Iungens Orientation, and Involvement in Innovation." *Administrative Science Quarterly* 50 (1): 100–130.

Olsen, T. 1965. "Silences: When Writers Don't Write." *Harper's Magazine* 231 (October supplement, "The Writer's Life"): 153–161.

Olson, M. 1965. *The Logic of Collective Action: Public Goods and the Theory of Group*. Cambridge, MA: Harvard University Press.

On The Media. 2013. "What Do You Broadcast When There's Nothing to Say? *On The Media,* April 19. http://www.onthemedia.org/story/288011-what-do-you-broadcast-when-theres-nothing-say.

Ong, W. 1982. *Orality and Literacy: The Technologizing of the Word*. London: Methuen Publishing.

Ornebring, H. 2008. "The Consumer as a Producer of What? User-generated Tabloid Content in *The Sun* (UK) and *Aftonbladet* (Sweden)." *Journalism Studies* 9 (5): 771–785.

Oudshoorn, N., and T. Pinch. 2003. "How Users and Non-users Matter." In *How Users Matter: The Co-construction of Users and Technology*, ed. N. Oudshoorn and T. Pinch, 1–25. Cambridge, MA: MIT Press.

Owen, L. H. 2015. "Soon, Publishers Will Be Able to Determine When Smartphone Users Are Bored and Push Content at Them." Nieman Lab, September 2, 2015. http://www.niemanlab.org/2015/09/soon-publishers-will-be-able-to-determine-when-smartphone-users-are-bored-and-push-content-at-them.

Palczewski, C. 2001. "Cybermovements, New Social Movements and Counterpublics." In *Counterpublics and the State*, ed. Robert Asen, 161–186. New York: SUNY Press.

Papacharissi, Z. 2014a. *Affective Publics: Sentiment, Technology, and Politics*. New York, Oxford: Oxford University Press.

Papacharissi, Z. 2014b. "Toward New Journalism(s): Affective News, Hybridity, and Liminal Spaces." *Journalism Studies*. published online March 2014.

Papacharissi, Zizi. 2015. "Toward New Journalism(s): Affective News, Hybridity, and Liminal Spaces." *Journalism Studies* 16 (1): 27–40.

Papacharissi, Z., and M. de Fatime Oliveira. 2012. "Affective News and Networked Publics: The Rhythms of News Storytelling on #Egypt." *Journal of Communication* 62 (2): 266–282.

Park, Robert E. 1940. "News as a Form of Knowledge: A Chapter in the Sociology of Knowledge." *American Journal of Sociology* 45 (5): 669–686.

Patten, D. A. 1986. *Newspapers and New Media*. White Plains, NY: Knowledge Industry Publications, Inc.

Patterson, Thomas E. 2013. *Informing the News: The Need for Knowledge-Based Journalism*. New York: Vintage.

Pavlik, J. 2001. *Journalism and New Media*. New York: Columbia University Press.

Pavlik, J. V. 2013. *Media in the Digital Age*. New York: Columbia University Press.

Pavlik, John V. 2013. "Innovation and the Future of Journalism." *Digital Journalism* 1 (2): 181–193.

Peng, T. Q., L. Zhang, Z.-J. Zhong, and J. Zhu. 2013. "Mapping the Landscape of Internet Studies: Text Mining of Social Science Journal Articles, 2000–2009." *New Media & Society* 15 (5): 644–664.

Persico, J. E. 1988. *Edward R. Murrow: An American Original*. New York: Dell Publishing Company.

Peters, J. D. 2015a. *The Marvelous Clouds: Towards a Philosophy of Elemental Media*. Chicago: University of Chicago Press.

Peters, J. D. 2015b. "The Anthropoid Condition: An Interview with John Durham Peters." *Los Angeles Review of Books,* July 10. http://lareviewofbooks.org/interview/the-anthropoid-condition-an-interview-with-john-durham-peters.

Petre, Caitlin. 2015. *The Traffic Factories: Metrics at Chartbeat, Gawker Media, and The New York Times*. New York: Columbia University Press; http://towcenter.org/research/traffic-factories.

Petre, Caitlin. 2015. "The Social Life of Metrics: The Production, Interpretation, and Use of Data Analytics in Online Journalism." Unpublished dissertation, NYU Department of Sociology.

Petrova, M. 2011. "Newspapers and Parties: How Advertising Revenues Created an Independent Press." *American Political Science Review* 105 (4): 790–808.

Pew Research Center. 2013. "Amid Criticism, Support for Media's 'Watchdog' Role Stands Out," August 8. http://www.people-press.org/2013/08/08/amid-criticism-support-for-medias-watchdog-role-stands-out.

Pew Research Center. 2014. "State of the News Media 2014." http://www.journalism.org/packages/state-of-the-news-media-2014.

Pew Research Center. 2014a. "State of the News Media 2014." http://www.journalism.org/packages/state-of-the-news-media-2014.

Pew Research Center. 2014b. "Paying for News: The Revenue Picture for American Journalism, and How It Is Changing." http://www.journalism.org/2014/03/26/the-revenue-picture-for-american-journalism-and-how-it-is-changing.

Pfau, M., M. Haigh, M. Gettle, M. Donnelly, G. Scott, D. Warr, and E. Wittenberg. 2004. "Embedding Journalists in Military Combat Units: Impact on Newspaper Story Frames and Tone." *Journalism & Mass Communication Quarterly* 81 (1): 74–88.

Pfister, D. S. 2014. *Networked Media, Networked Rhetoric: Attention and Deliberation in the Early Blogosphere*. University Park, PA: Penn State University Press.

Phelps, Andrew. 2012. "NPR Snags Brian Boyer to Launch a News Apps Team (and They're Hiring)." Nieman Lab, May 21. http://www.niemanlab.org/2012/05/npr-snags-brian-boyer-to-launch-a-news-apps-team-and-theyre-hiring.

Picard, Robert G. 2002. *The Economics and Financing of Media Companies*. New York: Fordham University Press.

Picard, Robert G. 2005. "Unique Characteristics and Business Dynamics of Media Products." *Journal of Media Business Studies* 2 (2): 61–69.

Picard, Robert G. 2010. *Value Creation and the Future of News Organizations: Why and How Journalism Must Change to Remain Relevant in the Twenty-first Century*. Lisbon: Media XXI.

Picard, Robert G. 2011. *The Economics and Financing of Media Companies*. New York: Fordham University Press.

Picard, Robert G. 2011. *Mapping Digital Media: Digitization and Media Business Models*. London: Open Society Foundations. http://www.opensocietyfoundations.org/sites/default/files/digitization-media-business-models-20110721.pdf.

Picard, Robert G. 2012. *Value Creation and the Future of News Organizations*. Porto, Portugal: MediaXXI.

Picard, Robert G. 2014. "Twilight or New Dawn of Journalism? Evidence from the Changing News Ecosystem." *Digital Journalism* 2 (3): 488–498.

Pickard, Victor. 2011. "Can Government Support the Press? Historicizing and Internationalizing a Policy Approach to the Journalism Crisis." *Communication Review* 14 (2): 73–95.

Pickard, Victor. 2014. *America's Battle for Media Democracy*. Cambridge: Cambridge University Press.

Pickard, V. 2014. "The Great Evasion: Confronting Market Failure in American Media Policy." *Critical Studies in Media Communication* 31 (2): 153–159.

Pickard, V. 2014b. "Laying Low the Shibboleth of a Free Press: Regulatory Threats against the American Newspaper Industry, 1938–1947." *Journalism Studies* 15 (4): 464–480.

Pickard, V. 2015a. *America's Battle for Media Democracy: The Triumph of Corporate Libertarianism and the Future of Media Reform*. New York: Cambridge University Press.

Pickard, V. 2015b. "The Return of the Nervous Liberals: Market Fundamentalism, Policy Failure, and Recurring Journalism Crises." *Communication Review* 18 (2): 82–97.

Pickard, Victor, Josh Stearns, and Craig Aaron. 2009. *Saving the News: Toward a National Journalism Strategy*. Washington, DC: Free Press.

Pickard, Victor, and Alex Williams. 2014. "Salvation or Folly? The Perils and Promises of Digital Paywalls." *Digital Journalism* 2 (2): 195–213.

Picker, J. 2012. "The Soundproof Study." In *The Sound Studies Reader*, ed. J. Sterne, 141–151. New York: Routledge.

Picone, I. 2011. "Produsage as a Form of Self-publication: A Qualitative Study of Casual News Produsage." *New Review of Hypermedia and Multimedia* 17 (1): 99–120.

Pilhofer, Aaron. 2010. "Programmer-Journalist? Hacker-Journalist? Our Identity Crisis." Mediashift, April 22. http://mediashift.org/idealab/2010/04/programmer-journalist-hacker-journalist-our-identity-crisis107.

Plaisance, P. L. 2013. *Media Ethics: Key Principles for Responsible Practice*. Thousand Oaks, CA: Sage Publications.

Plaut, E. R. 2015. "Technologies of Avoidance: The Swear Jar and the Cell Phone." *First Monday* 20 (11).

Plesner, Ursula. 2009. "An Actor-Network Perspective on Changing Work Practices: Communication Technologies as Actants in Newswork." *Journalism* 10 (5): 604–626.

Polgreen, Erin. 2014. "How Successful Media Startups Manage Growth." *Columbia Journalism Review*, December. http://www.cjr.org/business_of_news/startup_management_lessons.php.

Potter, Deborah. 2008. "Old News." *American Journalism Review*. http://ajrarchive.org/Article.asp?id=4448.

Potter, Deborah, and Katerina Eva Matsa. 2014. "A Boom in Acquisitions and Content Sharing Shapes Local TV News in 2013." Pew Research Center. http://www.journalism.org/2014/03/26/a-boom-in-acquisitions-and-content-sharing-shapes-local-tv-news-in-2013.

Portwood-Stacer, L. 2012. "Media Refusal and Conspicuous Non-consumption: The Performative and Political Dimensions of Facebook Abstention." *New Media & Society* 15 (7): 1041–1057.

Powers, Matt. 2012. "'In Forms That Are Familiar and Yet-to-Be Invented': American Journalism and the Discourse of Technologically Specific Work." *Journal of Communication Inquiry* 36 (1): 24–43.

Powers, Matthew, and Rodney Benson. 2014. "Is the Internet Homogenizing or Diversifying the News? External Pluralism in the U.S., Danish, and French Press." *International Journal of Press/Politics* 19 (2): 246–265.

Prenger, Mirjam. 2014. *Televisiejournalistiek in de jaren vijftig en zestig. Achter het nieuws en de geboorte van de actualiteitenrubriek*. Amsterdam: AMB.

Primo, Alex, and Gabriela Zago. 2015. "Who and What Do Journalism? An Actor-Network Perspective." *Digital Journalism* 3 (1): 38–52.

Prior, Markus. 2007. *Post-Broadcast Democracy: How Media Choice Increases Inequality in Political Involvement and Polarizes Elections*. New York: Cambridge University Press.

Prior, M. 2009. "The Immensely Inflated News Audience: Assessing Bias in Self-reported News Exposure." *Public Opinion Quarterly* 73 (1): 130–143.

Proctor, R. N. 2008. "Agnotology: A Missing Term to describe the Cultural Production of Ignorance (and Its Study)." In *Agnotology: The Making and Unmaking of Ignorance*, ed. R. Proctor and L. Schiebinger, 1–33. Stanford, CA: Stanford University Press.

ProPublica. 2015. "About Us." http://www.propublica.org/about.

Quandt, T. 2008. "News Tuning and Content Management: An Observation Study of Old and New Routines in German Online Newsrooms. In *Making Online News: The Ethnography of New Media Production*, ed. C. A. Paterson and D. Domingo, 77–97. New York: Peter Lang.

Quandt, T., M. Loffelholz, D. Weaver, T. Hanitzsch, and K.-D. Altmeppen. 2006. "American and German Online Journalists at the Beginning of the 21st Century." *Journalism Studies* 7:171–186.

Quinn, S. 2005. "Convergence's Fundamental Question." *Journalism Studies* 6 (1): 29–38.

Rainie, L., and M. Madden. 2015. "Americans' Privacy Strategies Post-Snowden." Pew Research Center, March 16. http://www.pewinternet.org/2015/03/16/americans-privacy-strategies-post-snowden.

Rand, P. 2014. *Thoughts on Design*. New York: Chronicle Books. (1970/1947).

Rao, S., and H. Wasserman. 2007. "Global Media Ethics Revisited: A Postcolonial Critique." *Global Media and Communication* 3 (1): 29–50.

Raviola, E., and M. Norbäck. 2013. "Bringing Technology and Meaning into Institutional Work: Making News at an Italian Business Newspaper." *Organization Studies* 34 (8): 1171–1194.

Razlogova, E. 2011. *The Listener's Voice: Early Radio and the American Public*. Philadelphia: University of Pennsylvania Press.

Reich, Z. 2008. "How Citizens Create News Stories." *Journalism Studies* 9:739–758.

Reich, Z., and Y. Godler. 2014. "A Time of Uncertainty: The Effects of Reporters' Time Schedule on Their Work." *Journalism Studies* 15 (5): 607–618.

Revers, M. 2014. "The Twitterization of News Making: Transparency and Journalistic Professionalism." *Journal of Communication* 64 (5): 806–826.

Rice, R. E., and G. A. Crawford. 1992. "Context and Content of Citations between Communication and Library and Information Science Articles." In *Between Communication and Information*. In *Information and Behavior*. vol. 4. Ed. J. Schement and B. Ruben, 189–218. New Brunswick, NJ: Transaction Publishers.

Rios, M. 2014. "Penalty Kicks, as Seen through Twitter Data." Twitter blog, July 3, 2014. https://blog.twitter.com/2014/penalty-kicks-as-seen-through-twitter-data.

Roberts, D. 2011. "Lessons from Our Open News Trial." *The Guardian*, October 17. http://www.theguardian.com/help/insideguardian/2011/oct/17/guardian-newslist.

Robinson, S. 2007. "Someone's Gotta Be in Control Here." *Journalism Practice* 1 (3): 305–321.

Robinson, Sue. 2011. "'Journalism as Process': The Organizational Implications of Participatory Online News." *Journalism & Communication Monographs* 12 (3): 137–210.

Robinson, S. 2011. "Convergence Crises: News Work and News Space in the Digitally Transforming Newsroom." *Journal of Communication* 61 (6): 1122–1141.

Robinson, S. 2012. "Experiencing Journalism: A New Model for Online Newspapers." In *The Handbook of Global Online Journalism*, ed. Eugenia Siapera and Andreas Vegas, 59–72. Malden, MA: Wiley-Blackwell.

Robinson, S., and C. DeShano. 2011. "'Anyone Can Know': Citizen Journalism and the Interpretive Community of the Mainstream Press." *Journalism: Theory, Practice, Criticism* 12 (8): 1–20.

Robinson, S., and M. Schwartz. 2014. "The Activist as Citizen Journalist." In *Citizen Journalism: Global Perspectives*, vol. 2, ed. Stuart Allan and Einar Thorsen, 377–390. New York: Peter Lang.

Rogers, Everett M. 1982. "The Empirical and the Critical Schools of Communication Research." In *Communication Yearbook*, vol. 5 ed. Michael Burgoon, 123–144. New Brunswick, NJ: Transaction Books.

Rogers, E. M., J. W. Dearing, and S. Chang. 1991. "AIDS in the 1980s: The Agenda-setting Process for a Public Issue." *Journalism Monographs (Austin, Tex.)* 126.

Romano, Carlin. 1986. "What? The Grisly Truth About Bare Facts." In *Reading the News*, ed. Robert Karl Manoff and Michael Schudson, 38–78. New York: Pantheon Press.

Rosen, Jay. 2014. "When Starting from Zero in Journalism, Go for a Niche Site Serving a Narrow News Interest Well." Jay Rosen's PressThink, March 26. http://pressthink.org/2014/03/when-starting-from-zero-in-journalism-go-for-a-niche-site-serving-a-narrow-news-interest-well.

Rosenstiel, T., M. Just, T. Belt, A. Pertilla, W. Dean, and D. Chinni. 2007. *We Interrupt This Newscast: How to Improve Local News and Win Ratings, Too*. Cambridge: Cambridge University Press.

Ross, Andrew. 2003. *No-Collar. The Humane Workplace and Its Hidden Costs*. Philadelphia: Temple University Press.

Rowland, Robert. 2000. "Panorama in the Sixties." In *Windows on the Sixties: Exploring Key Texts of Media and Culture,* ed. Anthony Aldgate, James Chapman, and Arthur Marwick, 154–182. London: I.B. Taurus.

Russell, A. 2001. "Chiapas and the New News: Internet and Newspaper Coverage of a Broken Cease-fire." *Journalism* 2 (2): 197–220.

Russell, Adrienne. 2011. *Networked: A Contemporary History of News in Transition.* Cambridge: Polity Press.

Rutigliano, L. 2007. "Emergent Communications Network as Civic Journalism." In *Blogging, Citizenship and the Future of Media,* ed. M. Tremayne, 225–237. New York: Routledge.

Ryfe, David. 2012. *Can Journalism Survive? An Inside Look at American Newsrooms.* Cambridge: Polity Press.

Ryfe, D. M. 2013. *Can Journalism Survive: An Inside Look at American Newsrooms.* Hoboken, NJ: John Wiley & Sons.

Saguy, Abigail. 2013. *What's Wrong with Fat?* New York: Oxford University Press.

Sall, M. 2013. "The Optimal Post is 7 Minutes." Medium, December 2. https://medium.com/data-lab/the-optimal-post-is-7-minutes-74b9f41509b

Saltzis, K. 2012. "Breaking News Online: How News Stories Are Updated and Maintained Around-the-Clock." *Journalism Practice* 6 (5–6): 702–710.

Salwen, M. B. 2005. "Online News Trends." In *Online News and the Public,* ed. M. B. Salwen, B. Garrison,and P. D. Driscoll, 47–77. Mahwah, NJ: Lawrence Erlbaum Associates.

Samuelson, Paul A. 1954. "The Pure Theory of Public Expenditure." *Review of Economics and Statistics* 36:387–389.

Santana, Arthur D. 2014. "Virtuous or Vitriolic: The Effect of Anonymity on Civility in Online Newspaper Reader Comment Boards." *Journalism Practice* 8 (1): 18–33.

Sayes, Edwin. 2014. "Actor-Network Theory and Methodology: Just What Does It Mean to Say That Nonhumans Have Agency?" *Social Studies of Science* 44 (1): 134–149.

Scacco, Joshua M., Alexander L. Curry, and Natalie Jomini Stroud. 2015. "Digital Divisions: Organizational Gatekeeping Practices in the Context of Online News," *#ISOJ, The Official Research Journal of the International Symposium on Online Journalism* 5 (1): 106–123.

Scacco, J. M., A. Muddiman, and N. J. Stroud. (in press). "The Influence of Online Quizzes on Learning from the News." *Journal of Information Technology & Politics.*

Schafer, R. M. 1993. *The Soundscape: Our Sonic Environment and the Tuning of the World.* Chicago: Destiny Books.

Scanlon, T. 1972. "A Theory of Freedom of Expression." *Philosophy & Public Affairs* 1 (2): 204–226.

Schauer, F. 1998. "Principles, Institutions and the First Amendment." *Harvard Law Review* 112:84–121.

Schiffer, A. J. 2006. "Blogswarms and Press Norms: News Coverage of the Downing Street Memo Controversy." *Journalism & Mass Communication Quarterly* 83 (3): 494–510.

Schiller, D. 1981. *Objectivity and the News: The Public and the Rise of Commercial Journalism*. Philadelphia: University of Pennsylvania Press.

Schlesinger, P. 1978. *Putting "Reality" Together: BBC News*. London: Constable.

Schlesinger, Philip, and Gillian Doyle. 2015. "From Organizational Crisis to Multi-Platform Salvation? Creative Destruction and the Recomposition of News Media." *Journalism* 16 (3): 305–323.

Schmitz Weiss, Amy, and David Domingo. 2010. "Innovation Processes in Online Newsrooms as Actor-networks and Communities of Practice." *New Media & Society* 12 (7): 1156–1171.

Schradie, J. 2011. "The Digital Production Gap: The Digital Divide and Web 2.0 Collide." *Poetics* 39:145–168.

Schradie, J. 2012. "The Trend of Class, Race, and Ethnicity in Social Media Inequality." *Information Communication and Society* 15:555–571.

Schudson, Michael. 1986. "When? Deadlines, Datelines, and History." In *Reading the News*, ed. Robert Karl Manoff and Michael Schudson, 79–108. New York: Pantheon Press.

Schudson, M. 1999. *The Good Citizen*. New York: Free Press.

Schudson, Michael. 2008. *Why Democracies Need an Unlovable Press*. Cambridge: Polity.

Schudson, M. 2010. "News in the United States: Panic—And Beyond." In *The Changing Business of Journalism and Its Implication for Democracy*, ed. David A. L. Levy and Rasmus Kleis Nielsen, 95–106. Oxford: Reuters Institute for the Study of Journalism/University of Oxford.

Schudson, Michael. 2013. "Would Journalism Please Hold Still!" In *Rethinking Journalism: Trust and Participation in a Transformed News Landscape*, ed. Chris Peters and Marcel Broersma, 191–199. London: Routledge.

Schulhofer-Wohl, Sam, and Miguel Garrido. 2013. "Do Newspapers Matter? Short-Run and Long-Run Evidence from the Closure of *The Cincinnati Post*." *Journal of Media Economics* 26 (2): 60–81.

Selwyn, N. 2003. "Apart from Technology: Understanding People's Non-use of Information and Communication Technologies in Everyday Life." *Technology in Society* 25 (1): 99–116.

Shah, H., and M. Thornton. 2004. *Newspaper Coverage of Interethnic Conflict: Competing Visions of America*. Thousand Oaks, CA: Sage.

Shah, H., and M. Thornton. 1994. "Racial Ideology in U.S. Mainstream News Magazine Coverage of Black-Latino Interaction, 1980–1992." *Critical Studies in Mass Communication* 11:141–161.

Shaker, L. 2014. "Dead Newspapers and Citizens' Civic Engagement." *Political Communication* 31 (1): 131–148.

Sherman, A., and B. Womack. 2012. "Twitter Blocks Reporter after He Posts NBC Executive E-mail." *Bloomberg Business*, July 30, 2012. http://www.bloomberg.com/news/articles/2012-07-30/twitter-blocks-journalist-after-he-posts-nbc-executive-s-e-mail.

Shibutani, T. 1966. *Improvised News: A Sociological Study of Rumor*. Indianapolis: Bobbs-Merrill.

Shirky, C. 2008. *Here Comes Everybody: The Power of Organizing without Organizations*. London: Penguin.

Shirky, C. 2010. *Cognitive Surplus: Creativity and Generosity in a Connected Age*. London: Penguin.

Shoemaker, P., and D. Stephen Reese. 2014. *Mediating the Message in the 21st Century: A Media Sociology Perspective*. New York: Routledge.

Sigal, L. 1986. "Who? Sources Make the News." In *Reading the News*, ed. Robert Karl Manoff and Michael Schudson, 9–37. New York: Pantheon Press.

Siles, Ignacio, and Pablo Boczkowski. 2012. "Making Sense of the Newspaper Crisis: A Critical Assessment of Existing Research and an Agenda for Future Work." *New Media & Society* 14 (8): 1375–1394.

Silverman, C., ed. 2014. *Verification Handbook: A Definitive Guide to Verifying Content for Emergency Coverage*. Maastricht, The Netherlands: European Journalism Centre.

Simmel, G. 1950. *The Sociology of Georg Simmel*. New York: Simon and Schuster.

Singer, J. 2003a. "Who Are These Guys?: The Online Challenge to the Notion of Journalistic Professionalism." *Journalism* 4 (2): 139–163.

Singer, J. 2003b. "Campaign Contributions: Online Newspaper Coverage of Election 2000." *Journalism & Mass Communication Quarterly* 80 (1): 39–56.

Singer, J. 2004. "Strange Bedfellows? The Diffusion of Convergence in Four News Organizations." *Journalism Studies* 5:3–18.

Singer, J. 2006a. "Partnerships and Public Service: Normative Issues for Journalists in Converged Newsrooms." *Journal of Mass Media Ethics* 21 (1): 30–53.

Singer, J. 2006b. "The Socially Responsible Existentialist: A Normative Emphasis for Journalists in a New Media Environment." *Journalism Studies* 7 (1): 2–18.

Singer, J. 2006c. "Stepping Back from the Gate: Online Newspaper Editors and the Co-production of Content in Campaign 2004." *Journalism & Mass Communication Quarterly* 83 (2): 265–280.

Singer, J. 2009. "Separate Spaces: Discourse About the 2007 Scottish Elections on a National Newspaper Web Site." *International Journal of Press/Politics* 14 (4): 477–496.

Singer, J. 2010. "Norms and the Network: Journalistic Ethics in a Shared Media Space." In *Journalism Ethics: A Philosophical Approach*, ed. C. Meyers, 117–129. New York: Oxford University Press.

Singer, J. 2014. "User-Generated Visibility: Secondary Gatekeeping in a Shared Media Space." *New Media & Society* 16 (1): 55–73.

Singer, J. 2015. "Out of Bounds: Professional Norms as Boundary Markers." In *Boundaries of Journalism: Professionalism, Practices and Participation*, ed. Matt Carlson and Seth C. Lewis, 21–36. New York: Routledge.

Singer, Jane B., and Ian Ashman. 2009. "'Comment Is Free, but Facts Are Sacred:' User-generated Content and Ethical Constructs at the *Guardian*." *Journal of Mass Media Ethics* 24 (1): 3–21.

Singer, J., Alfred Hermida, David Domingo, Ari Heinonen, Steve Paulussen, Thorsten Quandt, Zvi Reich, and Marina Vujnovic. 2011. *Participatory Journalism: Guarding Open Gates at Online Newspapers*. Malden, MA: Wiley-Blackwell.

Sirkkunen, E., and Clare Cook. 2012. "Chasing Sustainability on the Net." Tampere, Finland: Tampere Research Centre for Journalism, Media and Communication. http://www.submojour.net.

Sloboda, J. 1981. "The Uses of Space in Music Notation." *Visible Language* 15 (1): 86–110.

Soley, L., and Robert L. Craig. 1992. "Advertising Pressures on Newspapers: A Survey." *Journal of Advertising* 21 (4): 1–10.

Smit, G., Yael de Haan, and Laura Buijs. 2014. "Visualizing News: Make It Work." *Digital Journalism* 2 (3): 344–354.

Smith, A., ed. 1998. *Television. An International History*. New York: Oxford University Press.

Smith, B. 2014. "Why Buzzfeed Doesn't Do Clickbait." Buzzfeed, November 6. http://www.buzzfeed.com/bensmith/why-buzzfeed-doesnt-do-clickbait.

Snow, David A. 2001. "Extending and Broadening Blumer's Conceptualization of Symbolic Interactionism." *Symbolic Interaction* 24 (3): 367–377.

Snyder, J., and David Strömberg, 2008. "Press Coverage and Political Accountability." *Journal of Political Economy* 118 (2): 355–408.

Socolow, Michael J. 2010. "'We Should Make Money on Our News': The Problem of Profitability in Network Broadcast Journalism History." *Journalism* 11 (6): 675–691.

Soloski, John. 2013. "Collapse of the US Newspaper Industry: Goodwill, Leverage and Bankruptcy." *Journalism* 14 (3): 309–329.

Soper, T. 2013. "How to Bypass the New York Times Paywall: Go to Starbucks." Geekwire, February 28. http://www.geekwire.com/2013/bypass-york-times-paywall-starbucks.

Sousa, H. 2006. "Information Technologies, Social Change and the Future: The Case of Online Journalism in Portugal." *European Journal of Communication* 21 (3): 373–387.

Sparrow, B. H. 1999. *Uncertain Guardians: The News Media as a Political Institution*. Baltimore, MD: The Johns Hopkins University Press.

Squires, C. R. 2002. "Rethinking the Black Public Sphere: An Alternative Vocabulary for Multiple Public Spheres." *Communication Theory* 12:446–468.

Squires, C. R., and S. J. Jackson. 2010. "Reducing Race: News Themes in the 2008 Primaries." *International Journal of Press/Politics* 15:375–400.

Stack, S. 2000. "Media Impacts on Suicide: A Quantitative Review of 293 Findings." *Social Science Quarterly* 81 (4): 957–972.

Stanford Encyclopedia of Philosophy. 2014. "Scientific Explanation." http://plato.stanford.edu/entries/scientific-explanation.

Stark, D. 2011. *The Sense of Dissonance: Accounts of Worth in Economic Life*. Princeton, NJ: Princeton University Press.

Starkman, D. 2010. "The Hamster Wheel." *Columbia Journalism Review*. http://www.cjr.org/cover_story/the_hamster_wheel.php.

Starkman, D. 2014. *The Watchdog That Didn't Bark: The Financial Crisis and the Disappearance of Investigative Journalism*. New York: Columbia University Press.

Stavelin, E. 2014. "Computational Journalism: When Journalism Meets Programming." Doctoral thesis, University of Bergen, Bergen, Norway.

Steensen, S. 2009. "What's Stopping Them? Toward a Grounded Theory of Innovation in Online Journalism." *Journalism Studies* 10 (1): 1–16.

Steensen, S. 2011. "Online Journalism and the Promises of New Technology: A Critical Review and Look Ahead." *Journalism Studies* 12 (3): 311–327.

Steensen, S. 2014. "Newsroom Innovation Research and the Problem of Hidden Normativity." Presentation at the International Communication Association, Seattle, May 23–26.

Steinmetz, G. 2004. "Odious Comparisons: Incommensurability, the Case Study, and 'Small N's' in Sociology." *Sociological Theory* 22 (3): 371–400.

Sterne, J. 2012. "Sonic Imaginations." In *The Sound Studies Reader*, ed. J. Sterne, 1–17. New York: Routledge.

Stonebely, S. 2013. "The Social and Intellectual Contexts of the U.S. 'Newsroom Studies' and the Media Sociology of Today." *Journalism Studies* 16 (2): 259–274.

Storey, J., Salaman, G., Platman, K. 2005. "Living with Enterprise in an Enterprise Economy: Freelance and Contract Workers in the Media." *Human Relations* 58 (8): 1033–1054.

Stroud, N. 2011. *Niche News: The Politics of News Choice*. New York: Oxford University Press.

Stroud, S. 2008. "John Dewey and the Question of Artful Communication." *Philosophy & Rhetoric* 41 (2): 153–183.

Stroud, S. 2011. *John Dewey and the Artful Life: Pragmatism, Aesthetics, and Morality*. University Park, PA: Pennsylvania State University Press.

Sukumaran, Abhay, Stephanie Vezich, Melanie McHugh, and Clifford Nass. 2011. "Normative Influences on Thoughtful Online Participation." In *Proceedings of the 2011 Annual Conference on Human Factors in Computing Systems—CHI '11*, 3401–3410.

Sullivan, M. 2012. "Questions and Answers on How *The Times* Handles Online Comments from Readers." *New York Times*, October 15. http://publiceditor.blogs.nytimes.com/2012/10/15/questions-and-answers-on-how-the-times-handles-online-comments-from-readers.

Sullivan, M. 2013. "A Model of Restraint in the Race for News." *New York Times*, April 21. http://www.nytimes.com/2013/04/21/public-editor/a-model-of-restraint-in-the-race-for-news.html.

Sunstein, C. 1994. *Democracy and the Problem of Free Speech*. New York: Free Press.

Sunstein, Cass. 2007. *Republic 2.0*. Princeton, NJ: Princeton University Press.

Susskind, R. 2014. "The Internet in the Law: Transforming Problem-Solving and Education." In *Society and the Internet*, ed. M. Graham and W. H. Dutton, 272–284. Oxford: Oxford University Press.

Sydell, L. 2015. "Now You Can Sign Up to Keep Drones Away from Your Property. *NPR All Tech Considered*, February 23. http://www.npr.org/blogs/alltechconsidere

d/2015/02/23/388503640/now-you-can-sign-up-to-keep-drones-away-from-your-property.

Sylvie, George, and Peter Gade. 2009. "Changes in News Work: Implications for Newsroom Managers." *Journal of Media Business Studies* 6 (1): 113–148.

Szabo, G., and B. A. Huberman. 2010. "Predicting the Popularity of Online Content." *Communications of the ACM* 53 (8): 80–88.

Tandoc, E. C. 2014. "Journalism Is Twerking? How Web Analytics Is Changing the Process of Gatekeeping." *New Media & Society* 16 (4): 559–575.

Tandoc, Edson, and Ryan J. Thomas. 2015. "The Ethics of Web Analytics: Implications of Using Audience Metrics in News Construction." *Digital Journalism* 3 (2): 243–258.

Tanizaki, J. 1977. *In Praise of Shadows*. Stony Creek, CT: Leete's Island Books.

Taylor, C. 1979. "What's Wrong with Negative Liberty?" In *The Idea of Freedom: Essays in Honor of Isaiah Berlin*, ed. A. Ryan, 175–193. Oxford: Oxford University Press.

Taylor, John B. 2007. *Economics*. 5th ed. New York: Houghton Mifflin.

Tenenboim-Weinblatt, K., and M. Neiger. 2014. "Print Is Future, Online Is Past: Cross-Media Analysis of Temporal Orientations in the News." *Communication Research* 42 (8): 1047–1067.

Thelen, K. 1999. "Historical Institutionalism in Comparative Politics." *Annual Review of Political Science* 2:369–404.

Thompson, D. 2015. "The Unbearable Lightness of Tweeting." *Atlantic* 16 (February). http://www.theatlantic.com/business/archive/2015/02/the-unbearable-lightness-of-tweeting/385484/.

Thompson, E. 2002. *The Soundscape of Modernity*. Cambridge, MA: MIT Press.

Thornton, L. J., and S. M. Keith. 2009. "From Convergence to Webvergence: Tracking the Evolution of Broadcast-Print Partnerships through the Lens of Change Theory." *Journalism & Mass Communication Quarterly* 86 (2): 257–276.

Thorsen, E. 2013. "Routinisation of Audience Participation: BBC News Online, Citizenship and Democratic Debate." In *The Media, Political Participation and Empowerment*, ed. Richard Scullion, Roman Gerodimos, Daniel Jackson, and Darren Lilleker. New York: Routledge.

Thorson, E. 2016. "Belief Echoes: The Persistent Effects of Corrected Misinformation." *Political Communication* 33 (3): 460–480.

Thurman, N. 2008. "Forums for Citizen Journalists? Adoption of User Generated Content Initiatives by Online News Media." *New Media & Society* 10 (1): 139–157.

Thurman, N., and S. Schifferes. 2012. "The Future of Personalization at News Websites: Lessons from a Longitudinal Study." *Journalism Studies* 13 (5–6): 775–790.

Tilly, C. 1985. "Models and Realities of Popular Collective Action." *Social Research* 52 (4):717–747.

Tilly, C. 1999. "From Interactions to Outcomes in Social Movements." In *How Social Movements Matter*, ed. M. Giugni, D. McAdam, and C. Tilly, 253–270. Minneapolis: University of Minnesota Press.

Trammell, K. D., A. Tarkowski, and A. M. Sapp. 2006. "Rzeczpospolita blogów [Republic of Blog]: Examining Polish Bloggers through Content Analysis." *Journal of Computer-Mediated Communication* 11 (3): 702–722.

Tremayne, M. 2007. "Harnessing the Active Audience: Synthesing Blog Research and Lessons for the Future of Media." In *Blogging, Citizenship and the Future of Media*, ed. M. Tremayne, 261–272. New York: Routledge.

Trilling, D., and K. Schoenbach. 2013. "Skipping Current Affairs: The Non-Users of Online and Offline News." *European Journal of Communication* 28 (1): 35–51.

Truter, C. D. 1984. "First Amendment Rights of Prisoners: Freedom of the Prison Press. *University of San Francisco Law Review* 18 (3): 599–616.

Tuchman, G. 1972. "Objectivity as Strategic Ritual: An Examination of Newsmen's Notions of Objectivity." *American Journal of Sociology* 77 (4): 660–679.

Tuchman, G. 1978. *Making News: A Study in the Construction of Reality*. New York: Free Press.

Turner, F. 2005. "Actor-Networking the News." *Social Epistemology* 19 (4): 321–324.

Turow, J. 1994. "Hidden Conflicts and Journalistic Norms: The Case of Self-coverage." *Journal of Communication* 44 (2): 29–46.

Turow, J., and N. Draper. 2014. "Industry Conceptions of Audience in the Digital Space: A Research Agenda." *Cultural Studies* 28 (4): 643–656.

Upshaw, James, Gennadiy Chernov, and David Koranda. 2007. "Telling More Than News: Commercial Influence in Local Television Stations." *Electronic News* 1 (2): 67–87.

U.S. Supreme Court. 1969. *Red Lion Broadcasting Co. v. FCC*. 395 U.S. 367.

Ursell, G. D. 2001. "Dumbing Down or Shaping Up? New Technologies, New Media, New Journalism." *Journalism* 2 (2): 175–196.

Usher, N. 2014. *Making News at The New York Times*. Ann Arbor, MI: The University of Michigan Press.

Usher, N. 2016. *Interactive Journalism: Hackers, Data, and Code*. Champaign: University of Illinois Press.

Van der Haak, B., M. Parks, and M. Castells. 2012. "The Future of Journalism: Networked Journalism." *International Journal of Communication* 6:16.

van Dijck, J. 2013. *The Culture of Connectivity*. New York: Oxford University Press.

van Dijk, T. A. 1991. *Racism and the Press*. London: Routledge.

Van Leuven et al. 2014. "Networking or Not Working?" *Journalism Practice* 8:552–562.

van Maanen, H. 2009. *How to Study Art Worlds: On the Societal Functioning of Aesthetic Values*. Amsterdam: Amsterdam University Press.

Vergeer, M. 2014. "Peers and Sources as Social Capital in the Production of News Online Social Networks as Communities of Journalists." *Social Science Computer Review* 33 (3): 277–297.

Vobi, I. 2011. "Online Multimedia News in Print Media: A Lack of Vision in Slovenia." *Journalism* 12 (8): 946–962.

Vignelli, M. 2010. *The Vignelli Canon*. London, UK: Lars Muller.

Vis, Farida. 2013. "Twitter as a Reporting Tool for Breaking News: Journalists Tweeting the 2011 UK Riots." *Digital Journalism* 1 (1): 27–47.

Vos, Tim P., and Jane B. Singer. 2016. "Media Discourse about Entrepreneurial Journalism: Implications for Journalistic Capital." *Journalism Practice* 10 (2): 143–159.

Vu, H. T. 2014. "The Online Audience as Gatekeeper: The Influence of Reader Metrics on News Editorial Selection." *Journalism* 15 (8): 1094–1110.

Wacjman, J. 2015. *Pressed for Time: The Acceleration of Life in Digital Capitalism*. Chicago: University of Chicago Press.

Waisbord, S. 2013. *Reinventing Professionalism: Journalism and News in Global Perspective*. New York: John Wiley & Sons.

Waldman, S. 2011. *The Information Needs of Communities: The Changing Media Landscape in a Broadband Age*. Washington, DC: Federal Communications Commission. www.fcc.gov/infoneedsreport.

Wall, M. 2005. "'Blogs of War': Weblogs as News." *Journalism* 6 (2): 153–172.

Ward, S. J. 2005a. *Invention of Journalism Ethics: The Path to Objectivity and Beyond*. Kingston, ON: McGill-Queen's Press-MQUP.

Ward, S. J. 2005b. "Journalism Ethics from the Public's Point of View." *Journalism Studies* 6 (3): 315–330.

Ward, S. J. 2009. "Journalism Ethics." In *The Handbook of Journalism Studies*, ed. Karin Wahl-Jorgensen and Thomas Hanitzsch, 295–308.

Ward, S. J. 2010. *Global Journalism Ethics*. Kingston, ON: McGill-Queen's University Press.

Ward, S. J. 2011. *Ethics and the Media: An Introduction*. Cambridge: Cambridge University Press.

Ward, S. J. 2014. "The Magical Concept of Transparency." In *Ethics for Digital Journalists: Emerging Best Practices*, ed. Lawrie Zion and David Craig, 45–58. New York: Routledge.

Watt, I. 1957. *The Rise of the Novel: Studies in Defoe, Richardson, and Fielding*. Berkeley, CA: University of California Press.

Weber, M. 1978. *Economy and Society: An Outline of Interpretive Sociology*. Berkeley, CA: University of California Press.

Webster, J. 2014. *The Marketplace of Attention: How Audiences Take Shape in a Digital Age*. Cambridge, MA: MIT Press.

Wemple, E. 2014a. "Associated Press Polices Story Length." *Washington Post*, May 12. http://www.washingtonpost.com/blogs/erik-wemple/wp/2014/05/12/associated-press-polices-story-length.

Wemple, E. 2014b. "Reuters Polices Story Length Too." *Washington Post*, May 12. http://www.washingtonpost.com/blogs/erik-wemple/wp/2014/05/12/reuters-polices-story-length-too.

Westbrook, R. 1991. *John Dewey and American Democracy*. Ithaca, NY: Cornell University Press.

Westlund, O. 2011. *Cross-Media News Work: Sensemaking of the Mobile Media (r)evolution*. Gothenburg, Sweden: University of Gothenburg.

Westlund, O. 2013. "Mobile News: A Review and Model of Journalism in an Age of Mobile Media." *Digital Journalism* 1 (1): 6–26.

White, D. M. 1949. "The 'Gatekeeper': A Case Study in the Selection of News." *Journalism Quarterly* 27:383–390.

Williams, A., C. Wardle, and K. Wahl-Jorgensen. 2010. "'Have They Got News for Us?' Audience Revolution or Business as Usual at the BBC?" *Journalism Practice* 5 (1): 85–99.

Williams, B., and M. X. Delli Carpini. 2011. *After Broadcast News: Media Regimes, Democracy, and the New Information Environment (Communication, Society and Politics)*. New York: Cambridge University Press.

Williams, E. 2014. "A Less Long, More Connected Medium." Medium, February 24. https://medium.com/the-story/a-less-long-more-connected-medium-c345db2d6a56.

Williams, R. 1973. *Television: Technology and Cultural Form*. London: Routledge Classics.

Williams, W. 1915. "The Journalist's Creed." *The Editor and Publisher and the Journalist* 48 (1): 538.

Witschge, T. 2012. "Changing Audiences, Changing Journalism?" In *Changing Journalism*, ed. P. Lee-Wright, A. Phillips, and T. Witschge, 117–134. London: Routledge.

Willnat, L., and H. David Weaver. 2014. *The American Journalist in the Digital Age: Key Findings*. Bloomington: School of Journalism, Indiana University. http://news.indiana.edu/releases/iu/2014/05/2013-american-journalist-key-findings.pdf.

Witschge, T. 2013. "Transforming Journalistic Practice: A Profession Caught between Change and Transition." In *Rethinking Journalism: Trust and Participation in a Transformed News Landscape*, ed. Chris Peters and Marcel Broersma, 160–172. New York: Routledge.

Witschge, T., and Gunnar Nygren. 2009. "Journalism: A Profession under Pressure?" *Journal of Media Business Studies* 6 (1): 37–59.

Wodak, R., and M. Meyer, eds. 2009. *Methods of Critical Discourse Analysis*, 2nd ed. London; Thousand Oaks, CA: Sage Publications Ltd.

Woolgar, S. 1999. "Analytic Scepticism." In *Society on the Line*, ed. W. H. Dutton, 335–337. Oxford: Oxford University Press.

Wortham, J. 2012. "With Twitter, Blackouts and Demonstrations, Web Flexes Its Muscle." *New York Times*, January 18. http://www.nytimes.com/2012/01/19/technology/protests-of-antipiracy-bills-unite-web.html.

Wyndham, G. 1977. *Facing the Nation: Television and Politics, 1936–1976*. London: Bodley Head.

Xu, W. W., and M. Feng. 2014. "Talking to the Broadcasters on Twitter: Networked Gatekeeping in Twitter Conversations with Journalists." *Journal of Broadcasting & Electronic Media* 58:420–437.

Yglesias, M. 2015. "Refreshing the Evergreen." Vox, January 15. http://www.vox.com/2015/1/15/7546877/evergreen-experiment.

Yoneyama, M., J. i. Fujimoto, Y. Kawamo, and S. Sasabe. 1983. "The Audio Spotlight: An Application of Nonlinear Interaction of Sound Waves to a New Type of Loudspeaker Design." *Journal of the Acoustical Society of America* 73 (5): 1532–1536.

Young, Mary Lynn, and Alfred Hermida. 2015. "From Mr. and Mrs. Outlier to Central Tendencies." *Digital Journalism* 3 (3): 381–397.

Zamith, R. 2017. "On Metrics-Driven Homepages." *Journalism Studies*. Advance online publication. doi:10.1080/1461670X.2016.1262215.

Zamith, R., and Seth C. Lewis. 2014. "From Public Spaces to Public Sphere: Rethinking Systems for Reader Comments on Online News Sites." *Digital Journalism* 2 (4): 558–574.

Zelizer, B. 2004. *Taking Journalism Seriously: News and the Academy*. Thousand Oaks, CA: SAGE Publications.

Zelizer, B. 2015. "Terms of Choice: Uncertainty, Journalism, and Crisis." *Journal of Communication* 65 (5): 888–908.

Zickuhr, K. and Aaron Smith. 2013. "Home Broadband 2013." Pew Research Center. http://www.pewinternet.org/2013/08/26/home-broadband-2013.

Zion, Lawrie, and David Craig. 2015. *Ethics for Digital Journalists: Emerging Best Practices*. New York: Routledge.

Zuckerman, E. 2009. "Why We Fall for Fast News." http://www.ethanzuckerman.com/blog/2009/11/09/why-we-fall-for-fast-news.

Index

Absences/whitespace
democratic listening and, 131–132
as democratic self-governance, 129–132
historical meanings of press, 135–137
infrastructural holes and, 142–143
invisible audiences and, 140–141
as meaningful across domains, 132–135
media avoidance and, 139–140
organizational openings and, 141–142
professional norms and, 138–139
sources in the networked press, 137–143
storytelling and true, 149
See also Silence
Academia, 157–158, 174–176
collaboration between journalism industry and, 171, 211–212
contributions to news, 167–174
See also Business and democratic objectives
Achter het nieuws, 238
Actor-network theory (ANT), 30–36, 37, 284
constructionist conception of knowledge, 45n7
explanations offered by, 40–41, 45n8
problematization of normativity, 34
quality and, 35–36
systematic variations and, 39–41
Adams, Guy, 143
Administrative approach, critical orientation, 171–174
Administrative orientation, not critical approach, 169–170
Advertising
entrepreneurial journalism and, 196–197, 213
online, 16–17, 55
revenue decrease of newspaper, 177
Affective news, 153–154
Affective Publics, 257–258
Affordable Care Act, 100
Alexander, Jeffrey, 27, 33
Algorithmic journalism, 119, *120*, 120–121, 278
infrastructural holes and, 142–143
All the President's Men, 214
Ambient journalism, 118–119, *120*, 278
Analytic skepticism, 83
Ananny, Mike, 8, 18, 277, 285
commentary on, 148, 149, 151, 152, 153
See also Absences/whitespace
Anderson, C. W., 2, 6, 31, 41, 83, 220, 278, 280, 284, 285
on change as institution driven, 43
on comparative news ethnographies, 38–39
findings related to legacy inertia, 29

Anderson, C. W. (cont.)
on genealogical ethnography, 72–75, 76–79, 84
on journalists' understanding of their audiences, 178
on network analysis, 229
on primary work of journalism and struggle to be open/networked, 19–20
on "rebuilding the news," 260
See also Newsrooms
Antinoise campaigns, 133
AOL, 184, 214
Apple, 245
Arab Spring uprisings, 138
Arkin, William, 102
Armstrong, Jesse, 139, 140
Artful journalism, 212
Art worlds, 113–114, 124, 128n1, 278
Associated Press, *120*, 187
Associated Press Sports Editors (APSE), 116
Atlantic, 185, 188
Attention minutes metric, 186
Audiences
adjustment, 270
behavioral data, 165–166
data, 177–178
empowerment through citizen journalism, 255–259
invisible, 140–141
power and authority, 219–220, 275–276
relationship of online journalism to, 20–21
Authority, journalistic, 217–218
counterpublics and, 227, *230*
critical discourse analysis of, 226–227, *230*
critical framework of offline hierarchies and online networks, 220–224
defined, 218
different dimensions of, 224–226
field theory and, 228–229, *230*
institutional approach to, 218–220
network analysis of, 229, *230*
power and, 219–220, 231–232, 233n4, 268, 269–270
research frameworks, 226–231
science and technology studies (STS) and, 229–231
Autonomy, individual, 131

Baggi, Guia, 246
Balaji, M., 223–224
Balancing test, 179–180
Barnhurst, K. G., 17–18
BBC, 143, 238, 239–240
Beck, Ulrich, 244
Becker, Howard, 7, 28, 30, 112
on art worlds, 128n1, 278
on myth of artists working alone, 124
on sound mixing, 121
See also Worlds
Beckett, C., 231
Behind the News, 239–240, 241
Bell, Melissa, 104
Benson, Rodney, 6, 60, 64–65, 228, 233n4, 270
commentary on, 82, 83
"who, what, when, where, why and how" of journalism and, 278, 281, 284, 287
See also Descriptivism
Berry, D., 178
Big data analysis, 44n3, 142
"Black hole" shots, 137
Blair, Jayson, 268
Blind spots in journalism studies, 48–58, 85
Blogs, 21, 101, 120, 219
counterpublics, 227
critical framework of offline hierarchies and online networks of, 220–224

journalistic authority and, 220–224
live, 254
Bloomberg, 185
Boczkowski, Pablo J., 2, 5, 10–11, 81, 257
on collapse of daily news cycle into continuous publishing, 18–19, 279
on descriptive focus of research, 27–28
Digitizing the News, 64, 76, 261
on excitement of new technologies, 252–253
on gap between editors' and audiences' preferences, 21, 190, 248
on innovation in online journalism, 17, 274–275
on local factors in adoption of digital technologies, 43
on mainstream media response to online competition, 16
on systematic variation, 37
on types of news most often read by visitors, 165–166
on user-generated content, 20
"who, what, when, where, why and how" of journalism and, 278, 284, 285
Bosk, Charles, 268
Boston Marathon bombing, 2013, 138
Bourdieu, Pierre, 203, 228
Bowker, Geoff, 78
Brandpunt, 238
Braun, Joshua A., 34, 39, 124
Brennen, J. S., 49
Bridge decay, 22–23
Broadcast News, 214
Bruns, A., 259–260
Buijs, Laura, 124
Business and democratic objectives, 159
administrative approach, critical orientation, 171–174
administrative orientation, not critical approach, 169–170
harmful outcomes, 160–162
helpful outcomes, 166–167
nonadministrative quadrants, 170–171
off-diagonal quadrants, 162–166
BuzzFeed, 43, 101, 104, 106, 140, 181
clickbait and, 186
as commercial metric outlet, 184

Cage, John, 132–133
Can Journalism Survive?, 261
Capella, Joseph N., 165
Carey, James, 254, 282–283
Carlson, Matt, 123, 217, 225–226, 231
Carter, Jimmy, 98
Carvin, Andy, 119, *120*, 138
"Carving out quiet," 139
Cassidy, W. P., 28
Castells, Manuel, 18, 27, 32, 39, 219–220
network society theory, 31–32, 83
Causal mechanisms, 41
Cawley, A., 280
CBS, 240
Radio, 169
Censorship, 135–136, 137
Center for Public Integrity (CPI), 183, 188
Chadwick, Andrew, 260
Chaffee, Steven H., 40
Challenger Disaster, The, 268
Chartbeat, 37–38, 62, 186, 187, 189
Chicago Daily News, 97
Chicago Defender, 97
Chicago Sun, 97
Chicago Times, 97
Chicago Tribune, 97, 120
Christin, Angele, 38, 41
"Churnalism," 102
Cincinnati Post, 162
Citizen journalism, 255–259, 285
Citizen Kane, 214
Civic journalism, 197
Clarke, Debra, 263
"Click bait," 59, 186

Clinton, Bill, 98
CNN, 100–101, 104, 106
Coddington, Mark, 17
Collaborative nature of journalism, 123–125, 128n1
Collective activity, 112, 114–115, *120*, 128n1
Collective intelligence streams, 118
Columbia Journalism Review, 120
Comment sections, 171–173
Commercial metrics outlets, 184–185
Common Knowledge, 9, 212
Compaine, Ben, 207
Comparative case studies, 284
Competition, 249
Constructionist conception of knowledge, 45n7
Content management systems (CMSs), 74
Contextual journalism, 102
Conventions, 115, *120*, 125
Conversation, 246
Cook, Timothy, 219
Corporate libertarianism, 56
Correspondent, 246
Cottle, Simon, 258
Counterpublics, 227, *230*, 284
Creative destruction, 260
Crigler, Ann, 212
Crisis narratives for journalism, 47–49
Critical discourse analysis (CDA), 226–227, *230*, 284
Crowdfunding, 198, 202
Cultural pragmatics, 33
  systematic variations and, 42
Current affairs television, emergence of, 236–242
Curry, Alex, 173

Daily news cycle, 18–19, 279
Data Blog, 120
Data journalism, 48, 61, 116–117, 278
  ethics and, 189
  formal educational programs, 121
  technology and change in, 119–123
  world of, 118–119
Davis, Murray, 82–83, 87
Deciding What's News, 1, 45n5, 63, 85
Decluttering movement, 134
De Haan, Yael, 124
Delli Carpini, Michael X., 3–4, 5, 7, 11, 32
  on democratically relevant, politically useful media, 32
Democratic listening, 131–132
Democratic self-governance, meaningful absence as, 129–132
Descriptive data, 189
Descriptivism, 27, 82, 284
  actor-network theory (ANT) and, 30–31
  "big data" analysis, 44n3
  cultural pragmatics and, 33
  history of use in humanities and social sciences, 28–29
  network society theory (NS) and, 31–32
  old versus new, 27–34
  on systematic variation, 36–43
  unanswered questions in new, 34–43
  what's at stake in new, 34–36
  on "why" of systematic variation, 39–43
Deuze, Mark, 10, 256, 267, 271
  "who, what, when, where, why and how" of journalism and, 275, 278, 280, 283, 284, 285, 287
  See also Entrepreneurial journalism; Innovation
Dewey, John, 157–159, 175, 212, 280
  commonalities with Lippmann, 175–176
  on "great community," 163–164, 165
  on problems facing democracy, 166–167, 174
Digg, 143

Digital divide, 53
Digital exuberance, 48, 85
Digital journalism
  bridge decay, 22–23
  challenges to established professional dynamics, 19–21
  comments sections, 171–173
  deciding what's scholarship in, 85–88
  digital ad revenues, 16–17, 55
  digital divide and, 53
  digital exuberance, 48, 85
  empowerment of citizens and, 255–259, 272
  excitement about new technologies in, 252–255
  failure in, 251–265, 268, 286
  "hamster wheel," 187
  historical context and market environment, 16–17
  innovation processes, 17–18, 117–118
  legacy inertia in, 29
  mobile devices and, 200–201, 210, 248
  network analytic perspective of, 22
  normalization of, 19
  pathways ahead for study of, 21–25
  practices of, 18–19
  relationship with audiences and role of user-generated content in, 20–21
  "studying up" and the excitement of the new in, 251–252
  study of, 15–16, 26, 83–84
Digital Journalism, 30
Digital news, 91–93
  as *a* form of knowledge, 94–97
  as *forms* of knowledge, 97–105
  metrics, 50
  as new chapter in sociology of knowledge, 106–108
  news-about-relations, 92, 107–108
  news-as-impressions, 92, 277
  news-as-items, 92, 103, 107–108, 277
Digitizing the News, 64, 76, 261
Disconnection, 134
Documents and objects of evidence, 75–76
Domingo, David, 30, 34
Doudaki, V., 18
Doyle, Gillian, 260
Drucker, Peter, 202
Drudge Report, The, 43
Dutton, William H., 6–7, 276, 278, 280, 284

Economics of journalism, 54–58, 60
  entrepreneurial journalism and, 199–203
Economic theory, 203
Elizabeth, Queen, 240
Empowerment through new technologies, 255–259, 272
Encoding and decoding, 269
Engaging News Project, 163, 173
Entrepreneurial journalism, 235–236, 285
  defined, 195
  economic imperatives in, 199–203
  emergence of global start-up culture in, 242–247
  journalistic roles in, 203–210
  mobile devices and, 200–201, 210, 248
  normative boundaries and, 195–199, 204, 213
  value propositions in, 206
  See also Innovation
Esser, Frank, 38
Ethical metrics, 185–186
Ethics, journalism, 177–178, 192–193, 250, 285
  archetypes of, 182–188
  balancing test, 179–180
  commercial metrics outlets, 184–185
  individual journalists and, 188–190
  new research on, 190–192
  old values, new tensions in, 187–188
  as process driven, 178

Ethics, journalism (cont.)
professionalization of journalism and, 179
traditional ideal, 182–183
Ethnography. *See* Newsrooms
European Journalism Center, 138
Events turned into stories, 150–152
Explainer, The, 185
Explanatory research, 82–83

Facebook, 5, 39, 100, 101, 102, 245, 254, 270
comments sections, 173
journalistic authority and, 231
niche-community actors and, 223
"where" of journalism and, 281
Failure in digital journalism, 268, 286
approaches to understanding, 263–265
forms of, 263–264
inevitability of, 265, 271
studies of, 259–263
Fairness Doctrine, 51
Fallows, James, 98–99
Fenton, Natalie, 35
Ferree, Myra Marx, 35
Ferrier, Michelle Barrett, 208
Field theory, *230*, 284
constructionist conception of knowledge and, 45n7
journalistic authority and, 228–229
power relations and, 233n4
Fieldwork, 75
"Fireside chats," 97
First Amendment, U.S. Constitution, 51, 130, 196
FiveThirtyEight, 116, 120, *120*
as ethical metrics outlet, 185
Flow of impressions, 103
"Fly-in" news coverage, 281
Focus, 239, 240, 241
Ford, Franklin, 157, 176
Forgive and Remember: Managing Medical Failure, 268
Formal realism, 117
Foucault, Michel, 76–78
Fox, 41
Freedom of speech, 130–131
Freedom of the press, 50–51, 52
Frontline, 102

Galbraith, John Kenneth, 82
Gamson, William, 35
Gans, Herbert, 1, 28, 45n5, 63, 76, 85
Gap between editors' and readers' preferences, 21, 190–192, 248
Garrido, Miguel, 162
Gatekeeping function of journalists, 21, 208
Gauntlett, David, 256
Gawker, 38, 181
as commercial metric outlet, 184, 185
Geertz, Clifford, 4, 28, 30
Genealogical ethnography, 72–75, 76–79, 84
Generalizabiliy, 40
Gerhards, Jürgen, 35
Global Editors Network, 121
Global journalism, 282
Good Night and Good Luck, 214
Google, 105, 123, 142, 191, 220, 245
Fusion, 119
Goyanes, M., 17
Grand theory, 40
Grassroots media production, 256
Graves, Lucas, 38
Gross, Terry, 136
Guardian, 32, 119, 120, 122
media avoidance and, 139
"Open Newslist," 142
technological infrastructures, 141
"Gutenberg Parenthesis," 77

Hacker journalism, 48
Hahn, Kyu S., 164
Hall, Stuart, 269
Hallin, Dan, 38, 41, 281

Hamby, Chris, 183
"Hamster wheel" journalism, 187
Hanitzsch, Thomas, 38, 205
Hara, Kenya, 132
Harris, John F., 246
HBO, 41
Hemmingway, Emma, 30
Hermida, Alfred, 101, 254
Hindman, Matthew, 9, 212, 222, 229, 275, 284
 See also Ethical metrics
Historiography, 63–64, 84–85, 280, 284
Hoag, Anne, 207
Hogan, Hulk, 185
Horserace coverage, 165
"How and why" of journalism, 282–285
Huffington Post, 43, 98, 181
 as commercial metric outlet, 184
Huffington Post Investigative Fund, 25
Hybridity, 151–152
Hyperlinking, 73, 74

Independent, 143
Infographics, 167
Information as public good, 24, 55
Infrastructural holes, 142–143
Innovation, 235–236, 267–268, 285
 amnesia, 84
 competition and, 249
 emergence of current affairs television, 236–242
 emergence of global start-up culture and, 242–247
 processes in online journalism, 17–18, 117–118
 See also Entrepreneurial journalism
Innovation and Entrepreneurship, 202
Innovation labs, 62
In Praise of Shadows, 133
Instagram, 101, 254
Institutional approach to journalistic authority, 218–220
Intentional whitespace, 132
International Reporting Program, 25
Intrapreneuralism, 195
Introversion, 134
Investigative Reporters and Editors (IRE), 116
Invisible audiences, 140–141
IRPI, 246
Isolation, 134
 in mediated information flows of blogs, 222–223
Iyengar, Shanto, 164

Jacobs, Ron, 33
James, William, 91, 92–93, 94–95, 100, 277
Jamieson, Kathleen Hall, 165
Jansen, Sue Curry, 175
Jarvis, Jeff, 73
Jaspan, Andrew, 246
Journalism policy, 52–54, 60
Journalism Practice, 254
Journalism studies, 1, 282, 284–285
 blind spots in, 48–58, 85
 changes in modern journalism and, 1–2, 273–276
 conferences, 3–4
 deciding what's scholarship in, 85–88
 economics of journalism, 54–58, 60, 202–203
 establishing two-way street in with other disciplines and fields, 81–82, 284
 explanatory research in, 82–83
 on failed practices and institutions, 259–263
 focus on words, 1, 4
 journalism policy, 52–54
 journalistic authority, 226–231
 journalistic roles, 207–210
 market ontology, 52, 58–60
 multiplication of subfields in, 267–272
 new challenges for maturing field of, 81–85

Journalism studies (cont.)
  new descriptivism in (*See* Descriptivism)
  normative foundations of, 49–52, 60, 198–199
  ongoing journalism crisis and, 47–49
  public service journalism and, 24
  recent growth in, 47
  refocusing of, 60
  sense of destabilization in, 4–5
  studying up in, 251–252
  "why" of, 285–287
Journalistic Authority, 225
"Journalist's Creed, The," 177, 180
Just, Marion, 212

Kim, Ye Weon, 167
Kipling, Rudyard, 273
Kirk, Andy, 124
Kizen, 132
Klein, Ezra, 188
Klinenberg, Eric, 28
Knight News Challenge, 157
Knowledge, 91–93, 148, 279
  digital news as *forms* of, 97–105
  digital news as new chapter in sociology of, 106–108
  news as *a* form of, 94–97
  tacit, 95
Konieczna, Magda, 17, 38
Kreiss, Daniel, 3, 31, 38, 41, 64
  on normative theory, 49

Lang, G. E., 152–153
Lang, K., 152–153
La Silla Vacía, 246
Latour, Bruno, 27, 36, 39, 40–41
Lazarsfeld, Paul, 158, 168, 169, 171, 174
Lee, Eun-Ju, 167
Legacy intertia, 29
Legitimatized power, 219
Lehrer, Brian, 138
León, Juanita, 246
Levy, M. R., 202
Lewis, Paul, 119
Lewis, Seth C., 7, 19, 124, 125, 231
  commentary on, 148, 149, 151, 152, 154
  "who, what, when, where, why and how" of journalism and, 278, 283, 285
  See also Worlds
Life, 237
LinkedIn, 245
Lippmann, Walter, 147, 157–159, 189
  commonalities with Dewey, 175–176
  on "great community," 175
  on "perfect citizen," 163–164
  on problems facing democracy, 166–167
  on role of newspapers, 168
Liquid journalism, 256
Listening, democratic, 131–132
Literary journalism, 125–126
Little, 263
Liveblogging, 254
Local newspapers, 96–97, 264–265, 270–271
Los Angeles Times, 119, 120–121

Madrigal, Alexis, 107
Making News at the New York Times, 1, 45n5, 63–64
Management theory, 203
Mancini, Paolo, 38, 41
Manovich, Lev, 104–105
Mansell, R., 231
March of Time, The, 237
Marginalized groups, 219–220, 231–232, 268, 269–270
Market-driven journalism, 205–206
Market failure, 56–58
Market ontology, 52
  challenging, 58–60
Marketplace of Attention, The, 191

Marshall, John, 222
Marx, Leo, 85
Masip, P., 30, 34
Massing, Michael, 43–45
Matheson, Donald, 272
McCarthy, Joseph, 237, 239
McFadden Publications, 169
McHugh, Melanie, 172
Meadows, Laura, 38, 41
Media avoidance, 139–140
Mediality, 152–153
Mediapart, 246
Medium, 185, 186, 188
Meijer, I. Costera, 30, 34
Meiklejohn, Alexander, 51, 129, 130
Meraz, S., 21
Merit goods, 56
Merton, Robert, 39–40
Metrics, 37–38
  commercial, 184–185
  ethical, 185–186
  news, 50, 62
  return on investment, 163
Middle-range sociological theories, 39–40
Mieklejohn, Alexander, 130
Millennials, 200–201
Mitchelstein, Eugenia, 5, 11, 81, 257, 278, 284, 285
  on collapse of daily news cycle, 279
  on descriptive focus of research, 27–28
  on gap between editors' and audiences' preferences, 21, 190, 248
  on innovation in online journalism, 274–275
  on systematic variation, 37
  on types of news most often read by visitors, 165–166
  on user-generated content, 20
Mobile devices, 200–201, 210, 248
Mobile news, 254
MSNBC.com, 124
Muddiman, Ashley, 173
Murrow, Edward R., 237, 241
Mutz, Diana C., 165

Naldi, Lucia, 205
Napoli, P. M., 21
Narrative journalism, 116–117
Narrative Science, 4
Nass, Clifford, 172
National Institute for Computer-Assisted Reporting (NICAR), 116, 120
National Press Club, 180
NBC News, 34, 143
Negotiated meaning, 269
Neiger, M., 19
Network analysis, 229, *230*, 284
Networked journalism, 18, 22, 31–32, 48, 144–145
  affective news and, 153–154
  blogs and, 220–224
  sources of absence in, 137–143
Network society (NS) theory, 31–32, 83
Neuman, W. Russell, 9, 278
"News as a Form of Knowledge: A Chapter in the Sociology of Knowledge," 91
News Nerd Job, 122
Newspapers
  decline in circulation, 177, 273
  expected to turn a profit, 158–159
  harmful business and democratic outcomes, 160–162
  Lippmann on role of, 168
  local, 96–97, 264–265, 270–271
  online, 16
Newsrooms
  -centricity and failure, 262
  ethnographies and history, 1970s and 1980s, 65–72
  ethnography research projects, 61
  field research on newsroom culture and, 62–63

Newsrooms (cont.)
genealogical ethnography of, 72–75, 76–79
historicity of, 63–64, 84–85, 280
journalism, documents, and objects of evidence in, 75–76
journalistic roles in entrepreneurial, 203–210
online metrics and, 62
traces of the past in digital, 61–65
News values, 151
Newsvine, 34
New York Review of Books, The, 43
New York Times, 104, 106, 119, 120, 265, 278
bridge decay and, 23
continuous publishing, 18–19
descriptivist profiles of, 28–29
digital news as *forms* of knowledge, 98
as ethical metrics outlet, 185
failures, 268
influence of advertisers and audiences on, 197
limits on nonsubscribers, 141
metrics and, 38
ownership, management, and funding, 32
professional norms, 138–139
rewards for viral pieces with more traffic, 181
"right to be forgotten" and, 143
self-censoring by, 135
as traditional ideal, 182
visual guide to Iraq-ISIS conflict, 102
NFL Internet Group, 23
Niche-community actors, 223
Nielsen, Rasmus Kleis, 7, 31, 126, 202
commentary on, 148, 151–152, 154
"who, what, when, where, why and how" of journalism and, 277, 280, 287
See also Knowledge
Nieman Foundation, 44, 120
Nietzsche, Friedrich, 77
Nonadministrative quadrants, 170–171
Normative boundaries and entrepreneurial journalism, 195–199, 204, 213
Normative foundations of journalism studies, 49–52
Norpoth, Helmut, 164
NPR (National Public Radio), 120, 138
technological infrastructures, 141

Obama, Barack, 31, 33, 98, 99
Objectivity, 276
Objects of evidence, 75–76
Ochs, Adolph, 197
Offline hierarchies, 220–224
Online journalism. *See* Digital journalism
Operation Iraqi Freedom, 135
Opinion journalism, 33
Organizational openings and whitespaces, 141–142
Osborne, Peter, 36

Panorama, 238, 239–240, 241
Papacharissi, Zizi, 8, 257–258, 278
Park, Robert E., 7, 91–93, 106, 108, 277
on digital news as *forms* of knowledge, 97–105
on news as *a* form of knowledge, 94–97
Parks, M., 18, 32
Paywalls, 50, 142, 201
PBS, 102
Peretti, Jonah, 104
Performance and Power, 33
Personal Influence, 169
Peters, J. D., 78
Petre, Caitlin, 38
Pew Research Center, 44, 48, 160–161, 177
Pfister, D. S., 222, 232
Philadelphia Daily News, 74

Picard, Robert, 19, 118, 202, 205, 274
Pickard, Victor, 6, 85, 278, 280, 284, 285
Pilhofer, Aron, 122
Plain Dealer, The, 55
Plenel, Edwy, 246
Plesner, Ursula, 30–31
Policy, journalism, 52–54, 60
Political economy, 284
Politico, 43, 246
Popular Science, 172
Postindustrial phase of journalism, 47–48
Power relations and journalistic authority, 219–220, 231–232, 233n4, 268, 269–270
Powers, Matthew, 231
Pragmatism, 158, 168
Prenger, Mirjam, 10, 267, 271
  "who, what, when, where, why and how" of journalism and, 275, 277, 278, 280, 283, 284, 285, 287
  See also Entrepreneurial journalism; Innovation
Press Gazette, 119
Priest, Dana, 102
Primo, Alex, 30
Principles of Psychology, The, 94
Print journalism, subsidization of, 55
Problematization of normativity, 34
Professional norms and absence, 138–139
Profit expectations for news media, 158–159
Project for Excellence in Journalism, 17
ProPublica, 25, 43, 120, 183
"Prosaic Organizational Failure," 263
"Prosumers," 52
Public and Its Problems, The, 174
Public good
  information as, 24, 35–36, 55
  silence as, 133–134
Public Opinion, 166–167
Public service journalism, 24, 250

Quality of public goods, 35–36
QuickTake, 185

Radio, 97
"Random acts of journalism," 21
"Rational ignorance," 140
Rationalization of audience understanding, 21
Razlogova, E., 144
Razorfish, 247
Rebuilding the News, 72–75
Reddit, 153
Red Lion decision, 51
Reeves, Byron, 165
Relational whitespace, 132
Remensperger, John, 38, 41
Reputation, 115
Return on investment (ROI), 163
Reuters, 44, 172, 187
Reynolds, Glenn, 222
"Right to Be Forgotten," 105, 143
Robinson, Sue, 10, 17, 267, 269–270
  "who, what, when, where, why and how" of journalism and, 275, 278, 285, 287
  See also Authority, journalistic
"Robot Journalism: Don't Wait 'Til It's Too Late," 121
Rocky Mountain News, 161
Rogers, E. M., 168
Romano, Carlin, 276–277
Roosevelt, Franklin D., 97
Rosen, Jay, 255–256, 275
Ross, Andrew, 247
Rucht, Dieter, 35
Ryfe, David, 29, 38, 41–42, 261–262

Salon, 43
Santana, Arthur D., 172
Scaco, Josh, 173
Schlesinger, Philip, 252, 260
Schmitz Weiss, Amy, 30
Schonfeld, Eric, 101

Schudson, Michael, 11, 111, 264
  "who, what, when, where, why and how" of journalism and, 276, 278, 279, 285
Schulhofer-Wohl, Sam, 162
Schwencke, Ken, 120–121
Science and technology studies (STS), 229–231
"Scoop and shun" dynamics, 136
Seattle Post-Intelligencer, 161
Secondary reporting, 18
See It Now, 237–238, 239, 241
Segmentation, 166
Self-censorship, 135–136, 137
Self-governance, meaningful absence as democratic, 129–132
Self-help and motivational books, 134
Shaker, L., 161
Shared value systems in blogs, 223–224
Shirky, Clay, 256–257
Shorty Awards, 119
"Shovelware," 99
Silence
  as elite governance through unspoken norms, 134
  professional norms and, 138–139
  as public good, 133–134
  religious, 134–135
  See also Absences/whitespace
"Silent bargains and silent routines," 136
Silver, Nate, 116, 120, *120*
Simplicity movement, 134
Simpsons, The, 41
Singer, Jane B., 9, 213–214
  "who, what, when, where, why and how" of journalism and, 275, 277, 278, 280, 284, 285
  See also Entrepreneurial journalism
60 Minutes, 182–183
Skimm, the, 167
Slate, 43, 185
Sloan, Robin, 102–103, 104
"Slow news" movement, 138, 148
Smit, G., 124
Smith, Ben, 186
Snapchat, 101, 245
Snyder, J., 161
Social media, 5, 223–224, 254, 269–270
  absences and infrastructural holes, 143
  enabling events to be turned into stories, 151–152
  invisible audiences, 140–141
  networked journalism and, 18
  non-adopters, 139
  role in social movements, 257–258
  "where" of journalism and, 281
Social movements, 257–258, 269
Social power, 33
Society for News Design (SND), 116
Society of Professional Journalists (SPJ), 116
Sociology of knowledge, 106–108
Solitude, 134
Somaiya, Ravi, 119
Sound studies, 133
"Specious present," 93, 96
Spirals of silence, 134
Spyridou, L.-P., 18
Squires, Catherine, 227
SRC-CON, 121
Stanford Encyclopedia of Philosophy, 39
Starkman, Dean, 187–188
Starr, Susan Leigh, 78
Start-ups. *See* Entrepreneurial journalism
Status, 115–116, *120*
  changes, 119–123
Storytelling
  affective news and, 153–154
  events and, 150–152
  hybridity in, 151–152
  mediality in, 152–153
  technologies of, 147–150
Streams of collective intelligence, 118
Strömberg, David, 161

Stroud, Natalie Jomini, 8–9, 211–212
"who, what, when, where, why and how" of journalism and, 284, 287
See also Business and democratic objectives
Structured Stories, 75–76
Studying up, 251–252
Style and ethics, 189–190
Subaltern counterpublics, 227
Subscriptions, paid, 17, 141
Subsidization of journalism, 55
Sukumaran, A., 172
Symbolic interactionism, 113, 148
Systematic variation, 36–39
across fields, 42–43
ANT explanations for, 40–41
causal mechanisms, 41
cross-national studies of, 41
range of voices and, 43
reasons for, 39–43

Tableau, 119
Tacit knowledge, 95
"Talk back," 223
Talking Points Memo, 43
Tanizaki, J., 133, 145
Technologically oriented journalism, 116–123, 283
Technology
change in journalism, 117–118
empowerment through new, 255–259, 272
excitement about, 252–255
increasing role in journalism, 111–113
status changes with, 119–123
storytelling, 147–150
worlds of ambient, data, and algorithmic journalism, 118–119
Television Code, 239
Television news, 137
current affairs and, 236–242
Tenenboim-Weinblatt, K., 19
Thiel, Peter, 185
This American Life, 137
Thought News, 157, 174, 176, 280
Times-Picayune, 55
Times Wire, 101
Tow Center for Digital Journalism, 44, 120, 173, 202
Townsley, Eleanor, 33
Traditional ideal organizations, 182–183
TripAdvisor, 23
Tuchman, Gaye, 1, 2, 63–64
Turow, Joseph, 3
Tuskegee Institute, 92
Twitter, 39, 77, 153, 219, 245
digital news and, 100, 101, 102
domination of contemporary scholarship on online journalism, 253
infrastructural holes and visits to, 143
New York Times and, 106
privacy policy, 143
role in social movements, 257
"where" of journalism and, 281
whitespace and, 140–141

U. S. Geological Survey, 121
Universal Declaration of Human Rights, 51
Upshot blog, 120, 185
Upworthy, 186
USA Today, 23
User-generated content, 20–21
empowerment through, 255–259
power and authority in, 219–220
Usher, Nikki, 28–29, 45n5, 125, 181, 262–263

Vaca Valley Star, 135, 142
Vacaville Prison, 135
Valuation, 112, 126–127
Value propositions, 206
Values, news, 151, 250, 285
VandeHei, Jim, 246
Van Der Haak, B., 18, 32

Van Dijk, Jose, 39
Van Maanen, H., 113
Vaughan, Diane, 268
Vezich, Stephanie, 172
Videojournalism, 254
Vine, 107
Virtuous journalists, 181
Vox Media, 43, 104, 120, 167, 185, 188
Vu, 163

Wabi sabi, 132
Wahl-Jorgenson, Karin, 267, 270, 271, 275, 278, 284, 287
  See also Failure in digital journalism
Wall Street Journal, 167
Ward, 179, 180
Washington, Booker T., 92
Washington Post, 102, 182, 185
  failures, 268
Webster, James, 191
Westlund, Oscar, 124, 231
WeWork, 75
"What" of journalism, 276–278
"When" of journalism, 279–280
"Where" of journalism, 280–282
Whig interpretation of journalism history, 254–255, 258
Whitespaces. *See* Absences/whitespace
"Who" of journalism, 277
"Why" of journalism studies, 285–287
Wijnberg, Rob, 246
Wikipedia, 39
Williams, B., 32
Williams, Walter, 177, 180, 189
Wire, The, 41
Woolgar, Steve, 83
Worlds
  ambient, data, and algorithmic journalism, 118–119, 278
  art, 113–114, 124, 128n1, 278
  as framework, 113–114, 123–127
  key concepts for understanding, 114–115
  reputation and status in, 115–116
  of technologically oriented journalism, 116–123, 283
  technology in journalism, 111–113
World Wide Web, 2, 248
Wyndham Goldie, Grace, 238, 239–240, 241

Yglesias, Matthew, 188
YouTube, 39, 254, 281

Zago, Gabriela, 30
Zamith, Rodrigo, 7
  commentary on, 148, 149, 151, 152, 154
  "who, what, when, where, why and how" of journalism and, 278, 283, 285
  See also Worlds
"Zombie containers," 244

Inside Technology

edited by Wiebe E. Bijker, W. Bernard Carlson, and Trevor Pinch

Pablo J. Boczkowski and C. W. Anderson, editors, *Remaking the News: Essays on the Future of Journalism Scholarship in the Digital Age*

Jess Bier, *Mapping Israel, Mapping Palestine: Occupied Landscapes of International Technoscience*

Brice Laurent, *Democratic Experiments: Problematizing Nanotechnology and Democracy in Europe and the United States*

Stephen Hilgartner, *Reordering Life: Knowledge and Control in the Genomics Revolution*

Benoit Godîn, *Models of Innovation: History of an Idea*

Cyrus Mody, *The Long Arm of Moore's Law: Microelectronics and American Science*

Harry Collins, Robert Evans, and Christopher Higgins, *Bad Call: Technology's Attack on Umpires and Referees and How to Fix It*

Tiago Saraiva, *Fascist Pigs: Technoscientific Organisms and the History of Fascism*

Teun Zuiderent-Jerak, *Situated Intervention: Sociological Experiments in Health Care*

Basile Zimmermann, *Technology and Cultural Difference: Electronic Music Devices, Social Networking Sites, and Computer Encodings in Contemporary China*

Andrew J. Nelson, *The Sound of Innovation: Stanford and the Computer Music Revolution*

Sonja D. Schmid, *Producing Power: The Pre-Chernobyl History of the Soviet Nuclear Industry*

Casey O'Donnell, *Developer's Dilemma: The Secret World of Videogame Creators*

Christina Dunbar-Hester, *Low Power to the People: Pirates, Protest, and Politics in FM Radio Activism*

Eden Medina, Ivan da Costa Marques, and Christina Holmes, editors, *Beyond Imported Magic: Essays on Science, Technology, and Society in Latin America*

Anique Hommels, Jessica Mesman, and Wiebe E. Bijker, editors, *Vulnerability in Technological Cultures: New Directions in Research and Governance*

Amit Prasad, *Imperial Technoscience: Transnational Histories of MRI in the United States, Britain, and India*

Charis Thompson, *Good Science: The Ethical Choreography of Stem Cell Research*

Tarleton Gillespie, Pablo J. Boczkowski, and Kirsten A. Foot, editors, *Media Technologies: Essays on Communication, Materiality, and Society*

Catelijne Coopmans, Janet Vertesi, Michael Lynch, and Steve Woolgar, editors, *Representation in Scientific Practice Revisited*

Rebecca Slayton, *Arguments that Count: Physics, Computing, and Missile Defense, 1949–2012*

Stathis Arapostathis and Graeme Gooday, *Patently Contestable: Electrical Technologies and Inventor Identities on Trial in Britain*

Jens Lachmund, *Greening Berlin: The Co-Production of Science, Politics, and Urban Nature*

Chikako Takeshita, *The Global Biopolitics of the IUD: How Science Constructs Contraceptive Users and Women's Bodies*

Cyrus C. M. Mody, *Instrumental Community: Probe Microscopy and the Path to Nanotechnology*

Morana Alač, *Handling Digital Brains: A Laboratory Study of Multimodal Semiotic Interaction in the Age of Computers*

Gabrielle Hecht, editor, *Entangled Geographies: Empire and Technopolitics in the Global Cold War*

Michael E. Gorman, editor, *Trading Zones and Interactional Expertise: Creating New Kinds of Collaboration*

Matthias Gross, *Ignorance and Surprise: Science, Society, and Ecological Design*

Andrew Feenberg, *Between Reason and Experience: Essays in Technology and Modernity*

Wiebe E. Bijker, Roland Bal, and Ruud Hendricks, *The Paradox of Scientific Authority: The Role of Scientific Advice in Democracies*

Park Doing, *Velvet Revolution at the Synchrotron: Biology, Physics, and Change in Science*

Gabrielle Hecht, *The Radiance of France: Nuclear Power and National Identity after World War II*

Richard Rottenburg, *Far-Fetched Facts: A Parable of Development Aid*

Michel Callon, Pierre Lascoumes, and Yannick Barthe, *Acting in an Uncertain World: An Essay on Technical Democracy*

Ruth Oldenziel and Karin Zachmann, editors, *Cold War Kitchen: Americanization, Technology, and European Users*

Deborah G. Johnson and Jameson W. Wetmore, editors, *Technology and Society: Building Our Sociotechnical Future*

Trevor Pinch and Richard Swedberg, editors, *Living in a Material World: Economic Sociology Meets Science and Technology Studies*

Christopher R. Henke, *Cultivating Science, Harvesting Power: Science and Industrial Agriculture in California*

Helga Nowotny, *Insatiable Curiosity: Innovation in a Fragile Future*

Karin Bijsterveld, *Mechanical Sound: Technology, Culture, and Public Problems of Noise in the Twentieth Century*

Peter D. Norton, *Fighting Traffic: The Dawn of the Motor Age in the American City*

Joshua M. Greenberg, *From Betamax to Blockbuster: Video Stores tand the Invention of Movies on Video*

Mikael Hård and Thomas J. Misa, editors, *Urban Machinery: Inside Modern European Cities*

Christine Hine, *Systematics as Cyberscience: Computers, Change, and Continuity in Science*

Wesley Shrum, Joel Genuth, and Ivan Chompalov, *Structures of Scientific Collaboration*

Shobita Parthasarathy, *Building Genetic Medicine: Breast Cancer, Technology, and the Comparative Politics of Health Care*

Kristen Haring, *Ham Radio's Technical Culture*

Atsushi Akera, *Calculating a Natural World: Scientists, Engineers and Computers during the Rise of US Cold War Research*

Donald MacKenzie, *An Engine, Not a Camera: How Financial Models Shape Markets*

Geoffrey C. Bowker, *Memory Practices in the Sciences*

Christophe Lécuyer, *Making Silicon Valley: Innovation and the Growth of High Tech, 1930–1970*

Anique Hommels, *Unbuilding Cities: Obduracy in Urban Sociotechnical Change*

David Kaiser, editor, *Pedagogy and the Practice of Science: Historical and Contemporary Perspectives*

Charis Thompson, *Making Parents: The Ontological Choreography of Reproductive Technology*

Pablo J. Boczkowski, *Digitizing the News: Innovation in Online Newspapers*

Dominique Vinck, editor, *Everyday Engineering: An Ethnography of Design and Innovation*

Nelly Oudshoorn and Trevor Pinch, editors, *How Users Matter: The Co-Construction of Users and Technology*

Peter Keating and Alberto Cambrosio, *Biomedical Platforms: Realigning the Normal and the Pathological in Late-Twentieth-Century Medicine*

Paul Rosen, *Framing Production: Technology, Culture, and Change in the British Bicycle Industry*

Maggie Mort, *Building the Trident Network: A Study of the Enrollment of People, Knowledge, and Machines*

Donald MacKenzie, *Mechanizing Proof: Computing, Risk, and Trust*

Geoffrey C. Bowker and Susan Leigh Star, *Sorting Things Out: Classification and Its Consequences*

Charles Bazerman, *The Languages of Edison's Light*

Janet Abbate, *Inventing the Internet*

Herbert Gottweis, *Governing Molecules: The Discursive Politics of Genetic Engineering in Europe and the United States*

Kathryn Henderson, *On Line and On Paper: Visual Representation, Visual Culture, and Computer Graphics in Design Engineering*

Susanne K. Schmidt and Raymund Werle, *Coordinating Technology: Studies in the International Standardization of Telecommunications*

Marc Berg, *Rationalizing Medical Work: Decision-Support Techniques and Medical Practices*

Eda Kranakis, *Constructing a Bridge: An Exploration of Engineering Culture, Design, and Research in Nineteenth-Century France and America*

Paul N. Edwards, *The Closed World: Computers and the Politics of Discourse in Cold War America*

Donald MacKenzie, *Knowing Machines: Essays on Technical Change*

Wiebe E. Bijker, *Of Bicycles, Bakelites, and Bulbs: Toward a Theory of Sociotechnical Change*

Louis L. Bucciarelli, *Designing Engineers*

Geoffrey C. Bowker, *Science on the Run: Information Management and Industrial Geophysics at Schlumberger, 1920–1940*

Wiebe E. Bijker and John Law, editors, *Shaping Technology / Building Society: Studies in Sociotechnical Change*

Stuart Blume, *Insight and Industry: On the Dynamics of Technological Change in Medicine*

Donald MacKenzie, *Inventing Accuracy: A Historical Sociology of Nuclear Missile Guidance*

Pamela E. Mack, *Viewing the Earth: The Social Construction of the Landsat Satellite System*

H. M. Collins, *Artificial Experts: Social Knowledge and Intelligent Machines*

http://mitpress.mit.edu/books/series/inside-technology